D1203332

PRENTICE-HALL INTERNATIONAL, INC., *London*
PRENTICE-HALL OF AUSTRALIA, PTY. LTD., *Sydney*
PRENTICE-HALL OF CANADA, LTD., *Toronto*
PRENTICE-HALL OF INDIA PRIVATE LIMITED, *New Delhi*
PRENTICE-HALL OF JAPAN, INC., *Tokyo*

 FOUNDATIONS OF MODERN BIOCHEMISTRY SERIES

Lowell Hager and Finn Wold, editors

ORGANIC CHEMISTRY OF BIOLOGICAL COMPOUNDS*
Robert Barker

INTERMEDIARY METABOLISM AND ITS REGULATION
Joseph Larner

PHYSICAL BIOCHEMISTRY
Kensal Edward Van Holde

MACROMOLECULES: STRUCTURE AND FUNCTION
Finn Wold

SPECIAL TOPICS:

BIOCHEMICAL ENDOCRINOLOGY OF THE VERTEBRATES
Earl Frieden and Harry Lipner

* Published jointly in Prentice-Hall's *Foundations of Modern Organic Chemistry Series.*

ORGANIC CHEMISTRY OF BIOLOGICAL COMPOUNDS

ROBERT BARKER

Professor of Biochemistry
University of Iowa

Prentice-Hall, Inc., Englewood Cliffs, New Jersey

FOREWORD

Biochemistry has been and still is the major meeting place for biological and physical sciences, and most introductory courses in biochemistry have a rather unique and heterogeneous population of advanced students from all branches of the natural sciences. Such courses, therefore, have equally unique and heterogeneous requirements for background material and textbooks. The content of a first-year course in biochemistry based on two years of chemistry and biology would probably not be too difficult to define, but to write a single text which contains the needed material for all students becomes a much more controversial issue. As a solution to this dilemma, presenting the material in several packages offers some interesting possibilities. Without in any way compromising the basic content, such a multivolume text should above all allow for a great deal of flexibility: flexibility for the student to supplement previous experience with only the parts which represent new and unique features; flexibility for the instructor to offer a course covering an area of modern biochemistry with something less than the comprehensive text; and the flexibility to keep a text current by rewriting any outdated part without having to reject the whole text.

These thoughts were fundamental in formulating the Foundations of Modern Biochemistry series. So was the philosophy that a major purpose of a textbook is perhaps not so much to be encyclopedic as to select for the student the important principles and to illustrate and explain these in some depth. With this basic plan in mind, the next step was to divide the whole into logical subdivisions

v

or parts. The science of biochemistry seeks the answer to three basic questions:

1. What is the nature of the molecules and structures found in living cells?
2. What is the biological function of these molecules and structures?
3. How are they synthesized (and broken down) in the cell?

These questions were adopted as the basis for the division of this text. The first question, related to the qualitative and quantitative characterization of the biochemical world and to the methods available for structural analysis, is covered in two books: *Organic Chemistry of Biological Compounds* and *Physical Biochemistry*. The second question, concerning the elucidation of the biological function of these molecules, is discussed in *Macromolecules: Structure and Function*. The third question, covering all aspects of intermediate metabolism and metabolic regulation, is considered in *Intermediary Metabolism and Its Regulation*. As the work on the individual books progressed, it became apparent that this subdivision had one unexpected advantage; namely that the physical presentation of material as diversified as the mathematical derivations in physical chemistry, the many structures of organic chemistry, and the long and complicated road maps of intermediate metabolism can be handled much more rationally in the individual format of separate books than in a single volume.

Thus, the Foundations of Modern Biochemistry series came into being consisting of four *individual* books. The books can hopefully be used either separately or in any combination to meet the requirements of the individual student according to his particular background and goals. The absence of a numbered sequence represents a deliberate effort to emphasize that the books can be used in any order. The integration of the four parts into the total program is done by extensive cross-references between the individual books. Trying to retain the individuality and utility of the separate books and at the same time aiming for an integrated program required some compromises in the selection and distribution of the material to be covered. Quite expectedly, the main price paid for this dual purpose has turned out to be some duplication, which hopefully is not extensive enough to become a serious flaw.

In arriving at the definition of the Foundations of Modern Biochemistry content and at the philosophy represented by this set of books, the thinking of the "editorial board" (the four authors and Dr. Lowell P. Hager) underwent an extensive evolution. The evolutionary pressures were graciously, patiently, and sometimes even enthusiastically provided by colleagues, by the publishers, and most importantly by students, too many to mention individually. To all of these good people we express our sincere thanks.

FINN WOLD

CONTENTS

ONE | THE CHEMICAL REQUIREMENTS OF LIFE

The science of biochemistry encompasses all knowledge relevant to the understanding of biological systems at a molecular level. It is concerned with the description of the molecular constituents of living systems and the ways in which these constituents interact to allow and to regulate the operation of biological systems. Before outlining the approach to biochemistry used in this book, it is pertinent to consider some of the attributes of living systems and the ways in which they can be studied.

Biological systems are chemically highly organized, and "living" entails the maintenance, within fairly narrow limits, of this high degree of organization. If the ways in which organization is maintained in an animal are considered, it is readily apparent that it is an energy-requiring process and that the energy is derived from the environment in the form of food. It is also clear that the substances used to form the living system come from the diet and that a wide variety of foodstuffs can serve as precursors of the animal body. Compare, for example, the diets of man and the cow. Apparently foods are not incorporated into the body as eaten but are broken down into common constituents before being utilized.

1

The common constituents available to man (as a somewhat atypical animal) are oxygen, water, simple carbohydrates, amino acids, lipids, nucleosides and nucleotides in small proportion, and inorganic materials such as sodium, potassium, and chloride. To survive, he must possess mechanisms for the production of the substance of which he is made from these precursors. In fact, with few exceptions, he can do so with an even more limited diet. He can synthesize the carbohydrates from amino acids and vice versa with the exception of a few of the latter which he has lost the ability to synthesize during evolution. A small number of fatty acids which are abundant in plants is also required, as are very small amounts of a number of special organic compounds which are utilized catalytically, as components of enzyme systems (the vitamins). In addition, the inorganic components, which, in general, are carefully conserved, must be obtained. No biological system can synthesize sodium or phosphorus!

The synthetic biochemical processes in man, therefore, consist of pathways for the conversion of amino acids, carbohydrate, acetate (the common denominator of lipid and carbohydrate metabolism), and phosphate to the complex carbohydrate, protein, lipid, and nucleic acid structures which are required for life. The problem is very similar to that faced by the organic chemist who wishes to synthesize a complex structure. He must either build it up from very generally available small components or modify an already complex molecule made available by the efforts of other chemists or isolated from a biological system. Life would be much more precarious than it is at present if too great a dependence were placed upon the availability of complex compounds in the diet. The variations in eating patterns observable in our society would cease to be merely amazing—they would prove lethal. If man had to obtain many specific, highly complex compounds in large quantities in his diet, it would be entirely possible for him to "starve" in the presence of a plentiful supply of food. Such a possibility has been widely advertised in this country, where the vitamin deficiency has been equated to a "fate worse than death." Contrary to the tenor of this advertising, vitamin deficiencies and deficiencies in other dietary requirements are rare and the absence of the enzymes (catalysts) required to make these compounds is not normally a disadvantage. In fact, the lack of a requirement for the production of these enzymes may be an advantage. For example, the synthesis of a compound such as vitamin B_{12}, which is enormously complex, may require the organism to first synthesize a dozen or more specific proteins which are the catalysts for the various steps in the synthetic process. In addition, the information required for the synthesis of these proteins must be transmitted from generation to generation, which requires the presence and duplication of the nucleic acids which carry that information. Therefore, the protein and nucleic acid content of the organism able to synthesize this vitamin is greater than that of a similar organism lacking this ability, and the former has a greater dietary requirement for the precursors of protein and nucleic acid than the latter and requires an input of food for energy to accomplish their synthesis. Depending on the environment in which the organisms find themselves, the increased requirement for precursors of proteins and nucleic acids may be more of a disadvantage than the requirement for vitamin B_{12}. This is certainly true if the environment is abundant in the latter. There is a balance which can be achieved between dependence on the environment for specific dietary items and independence of it which requires an increase in cellular components and general dietary intake. This results in the greatest efficiency for each organism. At one extreme is the parasite which relies on a specific host for essential materials and is limited to that host. At the other extreme are the organisms which require only simple materials such as carbon dioxide, nitrogen, and inorganic elements. It is interesting that the most complex forms of life possess an intermediate dependence on their environment.

The discussion above draws attention to the fact that synthesis is an energy-requiring process and that man is a type of combustion apparatus in which some of his intake of foods is combusted (oxidized) to supply the energy for synthesis and function.

Let us now move back from the subject of man (as a representative animal) and inquire into the source of his foodstuff, which supplies energy and starting materials for his synthetic activity.

Tracing back from man to his food, to his food's food, and so forth, it can be established that the major ultimate source of energy is the sun and the primary source of food for the world is the green plant. The latter is capable of using carbon dioxide and nitrate to form carbohydrates, lipids, and proteins. The production of adequate nitrate for the plants is dependent upon the fixation of nitrogen by bacteria or by man in the form of commercial fertilizers.

The discussion above is intended to give a glimpse of the broad synthetic and degradative pathways operating in our world. The question may be asked, What is the relevance of this discussion to the purposes of this book?

The answer: In a sense, all chemistry can be considered as part of an attempt to understand the natural world, and the organic chemist and the biochemist share this common goal; this book represents an attempt to describe the areas of common interest. It would be possible to approach the topic of this text from the point of view of the types of reaction which occur in nature and a discussion of the present state of our knowledge of the detailed mechanisms of these reactions. Alternatively, the types of compounds which occur in nature can be described and an attempt made to show how their structures were established and how fundamental chemical approaches have been used to study them and their reactions. The latter approach has been taken. It is hoped that the reader will extrapolate from this limited treatment to the goal of understanding the mechanisms at work in the flow of energy and materials in the biological world. See Figure 1.1.

In the past few decades there has been an apparent separation of the biochemist from the chemist, although major efforts among the latter have been aimed at the study of natural products. An interesting difference in the approaches of the biochemist and the natural-products chemist is apparent. The former has focussed on the contents of the living cell, the dynamics of the living processes, and an understanding of the functions of the cell. His interest

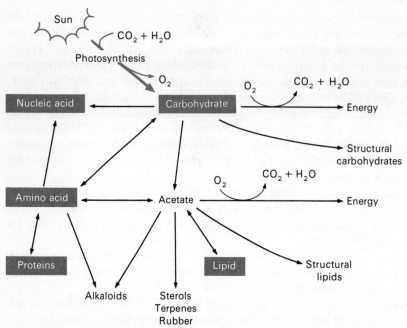

Figure 1.1   The major organic constituents of nature and the pathways for their inter-conversion.

is often in the transient metabolites which are never present in more than trace amounts. The natural-products chemist, on the other hand, has pursued the chemistry of accumulated metabolites—the carbohydrates, starch, and glycogen; and the lipids, carotenes, terpenes, sterols, and so forth.

Some of the accumulated compounds have been considered as "metabolic accidents," that is, of no value to the organism producing them. This point of view may eventually prove to be totally incorrect. Now that the "major biochemical pathways" have been established, investigators are focusing their attention on pathways and products whose role in the organism is more difficult to discern. As these are investigated, it becomes apparent that many of them serve exceedingly useful functions. It is also apparent, to even a casual observer, that nothing of natural origin is permanent and that all natural products must be destroyed by natural processes at a rate which approximates the rate of their formation. If this were not so, we would be knee-deep in accumulated natural products and natural-products chemistry would be much more fashionable than at present.

As stated above, this book is an attempt to describe the structures and reactivities of the major groups of compounds encountered by the biochemist. The systems he studies have many peculiarities and require a general knowledge of many areas of chemistry. He must be quite familiar with the chemical and physical properties of water and be aware of the attempts being made to understand its structure, since almost all the systems of interest involve water as the medium and/or as a reactant.

A second property of all living systems is their tendency toward outward symmetry and their possession of extreme internal molecular asymmetry. An appreciation of the levels of molecular symmetry and the possibilities for discrimination between chemically similar (or identical) groups is also essential "background" information for the biochemist. An attempt is made to supply it in Chapter 3.

The simple carbohydrates, amino acids, lipids, and nucleotides are the building blocks of nature and their chemistry serves as a basis for understanding the chemistry of the complex polymeric materials formed from them. Polymerization and depolymerization reactions are fundamental processes in living systems. In general they can be viewed as occurring by the removal or addition of the elements of water with the formation and scission of amide, acetal, and ester bonds—relatively simple chemical events. The chemistry of the monomeric materials can be quite complex, however, and an attempt has been made to focus on the ways in which these materials have been studied and the synthetic methods used in the preparation of biologically interesting compounds. The phosphate esters have been treated separately because of their relevance to the chemistry of the carbohydrates, lipids, and nucleic acids and because of their special role in metabolism.

It is usual for scientists to attempt to classify and arrange the knowledge they develop to facilitate communication and comprehension. Biological systems are sufficiently complex that classification schemes, applied rigorously, become restrictive rather than helpful. Even a relatively simple attempt to group the compounds of nature into broad categories by structure is difficult and results in a few being left over. These compounds are discussed separately in Chapter 9 because they deserve separate attention. Others have been omitted entirely from this book, and these omissions, when intentional, reflect only the prejudices of the author.

TWO | WATER

Almost all biological systems consist of dilute aqueous solutions in contact with membranes and organelles which are composed of lipid, protein, and/or carbohydrate. Life processes, therefore, involve many reactions and interactions occurring in dilute aqueous solution, at interfaces between water and semisolid components or inside these components. The chemistry of the membranes and organelles is as yet poorly understood and it is difficult to discuss the processes which go on inside them. Because of this, the majority of biochemical phenomena which have been investigated to date have involved the aqueous phase and the water-lipid or water-protein interface. Therefore it is pertinent to discuss some of the characteristics of water, of water interfaces, and of water-solute interactions.

Compared to liquids, gases and solids possess delightful simplicity. Gases are simpler because their molecules are almost independent of each other and thus can be thought of under some circumstances as behaving ideally. Solids are simpler because their molecules are in fixed relationships to one another and have definite intermolecular distances and interactions. In liquids the molecules have definite contacts and relatively strong intermolecular interactions, but they are constantly changing partners so that the kinds of interactions occurring are very difficult to discern. Thus, liquids possess "structures" which are time averages of many possible intermolecular arrays. The intermolecular arrays in water are held together by *hydrogen bonds*. Each water molecule can participate in the formation of a maximum of four such bonds, each contributing about 3 kcal/mole to the stability of the system.

In a sense, any body of water can be considered as a single giant molecule in which the bonds between monomers are hydrogen bonds rather than the more familiar covalent bonds. Because of their relevance to a discussion of water structure and their great importance to all biochemical systems, a short discussion of the characteristics of hydrogen bonds is given below.

2.1 HYDROGEN BONDS

A precise definition of the term *hydrogen bond* is difficult to achieve. Because hydrogen bonds are relatively weak, even their presence in a system is not always easy to establish. They involve the interaction of a covalently bound hydrogen with an electronegative atom or group. They can be considered to be present when a group AH which is capable of serving as a donor and an atom or group B which is capable of serving as an acceptor are present in a system and there is evidence for an interaction between AH and B which involves H. Hydrogen bonds are usually represented as a dashed or dotted line between H and the acceptor atom.

$$A—H\cdots B$$

There are many ways to establish that a hydrogen bond exists. For example, the infrared absorption maxima are quite different for hydrogen-bonded and nonhydrogen-bonded groups (RO—H absorbs at 3,650 to 3,590 cm^{-1}, RO—H$\cdots$B absorbs at 3,550 to 3,450 cm^{-1}). A second example is that hydrogen bonding between carboxylic acids in nonpolar media is sufficiently strong to cause dimer formation which can be detected by molecular weight determination. In most cases the existence of strong hydrogen bonds results in the donor and acceptor groups (A—H and B) being closer together than the sum of the van der Waals radii for A—H and B. The distance between A and B along the line of the bond is usually decreased by about 0.2 Å. This does not mean that there is no hydrogen bonding when this condition is not met; it simply implies that longer bonds will be weaker.

It is commonly held that the A—H$\cdots$B system must be linear for maximum stability. There is little evidence for this, however, and angles between the A—H vector and the A$\cdots$B vector as great as 30 deg have been observed in crystals. It is probable that the requirement for linearity in a given hydrogen-bonded system must be balanced against the other requirements of the system such as the formation of a maximum number of attractive interactions and the relief of a maximum number of repulsive interactions. It may be that nonlinear hydrogen bonds are the rule rather than the exception.

Two of the common hydrogen-bond acceptors are the carbonyl group and the water molecule. In the carbonyl group, the electrons in the sp^2 orbitals of the oxygen are involved in hydrogen bonding and it would be expected that the angle between the C=O vector and O$\cdots$H vector would be 120 deg.

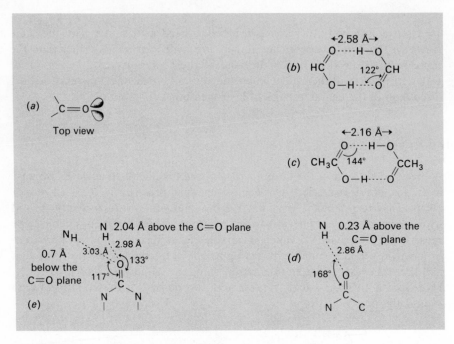

Figure 2.1   Hydrogen-bond formation involving the carbonyl group. (*a*) The carbonyl group showing the lone pairs of electrons on oxygen available for hydrogen bonding. (*b*) Bond angles and distances in crystalline formic acid. (*c*) Bond angles and distances in crystalline acetic acid. (*d*) The single hydrogen bond to the carbonyl group in the α helix of proteins. (*e*) Hydrogen bonds in crystalline thymidine. (Redrawn from *Structural Chemistry and Molecular Biology*, edited by Alexander Rich and Norman Davidson, W. H. Freeman and Company. Copyright © 1968.)

As shown in Figure 2.1, this is not usually the case. In crystalline formic acid, the angle is 122 deg; in crystalline acetic acid, it is 144 deg. In the important case of the carbonyl group, hydrogen bonding with an amide hydrogen atom in the α helix of proteins (see Chapter 4), the angle is 168 deg.

The water molecule is usually considered to have close to tetrahedral symmetry with two hydrogens and two sp^3 orbitals occupied by two pairs of unshared electrons as shown in Figure 2.2. In this case hydrogen bonding would be expected to be maximal when two donor groups and two acceptor groups surround each water molecule in a tetrahedral array. This situation is observed in ice I (Figure 2.4). In many other situations, a water molecule is hydrogen bonded between two donor and one acceptor group in a planar-trigonal array. Two examples are given [Figure 2.2(*b*) and (*c*)]. The water in these structures may possess other than pure sp^3 hybridization, thus changing the geometry of the central molecule.

The point to be gained from these considerations is that although hydrogen bonds may be most stable when the three participants (A—H···B) are in a linear array and when the orbital containing the electron pair of the donor is also in line, these are not absolute requirements. The energy of the bond

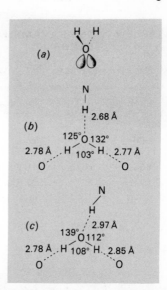

Figure 2.2

Hydrogen-bond formation involving water. (*a*) The water molecule showing lone pairs of electrons (and hydrogens) available for hydrogen bonding. (*b*) Water hydrogen bonded in crystalline glycyltryptophan dihydrate. (*c*) Water hydrogen bonded in crystalline cytosine monohydrate. (Redrawn from *Structural Chemistry and Molecular Biology*, edited by Alexander Rich and Norman Davidson, W. H. Freeman and Company. Copyright © 1968.)

may be only slightly affected by deviations from linearity of 30 deg in the former situation and 50 deg in the latter.

A further complication of the hydrogen-bond situation must be noted. It is possible for a hydrogen atom to bond simultaneously to two donor groups and to lie approximately equidistant from them. The H$\cdots$B distances are approximately 0.3 Å less than the sum of the van der Waals radii. This situation is referred to as the bifurcated hydrogen bond; it has been established in relatively few crystalline materials (Figure 2.3).

There have been claims for the occurrence of C—H$\cdots$O bonds in a number of systems. The occurrence of H$\cdots$O distances of less than 2.6 Å (the sum of the van der Waals radii) in crystalline materials have been interpreted as indicating the presence of hydrogen bonds with C—H acting as the donor. However, this cannot be considered an adequate single criterion. At present it seems safest to conclude that if this type of bond exists, it is rare. It will not be considered further in this text.

Some systems in which hydrogen bonds exist and which are important to biochemists are listed in Table 2.1 along with some values for ΔG^0. It is important not to overinterpret the values given for ΔG^0; they are valid only for specific systems under highly specified conditions. Values are given only to illustrate the general weak nature of the bond (5 ± 3 kcal/mole).

Figure 2.3

The bifurcated hydrogen bond in crystalline glycine (α form). (Redrawn from *Structural Chemistry and Molecular Biology*, edited by Alexander Rich and Norman Davidson, W. H. Freeman and Company. Copyright © 1968.)

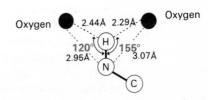

TABLE 2.1   SOME EXAMPLES OF HYDROGEN BONDS

Donors	Acceptors	System	ΔG^0 (kcal/mole)
−OH	−C=O	Formic acid dimers, 50°C	−7.0
HOH	H | O−H	Water (liquid)	−3.4 to 7.7
ROH	HOR	Methanol (liquid)	−4.7
N−H	N	Ammonia (gas)	−4.4
N−H	O=	Oxamide	−4.2
−S−H	| S−	Rubeanic acid	−3.8
F−H	| F−	Hydrofluoric acid	−6.8

Keeping in mind the flexibility of the hydrogen bond and the uncertainties which exist about it, let us consider the structure of water which is a resultant of the presence of an enormous number of hydrogen bonds. Most studies on liquids start from a consideration of the possible intermolecular interactions and these are most easily studied in the solid derived from that liquid: in this case, ice.

2.2   THE STRUCTURE OF ICE

Several kinds of ice are known; they are designated I, II, III, IV, V, VI, VII, and VIII; Ic; and vitreous ice. Of these, ice I is the familiar form; the others, ices II through VIII exist only at pressures exceeding 2,000 atm. Ice Ic is formed by heating ices II, III, or V or by deposition of water vapor at low pressures and low temperatures, and vitreous ice is formed by condensing vapors below −160°C. Although it may seem unlikely that these unusual ice structures will be important in connection with liquid water structure, they are important in considering possible arrangements for water molecules bound to solutes.

Ice I has the three-dimensional structure shown in Figure 2.4. This open semirigid structure does not represent the closest possible packing for water molecules and ice I has a density of 0.924 g/ml. The average distances between atoms and molecules are indicated in the figure; however, the structure is subject to thermal agitation and the atoms and molecules can vibrate about their equilibrium positions. At −10°C the vibrations of water molecules in ice I have a root-mean-square amplitude of 0.44 Å. In addition, the molecules can "rotate" in such a way that their dipole moments rapidly alternate between various equilibrium positions. The result of this rotational freedom is that ice has a very high dielectric constant.

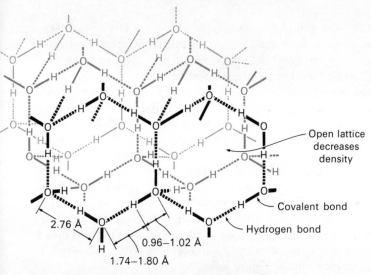

Figure 2.4   The structure of ice I. Note the open lattice which accounts for its low density.

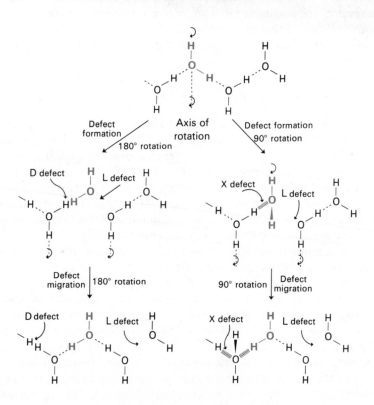

Figure 2.5   D, L, and X defects in the lattice of ice I.

The mechanism by which water molecules rotate in ice I is not known, but several proposals have been made. Two distinctly different principles have been invoked. One involves the migration of protons along the axis of the —O—H····O—bond so that a normal bond becomes a hydrogen bond and a hydrogen bond becomes a normal bond; the observed "rotation" is due to this rehybridization process. The second proposal is that rotation of intact water molecules does occur. To assist in explaining the ease with which rotation appears to occur, it has been necessary to propose that defects are present in the crystal structure which provide "holes" into which protons can move; such movements in turn regenerate the defect. Three types of defect have been proposed: D defects, in which there is no proton between oxygens; L defects, in which there are two protons between oxygens (both of these defects would be produced by rotation of one water molecule through 120 deg); and X defects, in which a water molecule has rotated 60 deg to bring one of its protons into a location such that it is not involved in hydrogen bonding, that is, does not lie on the axis between oxygen atoms (Figure 2.5).

It is obvious that these defects are accommodated by changes in the positions of other atoms in the crystal. For example, a D defect would be a very high energy arrangement (60 kcal/mole) if there was no relaxation of the lattice, but it has been calculated that small changes in bond angles and lengths in the vicinity of the defect decreases the strain energy to 5.2 kcal/mole of defects.

2.3 THEORIES CONCERNING WATER STRUCTURE

Most theories concerning the structure of liquid water use the known structure of ice I as a basis. It appears that at least a portion of liquid water is composed of three-dimensional arrays of molecules arranged in a tetrahedral fashion similar to that in ice I, but having slightly altered O—H····O bond, distances, or angles. The individual molecules in this structure are free to move as in the ice structure but do not, at low temperatures, rotate freely. (The rotation of molecules in ice I discussed above is not free rotation in the sense that relatively few molecules are involved and it is a highly ordered process).

Other regions in liquid water contain molecules which exist in other types of three-dimensional aggregate, and some molecules are essentially free from any structural restraint. The proportions of various kinds of water molecules are altered by changes in temperature, the presence of solutes, and at interfaces between water and other phases. The water system is never static, and individual water molecules at different times exist in different kinds of relations to their neighbors.

The description given above is one which is not intended to correspond exactly to any of the models which have been proposed for liquid water. The "vacant-lattice-point" model, the "water-hydrate" model, the "distorted-bond" model, the "flickering-cluster" model, and the "significant structure theory of water" have elements of similarity in that they propose that the system is dynamic, that it is rapidly fluctuating, and that individual molecules play changing roles. One of the major points of contention seems to be whether any of the water molecules are ever completely free from hydrogen bonding with their neighbors or whether some degree of bonding always exists. Since the transition from a dipole interaction to a covalent bond by way of a hydrogen bond could be an essentially continuous process, the argument may appear to involve detail rather than substance. The challenge of proposing a model which predicts all the anomolous behavior of water is an intriguing one, however, and it is vitally important to our understanding of the behavior of aqueous systems that it be met.

To illustrate the kind of approaches used in attempting to solve the problem of water structure, the model proposed by Eyring, based on his significant structure theory of liquids, is described briefly and illustrated in Figure 2.6. He assumes that water is composed of molecules in several different arrangements: two solidlike structures, one having a density and arrangement similar to that found in ice I and the other resembling ice III. The ice III-like structure is 20 percent denser than ice I. These solidlike structures are in a state of dynamic equilibrium with each other and with gaslike water molecules which are not bonded to each other. In addition, fluidized vacancies are present which can "migrate" through the structure in much the same way as defects move through ice structures (Figure 2.5).

The theory is applied by selecting a distribution of the molecules between different states and assigning parameters to the fluidized vacancies. The properties of the molecules in each state (for example, of ice I) are used to compute the average properties of the bulk water as the weighted sum of the contributions from the three forms and the vacancies. The theory has been utilized to compute values for viscosity, temperature of maximum density

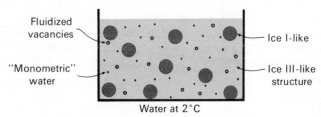

Water at 2°C

Figure 2.6   The equilibria involved in the significant structure theory of water. On melting of ordinary ice, the ice I structure (density 0.9249 g/ml) collapses to form ice III-like structure (density 1.14 g/ml) and a small number of free water molecules. A number of vacancies in the "lattice" is also formed. These vacancies can migrate rapidly by being occupied by adjacent molecules and are referred to as "fluidized." As the temperature increases from 0 to 4°C, the proportion of ice III-like structure increases to a maximum. Above 4°C the number of fluidized vacancies and free water molecules increases and the density decreases in a normal fashion. Ice III structure is more compact than that of ice I and has distorted hydrogen bonds.

(4°C), entropy of vaporization, vapor pressure, heat capacity, molar volume, and critical values which are in good agreement with experimental values.

As the temperature of water is increased from 0°C, the proportion of water molecules participating in "ice I-like" structure decreases. Between 0 and 4°C, hydrogen-bond bending allows closer packing to occur (ice III-like structure) and even though vacancies are formed the density increases over this range. Above 4°C thermal expansion takes over; that is, because of thermal agitation each molecule requires a larger volume as the temperature is increased.

2.4   THE INTERACTION OF WATER WITH SOLUTES

When solutes are present in water, the structure of the water is altered to some degree. There are several kinds of changes which can occur. In the case of slightly soluble nonpolar substances, solution probably involves formation or occupation of pre-existing cavities in the water structure which contain the solute molecule and which are surrounded by a layer of water molecules having an "ice like" or "semicrystalline" structure. In cases where extremely stable cage complexes (clathrate compounds) are formed, stable crystalline hydrates can be isolated. An example is shown in Figure 2.7 of a cage formed of 20 water molecules hydrogen bonded to form a regular pentagonal dode-cahedron with an unrestricted cavity 5 Å in diameter—sufficient to hold mole-cules the size of methane or acetone. Cages of this type can pack together in crystals or can be separated by other more random water molecules in solution. They form, break up, and reform with a rapidity depending on temperature, pressure, and the presence of other solutes. Large solute molecules or non-polar surfaces require increased cage sizes or large planes of structurally stabilized water. It seems clear that some water structures are stabilized in contact with nonpolar systems, but there is no definitive evidence on the thickness of the water layer which is affected. It is also clear that the con-formation (shape) that a large flexible nonpolar molecule or substituent will assume in solution will be strongly influenced by the stability of the cavity it occupies in the water structure.

Ionic materials in water influence water structure and have their own prop-erties modified as a consequence. Water, because of its strong dipole, tends

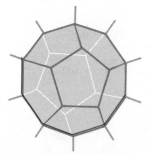

Figure 2.7

A regular pentagonal dodecahedral cage formed from 20 water molecules. The cavity can contain a free water molecule or another molecule of less than 5 Å diameter. The whole structure (that is, cage plus contents) is termed a *clathrate*.

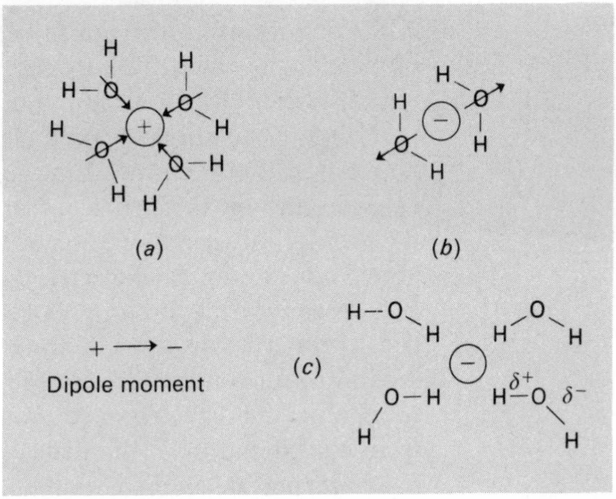

Figure 2.8

The hydration of cations and anions. (a) A relatively small cation with water molecule dipoles normal to the surface of the ion. (b) An anion with water dipoles oriented away from and normal to the surface of the ion. (c) An anion with water molecules oriented with hydrogens toward the ion.

to be oriented and more or less immobilized in the vicinity of ionic groups. The degree of orientation and immobility depends on the species carrying the charge. For example, Cr^{3+} forms a stable hexahydrate which in solution exchanges one half of its water molecules with those in the solvent during 40 hr; Cu^{2+}, although hydrated, exchanges water with solvent so rapidly that each water resides on the Cu^{2+} for only 5×10^{-7} sec. The orientation of water around a cation and an anion is diagrammed in Figure 2.8. It is probable that the angle which the dipoles of the nearest water molecules make with the surface of an ion will depend on the charge density at the surface (a function of valence, radius, and, in the case of complex ions, geometry). The closest water molecules also interact with water molecules in the next sphere and will be influenced by hydrogen bonding, dipole interactions, and thermal motions in that sphere. The interactions are complex and the resultant of them is not predictable.

In general, relatively small ions and multivalent ions are observed to increase the viscosity of water (F^-, Li^+, Na^+, H^+, Mg^{2+}, OH^-). This action is probably due to their producing a net increase in icelike structure by influencing water molecules beyond the first hydration shell. Large monovalent ions decrease the viscosity of water (K^+, NH_4^+, Cl^-, Br^-, ClO_4^-). They have a diffuse surface charge which tends to orient the nearest water molecules and decrease their ability to fit into ordered water structures. The effect is to increase the proportion of more randomly distributed water molecules. Anions in general are less hydrated than cations, presumably because the first hydration layer has the "broad end" of the water molecule toward the anion or because the water orients with only one proton toward the anion and is freer to interact with bulk water [Figure 2.8(b) and (c)].

An important effect of binding of water by ions is the marked decrease in dielectric constant of the bound water. The dielectric constant of bulk water is 78.56. The average value between two monovalent ions about 5 Å apart is approximately 19; it is lowest (3 to 6) at the surface of the ion and increases gradually to achieve the bulk value at 5 to 8 Å from the center of the ion.

It is important to distinguish between the bulk properties and the molecular properties of water. The former term refers to such properties as density, viscosity, dielectric constant, and so forth, which depend on the interactions of a large population of molecules and are weighted average values for that population. On the other hand, the properties of a specific set of water molecules depend on their environment and will generally be quite different from the average for the population as a whole. In a strict sense then the statement above, concerning the dielectric constant of water bound to an ion, is incorrect. It does convey the fact, however, that such water is very different in its electrical properties from bulk water. It is much poorer in maintaining charge separation and on the surface of a polyvalent macromolecule the force operating to separate similarly charged groups will be much greater than that predicted from consideration of the dielectric constant of bulk water. Since most proteins are polyvalent with both positive and negative charges, the attractive (between $+$ and $-$) and repulsive (between $-$ and $-$ or $+$ and $+$) forces on their surfaces may be much greater than those existing between similar groups in the bulk medium.

Polar compounds such as alcohols, unionized amines, sulfhydryls, carboxylic acids, ethers, esters, and phenols lie between nonpolar compounds and ions in their effect on water structure. They interact with the water by formation of hydrogen bonds and by dipole interactions. If they fit into water structures with relatively little distortion and can form stable hydrogen bonds with it, they will be soluble. If they disrupt the water structure and hydrogen bond poorly, they will be relatively insoluble. The strength of water-solute hydrogen bonds is similar to the strength of solute-solute hydrogen bonds and there is a competition between solvent and solute for hydrogen-bonding sites in those solutes having donor and acceptor groups for hydrogen-bond formation. Depending on the specific substance and the conditions used, solvation of polar molecules by water can strongly influence their physical, chemical, and biochemical properties.

2.5 HYDROGEN AND HYDROXYL IONS

The ions derivable from water deserve special mention. The hydrogen ion is very strongly solvated. In dilute solution it exists as at least a tetrahydrate, $H_9O_4^+$, although it is frequently referred to as though it were a monohydrate, H_3O^+ (the hydronium ion). The heat of hydration of a proton (H^+) is 276 kcal/mole, much greater than the heat of dissociation of water, 13.3 kcal/mole. It is probable that the bonding of all protons in the complex ion is equivalent and at least partly covalent. The hydroxyl ion is also strongly hydrated.

Pure water is only slightly ionized ($10^{-6.998}$ M of each ion at 25°C) but the rate at which the ions can be transported through the system makes even this slight degree of ionization important. The apparent velocity of protons in water is 36.2×10^{-4} cm/sec and that of hydroxyl ions, 19.8×10^{-4} cm/sec.

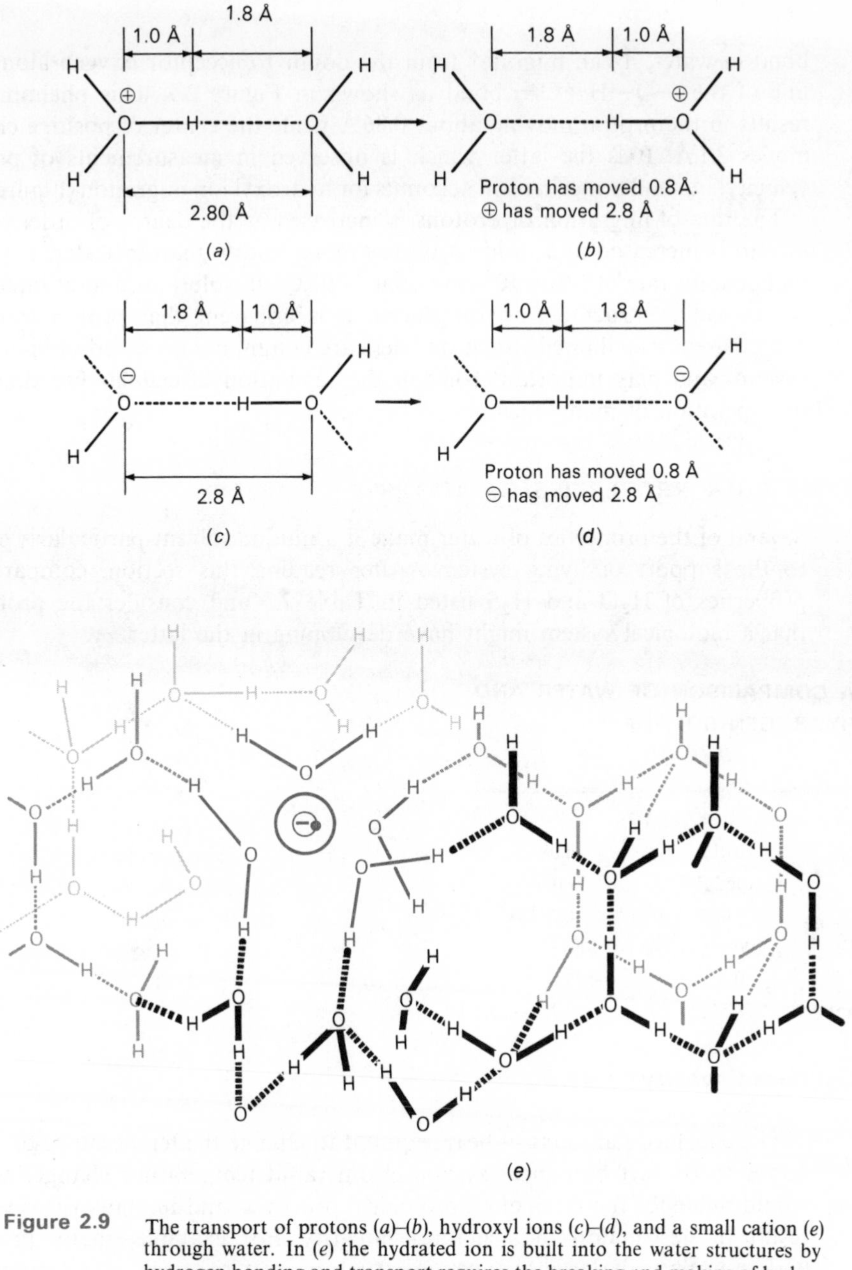

Figure 2.9   The transport of protons (a)–(b), hydroxyl ions (c)–(d), and a small cation (e) through water. In (e) the hydrated ion is built into the water structures by hydrogen bonding and transport requires the breaking and making of hydrogen bonds.

These rates are at least 2 to 4 times those of other small ions (4.0 to 8.0 × 10⁻⁴ cm/sec) even though the radius of the hydrated proton is at least as large (4.5 Å). The explanation lies in the fact that when a proton is part of a hydrogen bond in water, it can migrate† from the donor to acceptor oxygen along the line of the —O—H⋯O— bond as shown in Figure 2.9. This phenomenon results in the proton moving about 0.86Å while the center of positive charge moves 3.1Å. It is the latter which is observed in measurements of proton velocity. A similar mechanism accounts for hydroxyl ion migration (Figure 2.9).

The rate of migration of protons is increased as the degree of order of the system is increased and in ice it occurs more readily than in water, reaching an apparent rate of 1.9×10^2 cm/sec at $-10°C$. In solutions and at interfaces where icelike structures are produced, it is apparent that proton transfers will be greatly facilitated. Such transfers are commonly observed in biological systems and play important roles in the regulation of cellular function and the operation of many enzymes.

2.6 THE BULK PROPERTIES OF WATER

Several of the properties of water make it a unique solvent particularly suited to the support of living systems. After reading this section, compare the properties of H_2O and H_2S listed in Table 2.2 and consider the problems that a biological system might have developing in the latter.

TABLE 2.2 A COMPARISON OF WATER AND HYDROGEN SULFIDE

	H_2O	H_2S
Molecular weight (daltons)	18.015	34.08
Phase at 25°C	Liquid	Gas
Density (g/ml)	1.000	0.00154
Boiling point (°C)	100	-60.7
Freezing point (°C)	0	-85.5
Dielectric constant	78.54 (25°C)	9.26 (-85.5°C)

Heat Capacity

The enormous amount of heat required to change the temperature of water serves to protect biological systems from rapid temperature changes which would influence the rates of all biological processes and in many cases would result in destruction of structural integrity. For example, many proteins undergo relatively large changes in structure over relatively small ranges of temperature. These structural changes almost inevitably result in changes in function, such as an increase or decrease in catalytic ability. If water had a low heat capacity, the temperature of biological systems would fluctuate

† The migration of protons along the —OH⋯O band is referred to as "tunneling." The process takes place without the proton surmounting the energy barrier required for normal bond breaking-bond making. Tunneling is predicted by quantum mechanical considerations.

rapidly with changes in ambient temperature and the rates of many biological reactions (catalyzed by enzymes, all of which are proteins) would be affected.

Heat of Vaporization and Heat of Fusion

These are both very high for water. The high heat of vaporization tends to minimize loss of water by evaporation and to produce a significant cooling when it does occur. Protection from dehydration is an obvious requirement when it is considered that most biological processes occur in solution. The cooling effect of evaporation is also pertinent to the regulation of temperature discussed under heat capacity above. The high heat of fusion of water provides a similar protection against freezing; although in many biological systems subject to extreme cold, specialized antifreeze systems have been evolved.

Dielectric Constant

For processes involving ionization, charge separation, or the development of partial charge, the ability of water to reorient and solvate the charged centers plays an important regulatory role. For example, the force acting to separate two like unit charges is inversely proportional to the dielectric constant of the medium (see Van Holde in this series, Chapter 3). In bulk water this force is much smaller than in less polar solvents. This difference is very important in determining the fine structure of proteins since it is energetically unfavorable for like-charged groups to be "on the inside" where the dielectric constant is probably quite low. Similarly, the activity of all ions decreases as the dielectric constant of the medium increases. For this reason the solubility of ionic compounds increases as the dielectric constant of the medium increases and water is almost unique in its ability to dissolve ionic compounds (see Van Holde, this series). It is the almost universal solvent.

High Viscosity

The viscosity of water is due to its being a hydrogen-bonded continuum. Many biological systems use water as a transport medium (blood, lymph, cell sap), and its viscosity, which of course is influenced by the solutes in it, allows it to flow and be pumped in a controlled way.

Surface Tension and Adhesion

All liquids tend to exhibit the smallest possible surface because their surface tension acts against the force of gravity which is acting to spread them out. Water has a relatively high surface tension, but it also has a great capacity to wet surfaces, that is, to solvate or bond to them). This is particularly obvious when the rise of water in a capillary is considered. The strong adhesive force between water and the material of the capillary and the strong cohesive force between water molecules at the surface act to produce the upward flow of water through capillaries of soils and plants. The presence in all membranes of lipid materials, which are not wetted by water (low adhesive force), is probably important in the control of water flow across membranes.

2.7 THE CHEMICAL PROPERTIES OF WATER

Water is involved as a reactant or as a product in many biologically important reactions. Some of them are listed in Table 2.3 along with approximate values of ΔG^0 for the process in water at 25°C. When it is recognized that these values of ΔG^0 reflect the assumption that the concentration of water is $1.0M$, when in fact it is $55.6M$, its role in determining the position of the equilibria is seen to be very important.

The role of water in determining the direction in which many biological reactions *tend* to go is very important. For example, the formation of ethyl acetate from ethyl alcohol and acetic acid at 25°C has $K_{eq} \cong 1$ and

$$CH_3CH_2OH + CH_3C \overset{O}{\diagup}_{\diagdown OH} \rightleftharpoons CHC \overset{O}{\diagup}_{\diagdown OCH_2CH_3} + H_2O$$

liquid   liquid   liquid   liquid

$\Delta G^0 = -RT \ln K = 0$ kcal/mole. If ethyl alcohol, acetic acid, ethyl acetate, and water are mixed in equimolar proportions, the system will be quite stable. If, on the other hand, an aqueous solution $1 \times 10^{-3} M$ in ethyl alcohol, $1 \times 10^{-3} M$ in acetic acid, and $1 \times 10^{-3} M$ in ethyl acetate is considered, ΔG for the formation of ethyl acetate is

$$\Delta G^0 + RT \ln \frac{[10^{-3}][10^{-3}]}{[10^{-3}][55.6]} = +6.47 \text{ kcal/mole}$$

In this system ethyl acetate will hydrolyze almost completely. Its final concentration will be approximately $8 \times 10^{-7} M$.

Water is amphoteric; it can act as an acid and as a base and plays both roles in chemical reactions. For example, it serves as a weak base in the ionization of weak acids such as acetic acid

$$RCOOH + H_2O \rightleftharpoons RCOO^- + H_3^+O$$

acid   base   base   acid

and in the removal of protons in the general acid catalysis of enolization.

$$CH_3-\overset{O}{\overset{\|}{C}}-CH_3 + H^+ \rightleftharpoons CH_3-\overset{^+OH}{\overset{\|}{C}}-CH_2-H:O\overset{H}{\diagup}_{\diagdown H} \rightleftharpoons CH_3-\overset{OH}{\overset{|}{C}}=CH_2 + H_3O^+$$

TABLE 2.3 REACTIONS INVOLVING WATER

H_2O	Approximate $-\Delta G^{0\prime}$ at pH 7.0 kcal/mole	Catalyst
$+\ RC(=O)(NR'H) \rightleftharpoons RC(=O)(OH) + H_2NR'$	3	H^+, OH^-
$+\ RC(=O)(OR') \rightleftharpoons RC(=O)(OH) + HOR'$	3	H^+, OH^-
$+\ RC(=O)(SR') \rightleftharpoons RC(=O)(OH) + HSR'$	7	H^+, OH^-
$+\ ROP(=O)(OH)(OH) \rightleftharpoons ROH + HOP(=O)(OH)(OH)$	3	H^+
$+\ ROP(=O)(OH)OP(=O)(OH)(HO) \rightleftharpoons ROP(=O)(OH)(OH) + HOP(=O)(OH)(OH)$	7	H^+
$+\ \underset{}{}C(\!-\!O\!-\!)(\!-\!O\!-\!)(OR') \rightleftharpoons C(\!-\!O\!-\!)(\!-\!O\!-\!)(OH) + HOR'$	3	H^+
$+\ \ C{=}O \rightleftharpoons C(OH)(OH)$	3	H^+, OH^-
$+\ \ C{=}NR \rightleftharpoons C{=}O + H_2NR$	3	H^+

Water serves as an acid in the protonation of ammonia and in the catalysis of mutarotation it serves as both an acid and a base.

$$\underset{\text{base}}{NH_3} + \underset{\text{acid}}{H_2O} \rightleftharpoons \underset{\text{acid}}{NH_4^+} + \underset{\text{base}}{OH^-}$$

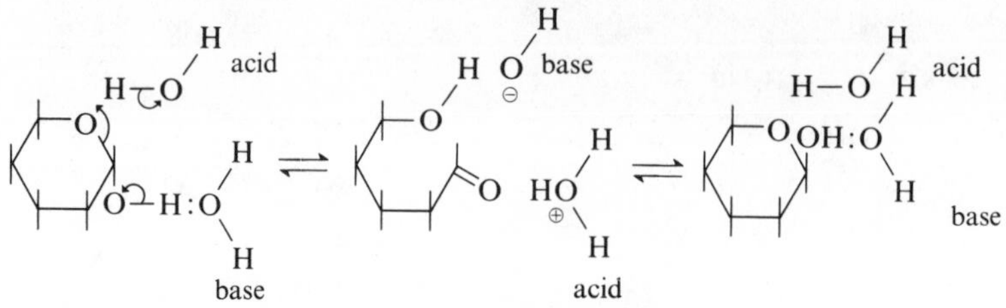
acid
base
acid
base
acid
base

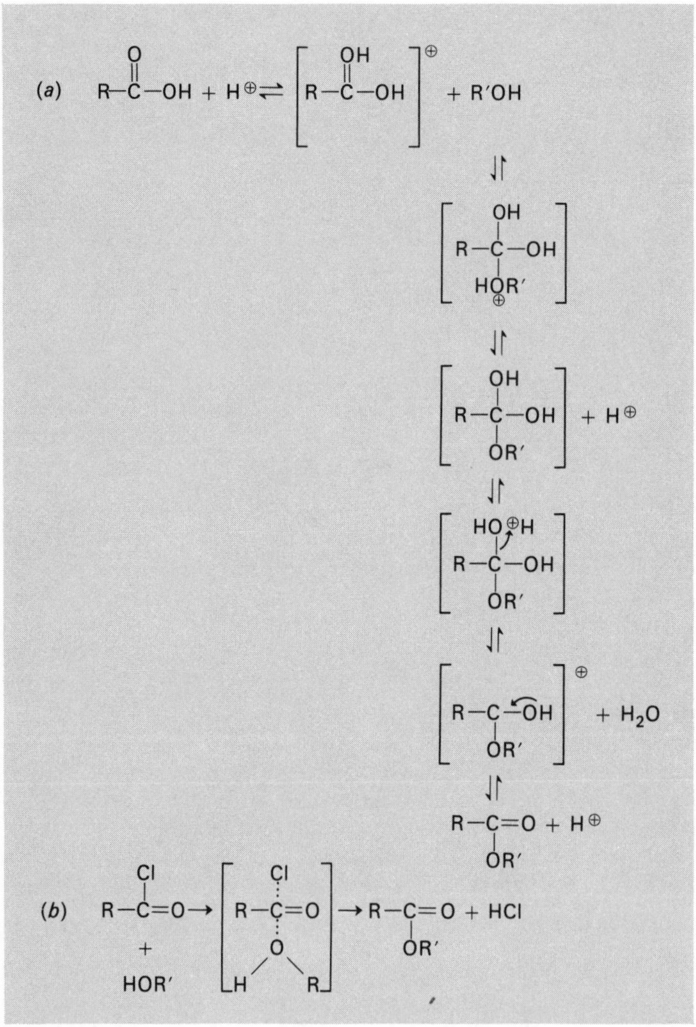
(a)
(b)

Figure 2.10

Ester formation. (*a*) The acid-catalyzed reaction of an alcohol with an acid in which water is a leaving group; the reaction is quite sluggish. (*b*) The reaction of an alcohol with an acid anhydride in which chloride ion is the leaving group; the reaction is usually very rapid.

That water is important as a nucleophile is implied by the roles it plays in the reactions listed in Table 2.3. It is not as potent a nucleophile as many of those known, but this is compensated for by its suitability as a solvent. Some common nucleophiles arranged in order of increasing effectiveness are ClO_4^-, H_2O, F^-, Cl^-, HPO_4^{2-}, Br^-, and CN^- and a typical nucleophilic reaction is shown below.

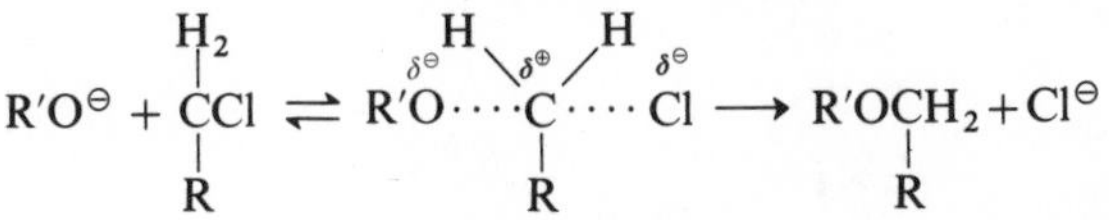

$$R'O^\ominus + \overset{\overset{H_2}{|}}{\underset{\underset{R}{|}}{C}}Cl \rightleftharpoons R'\overset{\delta\ominus}{O}\cdots\overset{H}{\underset{\underset{R}{|}}{\overset{\delta\oplus}{C}}}\overset{H}{\cdots}\overset{\delta\ominus}{Cl} \longrightarrow R'O\underset{\underset{R}{|}}{CH_2} + Cl^\ominus$$

Water is not a very good leaving group and reactions which could occur by the protonation of a hydroxyl group to form a water molecule usually are expedited by replacing the hydroxyl group with a better leaving group. For example, the formation of esters can be accomplished by an acid-catalyzed reaction involving water as a leaving group. It is much more readily achieved, using an acyl halide or, in biochemical systems, an acyl phosphate or a thioester (Figure 2.10). These few examples are given to focus on the importance of water as a reactant in biological systems and to illustrate the relevance of knowledge of its properties, structure, and chemistry to a full understanding of biological systems. There is a lot to be done.

REFERENCES

Eisenberg, D., and W. Kauzman, *The Structure and Properties of Water*, Oxford University Press, 1969. A physical–chemical approach to the water problem in which sufficient background material is presented to allow the presentation to be followed by those who have had a first course in physical chemistry.

Kavanau, J. Lee, *Water and Solute-Water Interactions*, Holden-Day, Inc., San Francisco, Calif. 1964. This short monograph gives a good coverage of the theories of water structure and their experimental basis. Many references to the original literature are given. The author is interested primarily in biological membranes and this text reflects his interest.

Lippincott, E. R., R. R. Stromberg, W. H. Graut, and G. L. Cessac, "Polywater," *Science*, **164,** 1482 (1969). An interesting new type of water is described which has a very low vapor pressure, solidifies at −40°C, is stable to temperatures of approximately 500°C, and has a density of 1.4 g/cm^3. The material behaves like a polymer. Whether or not this type of water plays a role in biological systems remains to be seen, but its description presents an entirely new dimension for consideration in studies of water structure. There is some disagreement concerning the existence of polywater, and the reader should consult also, Roussea, D. I., and S. P. S. Porto, "Polywater: Polymer or Artifact," *Science*, **167,** 1715 (1970).

Symposium on Cryobiology, *Federation Proceedings*, **24,** No. 2, Part 3, Supplement No. 15 (1965). A collection of papers describing water models and the ways in which water structure influences biological activity.

THREE | SYMMETRY AND ASYMMETRY

A casual observer of nature might quickly conclude that there is a tendency for almost everything to develop symmetrically. Most animals appear to possess bilateral symmetry (that is, are symmetrical about their midline) and most plants when not deformed by competition with other plants or their environment develop almost cylindrical symmetry with only a slight bias in the north–south direction. The tendency to be symmetrical is even more apparent in simpler organisms. This outward symmetry might lead to the conclusion that the building blocks of nature are symmetrical. Although there are some tendencies for symmetry to be maintained, however, the majority of molecular biological processes involve interactions between asymmetric molecules. The sugars, amino acids, lipids, and nucleic acids all possess varying degrees of asymmetry. It is important to recognize these and to be alert to their role in chemical processes.

In this chapter the types of asymmetry encountered in biological systems are discussed. In addition, the possibilities for discriminating between similar groups in symmetrical molecules are considered along with examples of the role that such discrimination plays in the conversions which occur in cells. The question of identity of similar groups is also important in nuclear magnetic resonance spectroscopy since like groups in identical environments have identical chemical shifts whereas like groups in magnetically nonidentical environments have different chemical shifts. Finally, a very brief treatment is given of the symmetry elements and point groups, a classification into which all objects can be fitted. This last section is presented not because it is commonly used by biochemists but in the hope that the student will spend about an hour working through the presentation and will recognize that it offers a completely logical approach to the consideration of questions of symmetry (or lack of it). Since such questions are components of almost all biochemical problems, an awareness of the existence of the system and some facility in using it may be worthwhile.

3.1 THE ASYMMETRIC CARBON ATOM

To facilitate communication of chemical information, formulations have been developed which allow chemical structures to be represented in two dimensions (as on a sheet of paper or a chalkboard). Frequently the student develops such facility and familiarity with these representations that he does not need to translate them back into the three-dimensional objects which they stand for. The development of this skill is perhaps advantageous in the context of examinations but can be detrimental in the context of the research laboratory or in those areas of chemistry where asymmetry is common. Facility in thinking of molecules—as they are—in three dimensions should be pursued. This pursuit is greatly aided by the use of models. Both space-filling and framework models are very useful in helping "see it like it is" and the use of the latter will be very helpful in understanding this section. In addition, questions concerning molecular geometry and interaction are often most readily answered by model building and visual comparison.

The usual discussion of symmetry in biochemistry centers around the tetrahedral carbon atom and the fact that when attached to four different substituents it is totally lacking in symmetry and is termed *asymmetric*. An asymmetric carbon atom is not superposable on its mirror image. As shown in Figure 3.1, mirror images can be constructed by drawing a line to represent the reflecting surface and by projecting each group along a line perpendicular to the surface up to it and an equal distance out the other side; for example, a is moved as shown by the dotted lines in the figure. Isomers which are mirror images of each other are *enantiomers*.

In any specific case, such as glyceraldehyde [Figure 3.1(b)], the two isomers have identical physical and chemical properties with two notable exceptions. First, they rotate plane-polarized light equally but in opposite directions (see Van Holde this series, Chapter 10); second, they will react at different rates with asymmetric reagents. This second point is considered in more detail later in this chapter. It should be pointed out now, however, that essentially all biochemical transformations are catalyzed by a group of substances termed *enzymes* which are proteins and highly asymmetric (see Chapter 4). They are usually capable of interacting with only one of a mirror-image pair.

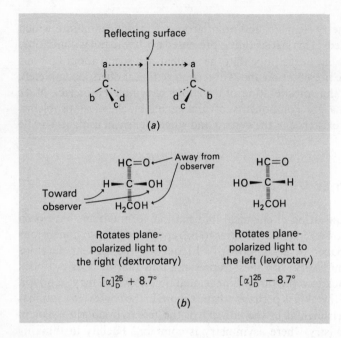

Figure 3.1

The asymmetric carbon atom. (*a*) An asymmetric carbon atom and its mirror image. (*b*) Enantiomers of glyceraldehyde showing the symbols used to represent groups toward and away from the observer. $[\alpha]_D^{25}$ is the specific rotation at 25° using light at the D line of sodium.

The usual shorthand representation for asymmetric carbons originated by Fischer is shown in Figure 3.2. When this type of representation is used, the three-dimensional implications must be clearly recognized. Note that the Fischer representation can be rotated only in the plane of the paper. Rotation out of the plane is equivalent to conversion to the enantiomer (Figure 3.2). Exchanging two groups is also equivalent to converting to the enantiomer.

There are several ways of specifying arrangements about an asymmetric carbon atom.

1. D *and* L

The reference substance is D-glyceraldehyde represented below in the Fischer convention.

$$
\begin{array}{ccc}
\text{HC=O} & & \text{HC=O} \\
\text{H-C-OH} & & \text{HO-C-H} \\
\text{H}_2\text{COH} & & \text{H}_2\text{COH} \\
\text{D-glyceraldehyde} & & \text{L-glyceraldehyde}
\end{array}
$$

With the hydroxyl group to the right, the compound is D; its enantiomer is L-glyceraldehyde. These configurations have been established by X-ray crystallographic analysis of compounds whose relationship to the glyceraldehydes has been unequivocally established.

It is difficult to apply the D-L system to molecules having more than one asymmetric center. The carbon chain is numbered according to the usual

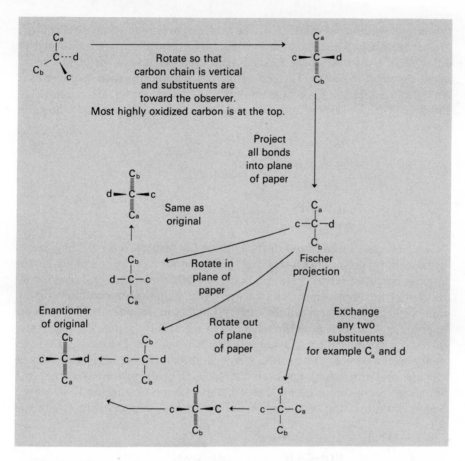

Figure 3.2   The Fischer convention for representing asymmetric carbon atoms. Note that since the convention requires the carbon chain to be away from the observer, structures can be written with the carbon chain horizontal.

conventions and arranged vertically with the lowest numbered carbon up. The configuration at each center is then specified. Some examples are given below. The monosaccharide series (which is discussed in Chapter 6) consists

2L,3D-dibromobutane

3D,4L-dihydroxy-
4D-methyl tetrahydrofuran

3L-methylhexan-3D-ol

of a homologous series of which D- and L-glyceralde are the simplest members. Higher members which may have several asymmetric centers are related to either D- or L-glyceraldehyde through their configuration at the highest numbered asymmetric center. The rest of the configuration is then specified by a trivial name and a rather large number of trivial names are required. D-Erythrose is shown below and the complete set of trivial names are shown in Figure 6.2.

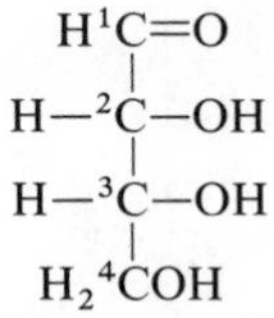

$$H^1C{=}O$$
$$H{-}^2C{-}OH$$
$$H{-}^3C{-}OH$$
$$H_2{}^4COH$$

D-erythrose

In some cases, application of the D-L system developed for naming carbohydrates runs afoul of the D-L system developed for naming amino acids. The latter is based on serine (Figure 3.3) and the other members of the family for which it is the simplest member are named from it on the basis of the configuration of the group adjacent to the carboxyl function. Thus the two systems use opposite ends of the molecule to establish the family relationship. Consider threonine, by convention it is designated L, but it could be named as a D compound in the carbohydrate series. There is no contention at present over the naming of L-threonine; however, the tartaric acids present a similar but presently unresolved problem. The structures for the various forms are shown below. Note that the two forms in the center are identical since if either is

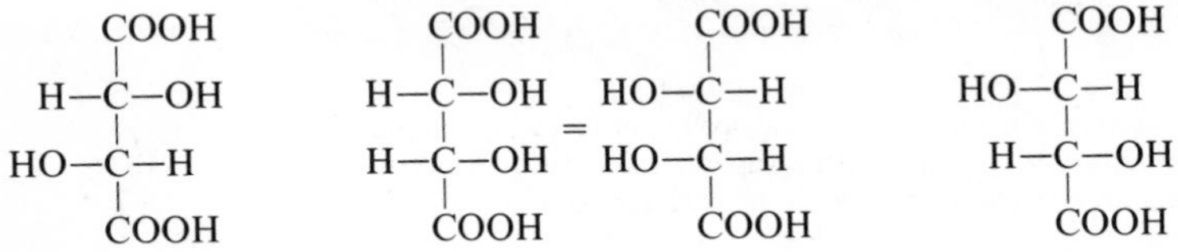

```
        COOH          COOH     COOH              COOH
         |             |        |                 |
     H—C—OH        H—C—OH   HO—C—H           HO—C—H
         |             |   =    |                 |
    HO—C—H         H—C—OH   HO—C—H            H—C—OH
         |             |        |                 |
        COOH          COOH     COOH              COOH

[α]²⁰_D    +12 deg           0 deg              −12 deg
```

rotated by 180 deg *in the plane of the paper*, the other form is obtained. This form is termed *mesotartaric acid* and has no optical activity since it is superposable on its mirror image. The other tartaric acids present a problem. They can be designated + and − on the basis of their observed optical activity. If they are named by the carbohydrate system, they are L (on the left) and D, respectively. If they are named by a system similar to that used for the amino acids, however, they will bear the opposite designation. Both systems have been used and since tartaric acids were studied more than 100 yr ago by Pasteur and have played an important role in the establishment of the absolute configuration of asymmetric compounds, the number of references in the literature to the tartaric acids is large. It would be best to draw the structure or specify

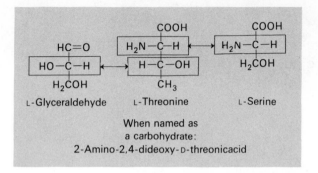

Figure 3.3

The relationship between the D–L systems for naming carbohydrates and amino acids.

the sign of the rotation in discussing these compounds. The subscripts G for glyceraldehyde and S for serine have also been suggested. For example, (+) tartaric acid is L_G- or D_S- tartaric acid.

Although the D-L system is difficult to apply and requires acceptance of trivial names and arbitrary relationships, it is used in dealing with most of the stereochemistry of the simple carbohydrates, amino acids, lipids, and nucleic acids. Its application to these compounds is described in the chapters devoted to their chemistry.

2. R and S

This is a much more generally applicable method for designating configuration. The major features are described here but its application to more complex systems requires the use of further conventions. The method is described in detail in Eliel's book referenced at the end of this chapter.

The substituents about an asymmetric center are arranged in order of decreasing atomic number of the atom directly attached to the asymmetric center. If two or more of these atoms have the same atomic number, their substituents are considered. The group with substituents of highest atomic number takes precedence. When a multiple bond to a substituent is encountered, the substituent is arbitrarily considered to occur twice or thrice; for example,

$$C=O \text{ is taken as } C\overset{O}{\underset{O}{\diagup}} \quad , \quad C=N \text{ is taken as } C\overset{N}{\underset{N}{-}}N.$$

If, in applying this rule, a "natural" doubly or triply substituted atom is being compared with one which is generated by the rule, the "natural" one takes precedence. Higher mass number isotopes take precedence; for example, ^2H precedes ^1H.

When the substituents have been ranked in decreasing order, the smallest is placed away from the observer (behind the asymmetric carbon) and the other

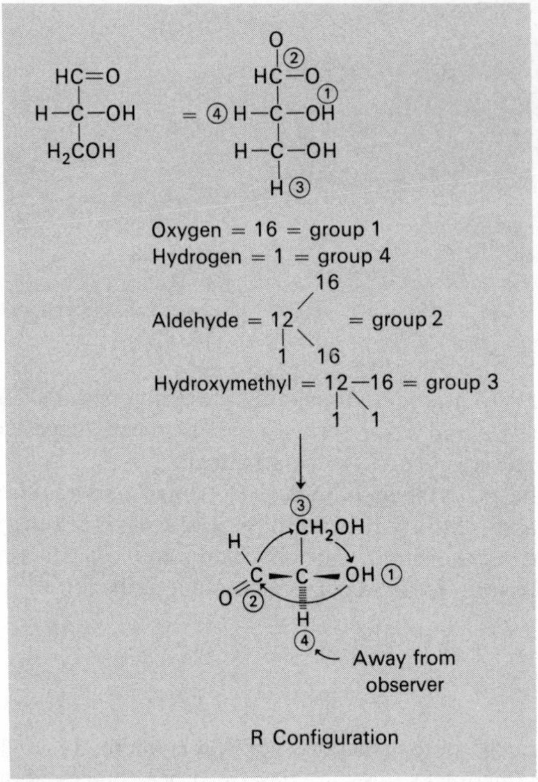

Figure 3.4   Application of the R-S system for specifying asymmetry to glyceraldehyde.

three project toward the observer. If the groups numbered as described above lie in a clockwise (right-handed) array of decreasing atomic number, the center is designated R (rectus). The left-handed array is designed S (sinister).

The application of the R-S system to D-glyceraldehyde is shown in Figure 3.4. To apply the system, carry out the following manipulations. *After* the substituents have been assigned numbers, exchange *two* pairs of substituents so that the highest numbered one occupies the bottom position as shown below.

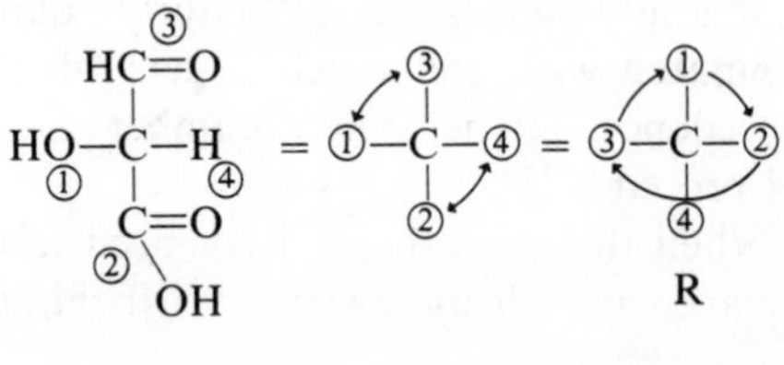

The R-S system works very well and should be used where systems without firmly established sets of conventions and trivial names are encountered. Apply it to the following examples.

$$
\begin{array}{c}
H \diagdown \diagup H \\
C \\
\parallel \\
C \\
| \\
CH_3-C-H \\
| \\
H_2COH
\end{array}
\qquad
\begin{array}{c}
H-C{=}O \\
| \\
H-C-OH \\
| \\
H-COH \\
| \\
H_2COH
\end{array}
\qquad
\begin{array}{c}
COOH \\
| \\
H-C-NH_2 \\
| \\
H_2COH
\end{array}
\qquad
\begin{array}{c}
COOH \\
| \\
H-C-OH \\
| \\
HO-C-H \\
| \\
COOH
\end{array}
$$

3. Plus (+) and Minus (−)

When the absolute configuration of an asymmetric center is not known and when its relationship to another center of unknown absolute configuration is discussed, it is usual to designate them using + and −. These symbols are also used to designate the direction in which an isomer rotates plane-polarized light; + is dextrorotary; − is levorotatory.

Note: Neither D-L nor R-S designations carry any implication concerning the direction of rotation of polarized light. In the older literature the designations *d* or *l* may be found. These usually referred to the dextro- or levorotatory properties of the material.

3.2   MOLECULAR ASYMMETRY

Although the asymmetric carbon atom is of fundamental importance in the production of asymmetric structures, it is also important to recognize that molecules having asymmetric carbon atoms may still possess symmetry and that molecules having no asymmetric carbon atoms may be asymmetric.

Let us first consider compounds which contain more than one asymmetric carbon atom. Since each such atom can have one of two possible arrangements, it follows that for a compound with n asymmetric centers, 2^n different isomeric forms can be written. For example, for 2,3-dihydroxy butyric acid, which has two asymmetric carbons, four structures can be written. They are shown in Figure 3.5. There are two sets of enantiomers (nonsuperposable mirror images); in addition, the relationship between either member of one pair of enantiomers and the members of the other pair can be termed a *diastereomeric* one; that is, A and C, A and D, B and C, and B and D are diastereomers. The term *diastereomer* has usually been applied only to those compounds which are stereoisomers but *not* enantiomers and owe their differences to asymmetric centers. The broader definition, however, of diastereomers as any set of nonenantiomeric stereoisomers is now usually accepted. This definition includes *cis* and *trans*

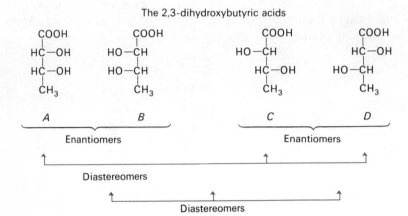

Figure 3.5   The dihydroxybotyric acids showing the four possible isomers, the enantiomeric pairs, and the diastereomers.

compounds as diastereomers, for example, the crotonic acids shown on page 35. The term *diastereoisomers* can be used also.

Turning again to Figure 3.5, if a mixture containing equal amounts of A and B is prepared, it will have no optical activity since the contribution from A exactly counteracts the contribution from B. Such a mixture is termed a *racemic* mixture. It must be clearly distinguished from the *meso* compounds discussed below.

An important "special case" of the generalization concerning the number of isomers which can exist arises when a molecule having asymmetric centers possesses end-to-end symmetry. If an isomer has a plane of symmetry in it such that one half is the mirror image of the other, it must be superimposable on its mirror image. One such case has already been described—the tartaric acids. Some other examples are shown below:

$$
\begin{array}{l}
H_2COH \\
| \\
HCOH \\
| \\
HCOH \\
| \\
H_2COH
\end{array}
$$

3 diastereoisomers can exist

$$
\begin{array}{l}
H_2COH \\
| \\
HCOH \\
| \\
HCOH \\
| \\
HCOH \\
| \\
H_2COH
\end{array}
$$

4 diastereoisomers can exist

$$H_2COH$$
$$|$$
$$HCOH$$
$$|$$
$$HOCH$$
$$|$$
$$HOCH$$
$$|$$
$$HCOH$$
$$|$$
$$H_2COH$$

10 diastereoisomers can exist

One of the isomers in which compensation is observed is shown and the number of isomers possible is stated. Write the structures of the other diastereomers including any others which have a plane of symmetry. Although all the examples are polyhydric alcohols, the same sort of situation will exist regardless of substitution. Compounds which have asymmetric carbon atoms but which are internally compensated are termed *meso* compounds.

The *meso* isomers are cases of compounds which have asymmetric elements but overall symmetry. If we consider their diastereoisomers, however, we find that although they have overall asymmetry as well as asymmetric carbon atoms, they are not totally lacking in symmetry. A three-dimensional representation of D-threitol is shown in Figure 3.6(a) along with the Fischer projection. Neither is superposable on its mirror image and the compound is not symmetrical. If an axis is drawn between carbons 2 and 3 in the plane of the paper, however, the compound can be rotated about the axis by 180 deg to produce a form indistinguishable from the original. This axis is termed a C_2 axis since rotation by 360 deg/2 (180 deg) produces an identical arrangement. It is one of the symmetry elements discussed at the end of this chapter. Its only importance in this context is that it demonstrates the presence of symmetry in a molecule with overall asymmetry as judged by having nonsuperposable mirror images (that is, enantiomers). Other conformations† of threitol are shown in Figure 3.6(b), (c), and (d) in stereo representation and in Newman projection. Each has an axis of symmetry. It is therefore incorrect to say that D-threitol is *asymmetric*; rather, it is *dissymmetric*. The two terms should be used as follows. Asymmetric means totally lacking in symmetry. Dissymmetric means having nonsuperposable mirror images but possessing elements of symmetry.

When the dissymmetric compound D-threitol and its diastereomer erythritol (a *meso* compound) are examined using the R-S convention, D-threitol is seen to be composed of two R arrangements whereas erythritol is composed of an R and an S arrangement. On this basis, threitol would be expected to possess

---

† A conformation is an arrangement of the atoms of a molecule in space which can be achieved by rotation about single bonds. An infinite number of conformations exist for all polyatomic molecules but for most a few conformations are more stable and are dominant. Changes in conformation do not affect configurations. The latter can only be changed by breaking bonds.

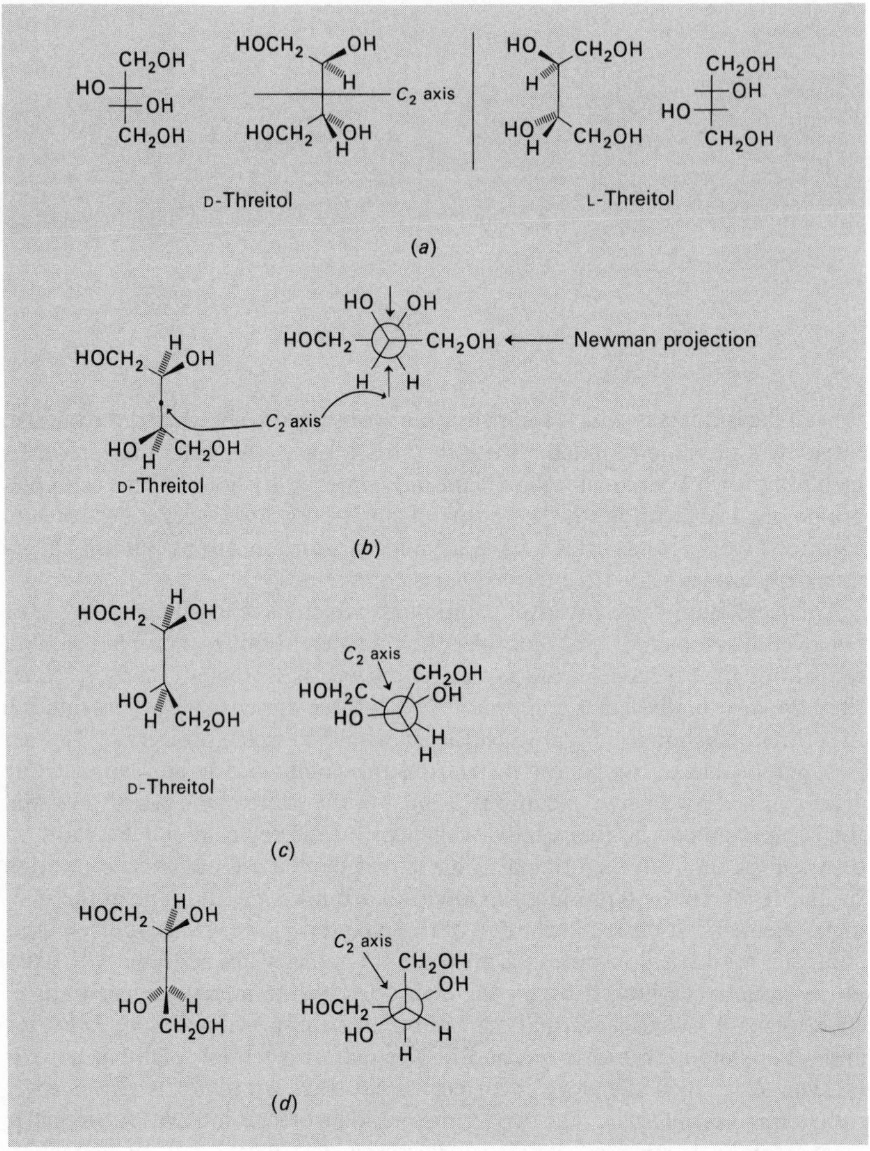

Figure 3.6   D-Threitol, a compound with overall asymmetry but having some elements of symmetry. (a), (b), (c), and (d) show different conformations of D-threitol, each possessing an axis of symmetry.

optical activity (R + R = 2R) and erythritol none (R + S = 0, since R = −S) (Figure 3.7). This approach will prove useful in considering the possibility of discriminating between similar groups in a molecule and will help in deciding if a given isomer is a *meso* compound.

In addition to molecular asymmetry arising from the presence of asymmetric carbon atoms, asymmetry can arise because of the presence of double bonds or rings. The familiar *cis-trans* isomerism possible in ethylenic compounds is illustrated by the crotonic acids. No optical activity is associated with this

$$CH_3\quad COOH \qquad H\quad COOH$$

$$C=C \qquad\qquad C=C$$

$$H\qquad H \qquad\qquad CH_3\quad H$$

cis   *trans*

crotonic acids

type of isomerism. On the other hand, the presence of a ring rather than a double bond produces a type of *cis-trans* isomerism in which enantiomers can be present. The four 2-methylcyclopropanecarboxylic acids are shown below. Although these compounds are termed *cis* and *trans* and seem because

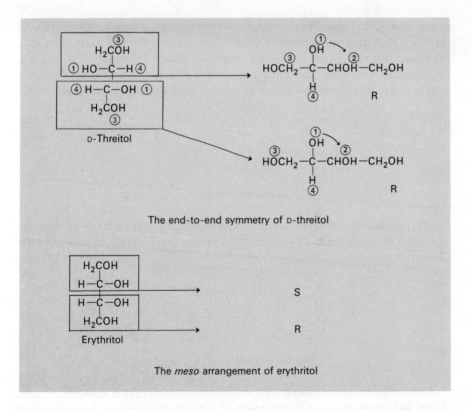

The end-to-end symmetry of D-threitol

The *meso* arrangement of erythritol

Figure 3.7 Application of the R-S system to D-threitol and erythritol showing the end-to-end symmetry of D-threitol and the internal compensation of asymmetric centers in erythritol.

of this terminology to be related to the ethylenic compounds, they are more properly related to the acyclic 2,3-dihydroxybutyric acids discussed above. In fact, they have two asymmetric carbons and no end-to-end symmetry and thus have 2^2 (four) stereoisomers.

Substitution at two positions is necessary to produce an asymmetric carbon

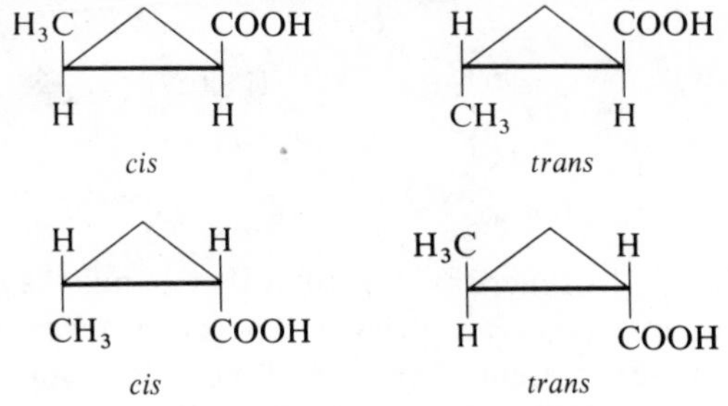

cis   *trans*

cis   *trans*

2-methylcyclopropanecarboxylic acids

in any homocyclic ring. In ring systems having an even number of carbons, if there are only two substituents and they are located directly across the ring from each other, it is always possible to draw a plane of symmetry through the molecule. In this case, the molecule is optically inactive and enantiomers cannot exist. As shown below, however, *cis-trans* isomers can exist.

trans   *cis*

4-methylcyclohexan-1-ol

When substituents are not located symmetrically, the number of isomers predicted by the number of asymmetric carbons are possible as shown by the 2-methylcyclohexan-1-ols.

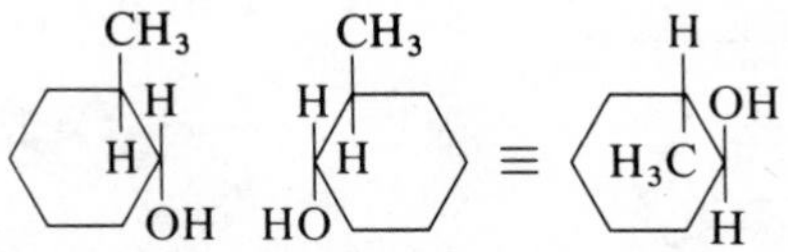

2-methylcyclohexan-1-ol

Another important example of a dissymmetric structure is shown in Figure 3.8(*a*) and (*b*). It is a spiral composed of some homogeneous substance such as a piece of wire. It can exist as nonsuperposable mirror images, but each has an infinite number of C_2 axes through the midpoint of the coil. Note that

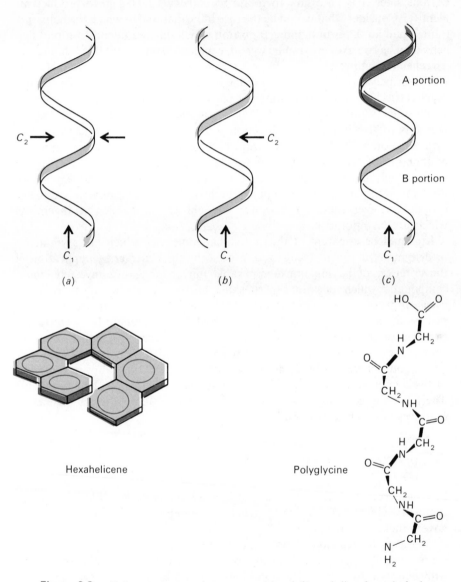

$C_2 \rightarrow$ $\leftarrow$

$\uparrow$
C_1

(a)

$\leftarrow C_2$

$\uparrow$
C_1

(b)

A portion

B portion

$\uparrow$
C_1

(c)

Hexahelicene

Polyglycine

Figure 3.8 Helices as asymmetric structures. (a) and (b) are helices formed of a homogeneous material such as a piece of wire showing the possibility for non-superposable mirror-image structures. (c) is a helix formed of a strand of material with differences in composition at the ends. Hexahelicene is an example of the homogeneous (palindromic) helix; polyglycine, of the non-palindromic helix.

what might be considered the major axis, that is the longitudinal axis, is a C_1 axis since it is necessary to rotate by $360 \deg/1$ ($360 \deg$) to produce an identity. Consider what would be the case if the material of which the helix was formed had an A portion and a B portion which differed in constitution. The helix then has no axes of symmetry greater than C_1, and it is totally asymmetric. The chemical equivalent of a helix with C_2 axes is known to exist in hexahelicene. It is described as palindromic, that is, having end-to-end symmetry. The totally asymmetric helix (nonpalindromic) is present in many biopolymers [Figures 4.16(b) and 8.37(a)]. All that is necessary to cause a coil to become totally asymmetric is to give it a sense of direction so that the ends can be distinguished from each other. Helices formed from polymers containing unique sequences of amino acids and nucleotides are totally asymmetric, but so is a helix made from polyglycine or polyadenylic acid since each monomer has a sense of direction. Only helices with complete uniformity in the monomers are palindromic and even these can be coiled in right- or left-handed arrangements which are enantiomeric.

It should be apparent at this point that molecules which are asymmetric or dissymmetric can be recognized to be so by the observer by application of the criterion of having nonsuperposable mirror images. Conversely, those compounds which possess superposable mirror images are symmetric. In *chemical* processes occurring in biological systems, however, the *reagent* must possess the ability to discern asymmetry and to discriminate between optical isomers. In addition, the fact that there are differences in reactivity between similar groups such as primary and secondary hydroxyl groups is clear, but what about differences between the two primary hydroxyl groups in D-threitol or those in erythritol (Figure 3.7)? Is it possible to distinguish between them? The following section deals briefly with these questions, using a number of examples of significance in biological systems.

3.3   REACTIVITY AND SYMMETRY

In considering interactions of all kinds between molecules, the changes in symmetry which occur during the reaction are important since they frequently govern the ease of reaction.

Consider the reaction of both enantiomers of a tetrahedrally asymmetric compound, for example, 2-*O*-methyl-D- and L-glyceraldehyde dimethyl acetal with benzoyl chloride under conditions which will facilitate reaction of the primary hydroxyl group (Figure 3.9). Both compounds will react equally well and if a mixture were partially benzoylated, the product would contain equal amounts of the D and L esters, which would be identical except for the direction in which they rotate plane-polarized light. If the same D + L mixture well and if a mixture were partially benzoylated, the product would contain however, the two compounds formed would possess distinctly different properties; they would be diastereomers (Figure 3.9). Not only can the products

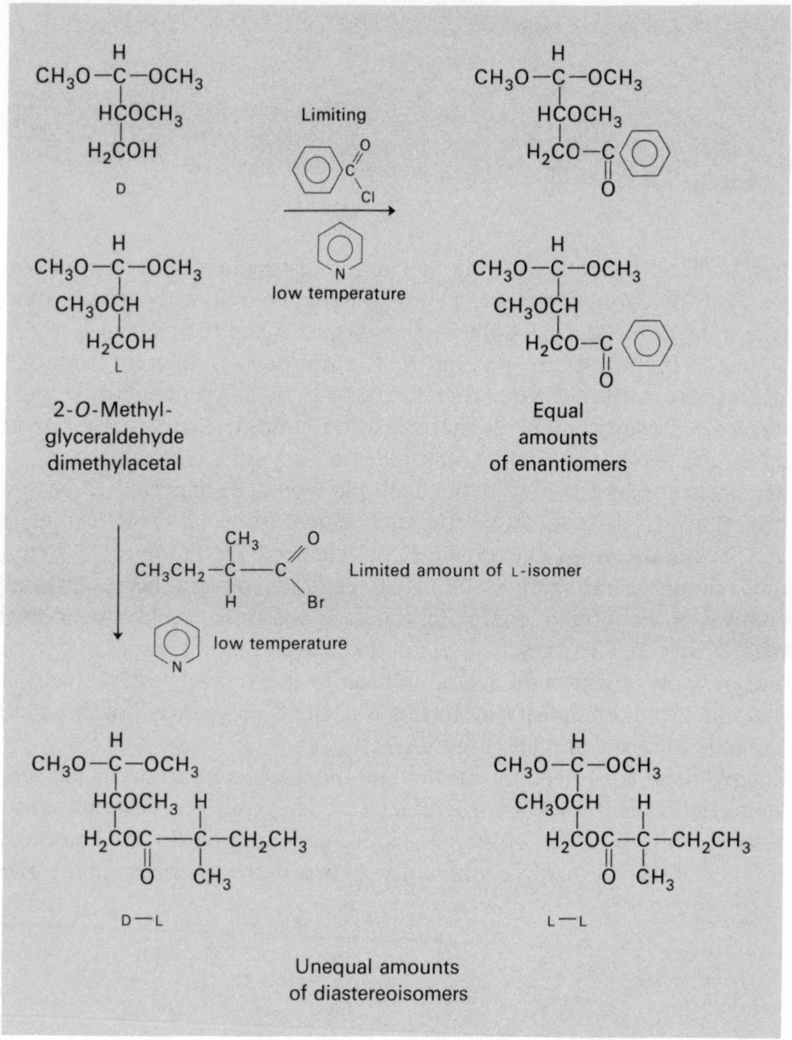

Figure 3.9 The reaction of a racemic mixture with symmetric and asymmetric reagents.

of this type of reaction be distinguished, and in some cases separated, the rate at which the reaction occurs may well differ for each enantiomer. This can be explained as follows: As a reagent approaches the hydroxyl group of the D-glyceraldehyde derivative, it enters an environment which is asymmetric and which is the enantiomer of the environment met on approaching the L compound from the same direction as shown below. If the reagent is itself

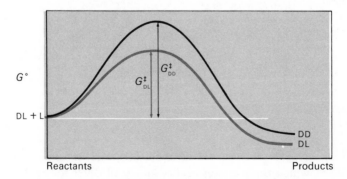

symmetric, such as benzoyl chloride, it cannot distinguish between the two situations. The reaction, therefore, proceeds equally well with either enantiomer (the transition states have the same energy). When the reagent is asymmetric, however, the transition state for the D enantiomer is different from that formed with the L enantiomer and, therefore, possesses different energy (Figure 3.10). Since the energy content of the transition state determines the rate of the reaction, the two enantiomers could react at very different rates.

The asymmetric reagent used in the example would be expected to have a fairly low ability to discriminate between D and L forms. Its own center of asymmetry is remote from the position of reaction. In biological systems, even when simple compounds such as glyceraldehyde are being utilized, discrimination is essentially complete. This is possible because biological catalysts (enzymes) are extremely asymmetric. Although enzymes catalyze a reaction at only one position on a small molecule, they usually interact with it (bind it) at more than one position. In this way the "recognition" of the small molecule can be almost complete (Figure 3.11).

In principle there is no need for more than one interaction to occur between an asymmetric reagent and an asymmetric compound for discrimination between enantiomers to be observed. For discrimination to be complete, however, there must be a large difference between the energies of the two

Figure 3.10   A hypothetical activation diagram for the reaction of an asymmetric reagent (L) with enantiomers (DL). $\Delta G^{\ddagger}_{DD}$, the activation energy for the formation of product DD, is significantly greater than $\Delta G^{\ddagger}_{DL}$ and discrimination will be significant. The *rate* of formation of DL product will be greater than the *rate* of formation of DD product.

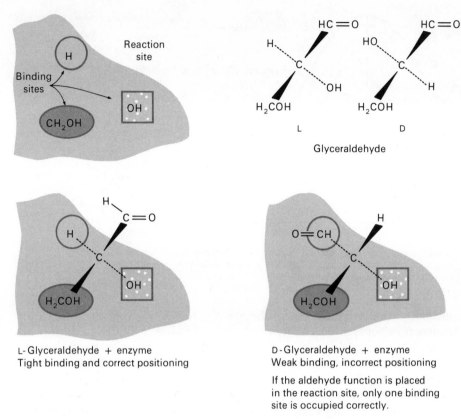

Glyceraldehyde

L-Glyceraldehyde + enzyme
Tight binding and correct positioning

D-Glyceraldehyde + enzyme
Weak binding, incorrect positioning

If the aldehyde function is placed
in the reaction site, only one binding
site is occupied correctly.

Figure 3.11   The reaction of an asymmetric substance with an enzyme having multiple
binding sites.

possible transition states. This energy difference is significantly enhanced
if more than one interaction with the reagent is required before, or for, reaction.
It is possible that discrimination can occur without physical contact between
the asymmetric reagent and the asymmetric reactant since the energy developed
in the approach of the reactants to each other should be different at distances
greater than bond lengths.

In the case of enzyme-catalyzed reactions, the asymmetric reagent is the
enzyme. It is involved only transiently in the overall process and the reaction
itself may appear to involve a symmetrical agent such as H_2. In these cases,
however, the hydrogen (or symmetric agent) is also incorporated into the
asymmetric reagent and can be added asymmetrically.

The enzymes, the biochemical catalysts, are polymers of amino acids. They
range in molecular weight upward from several thousand daltons. Of importance to us here is that they are highly specific because they provide binding
and catalytic sites which are precisely positioned to accept the molecule(s)
with which they react (the substrates) (Figure 3.11). In doing this, they

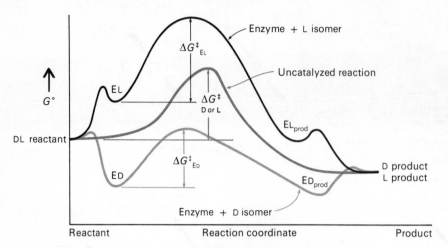

Figure 3.12   A hypothetical activation diagram for an enzyme-catalyzed transformation. Enzyme E binds D much more tightly than it does L. In addition, the $\Delta G^{\ddagger}$ for ED is much lower than that either for EL or for L or D alone.

effectively eliminate the tendency for side reactions to occur and allow for discrimination between very closely related compounds (see Figure 3.12). For example, almost all naturally occurring carbohydrates are related to D-glyceraldehyde and almost all amino acids are related to L-serine. When enzymes which operate on these forms are examined, they are usually found to be incapable of reaction with enantiomeric forms and often bind those forms only very poorly. They may even fail to react with diastereomers in which the differing asymmetric site is remote from the site of reaction. The enzymes owe their specificity to the presence of charged groups, crevices, surfaces and attractive interactions which provide an environment complimentary to the shape and chemical composition of the substrate molecule. In addition, when the substrate molecule is bound, acidic and basic residues and other catalytic residues are precisely located to provide for an easy passage of starting material to product. They also possess another property, flexibility, which may be very important in the binding and catalytic processes but usually they are not sufficiently flexible to allow for the "wrong" chemical even to occur.

We have established that asymmetric compounds can be distinguished by asymmetric reagents. Is it possible to distinguish between similar groups in compounds which have overall symmetry?

3.4   INTRAMOLECULAR SYMMETRY

The fact that a whole molecule is symmetrical does not mean that all of its components are also symmetrical. This is well illustrated in the case of erythritol which has been shown (Figure 3.7) to be composed of two asymmetric

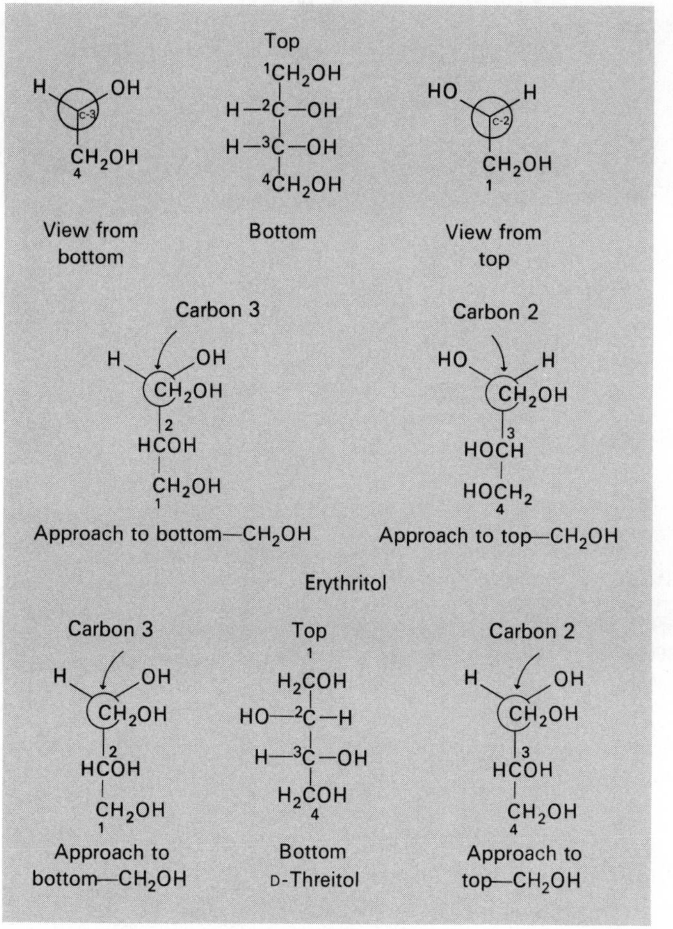

Figure 3.13

The views from opposite ends of erythritol and D-threitol showing the possibility for discrimination in the former and the lack of it in the latter.

parts. As shown in Figure 3.13, an observer viewing an erythritol molecule from one end or the other sees two different arrangements of the groups (H, OH, and CH_2OH). He can, therefore, distinguish between the two ends of the molecule, and if he approaches any of the groups, he will have a different group to his right or left. For example, if he approaches the primary hydroxyl groups, he will see the two arrays shown in Figure 3.13. Similarly, an asymmetric reagent can discriminate between these two arrangements. There are three groups available for interaction in this case, just as there were in the case of the fully asymmetric compound D-glyceraldehyde. It is interesting to compare the possibility for discrimination by an asymmetric reagent between the ends of erythritol with the possibility for a similar reagent to distinguish between the ends of D-threitol. Recall that D-threitol is dissymmetric and nonsuperimposable on its mirror image, L-threitol, and that an asymmetric

Point group	Compound	Distinguishable groups	Isotopically substituted derivative	Point group
C_s	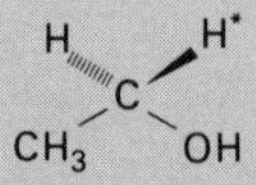 (structure: H, H*, C, CH₃, OH)	H and H*	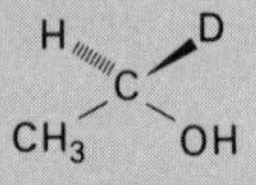 (structure: H, D, C, CH₃, OH)	C_1
C_s	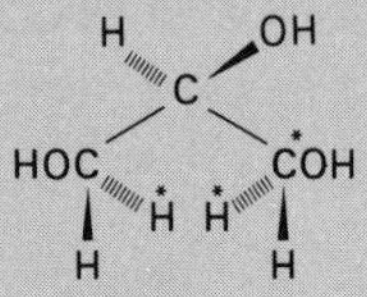 (structure: H, OH, C, HOC, COH, H, H*, H*, H)	C and C* or H and OH groups on them H and H*	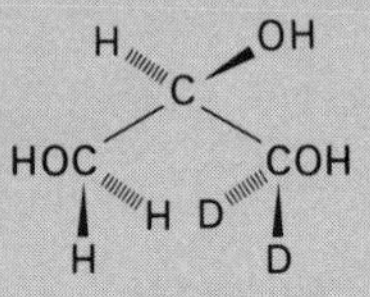 (structure: H, OH, C, HOC, COH, H, D, D)	C_1
C_s	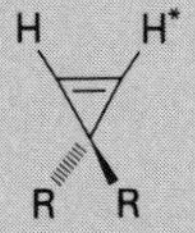 (cyclopropene structure: H, H*, R, R)	H and H*	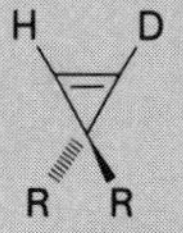 (cyclopropene structure: H, D, R, R)	C_1
C_{2v}	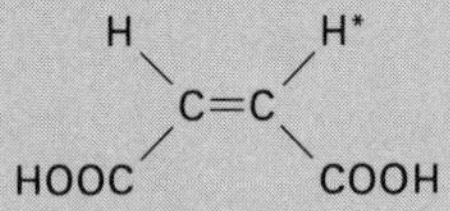 (structure: H, H*, C=C, HOOC, COOH)	None	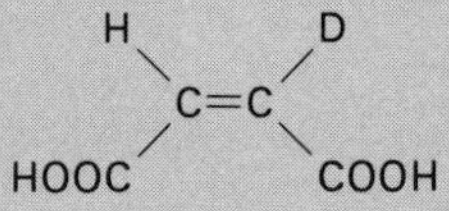 (structure: H, D, C=C, HOOC, COOH)	C_s
C_s	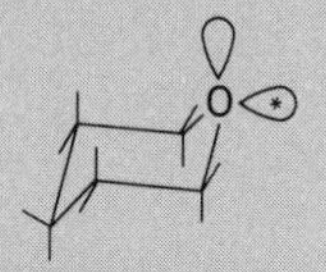 (ketone/cyclohexanone structure: O and O⊙)	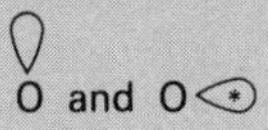 O and O⊙		
C_{2h}	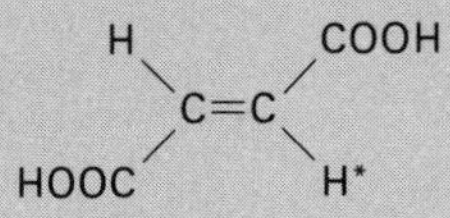 (structure: H, COOH, C=C, HOOC, H*)	None	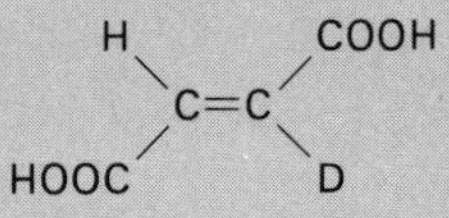 (structure: H, COOH, C=C, HOOC, D)	C_s

Figure 3.14 Examples of compounds possessing overall symmetry; some of these contain similar groups that can be distinguished from each other.

reagent can discriminate between them; however, there is *no* possibility for such a reagent to distinguish between the ends, or any of the pairs of similar groups, of the threitol molecule. The projections in Figure 3.13 show that the ends are identical. A number of molecules which possess groups or surfaces which can be involved in asymmetric synthesis are shown in Figure 3.14. It is sometimes helpful in considering this question of identity between groups to convert one of them (figuratively) to an isotopic or marked group to see if this produces an asymmetric group. If it does, then discrimination by an asymmetric reagent is possible.

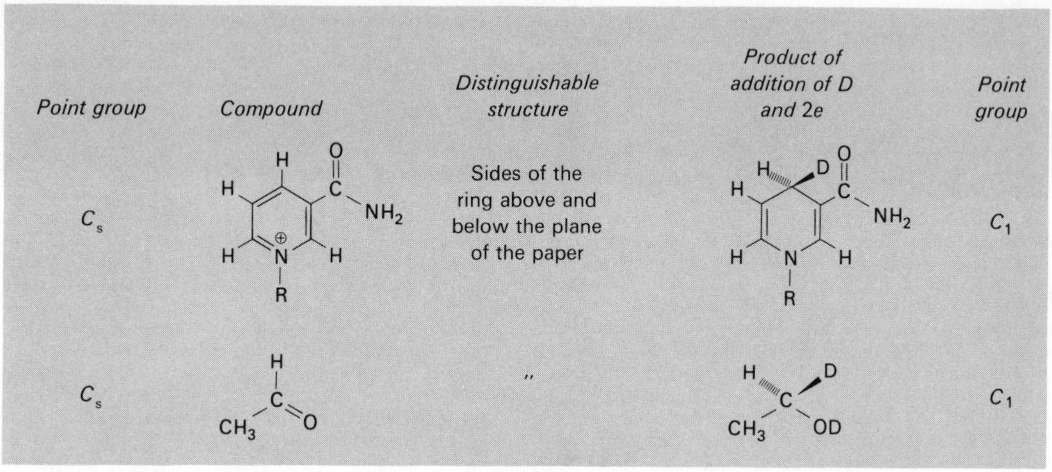

Point group	Compound	Distinguishable structure	Product of addition of D and 2e	Point group
C_s		Sides of the ring above and below the plane of the paper		C_1
C_s		,,		C_1

Figure 3.15 The addition of deuterium and electrons to accomplish the stereospecific reduction of nicotinamide and acetaldehyde.

The possibility for distinguishing between groups in a molecule does not require tetracovalent carbon. Only two groups in addition to the reaction site are necessary and the two sides of planar molecules can often be distinguished. Two important biochemical examples are given in Figure 3.15. The reduction of nicotinamide derivatives is always stereospecific in biological systems, as is the reoxidation. In addition, the reduction of aldehydes to alcohols is stereospecific.

There are few cases in which compounds are handled symmetrically in biological systems and when they are, as in the conversion of fumaric acid to L-malic acid, there is good reason. In this case, discrimination of the two

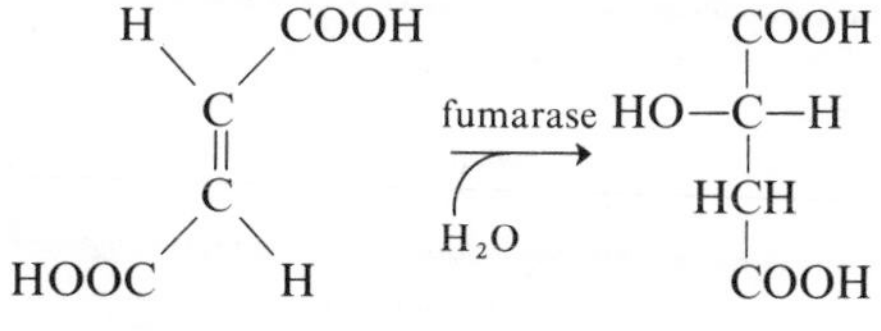

ends of fumaric acid even by a highly asymmetric enzyme is not possible. No matter which way the fumaric acid fits on the enzyme surface, however, the subsequent reaction with water is totally stereospecific and only L-malate is formed!

The term *enantiotopic* has been suggested for atoms or groups in a molecule which reside in enantiomeric environments, for example, the two α hydrogens of ethanol (Figure 3.14).

A simple rule has been proposed by Hirschmann for deciding whether it is possible to discriminate between "identical" groups in a molecule. The rule

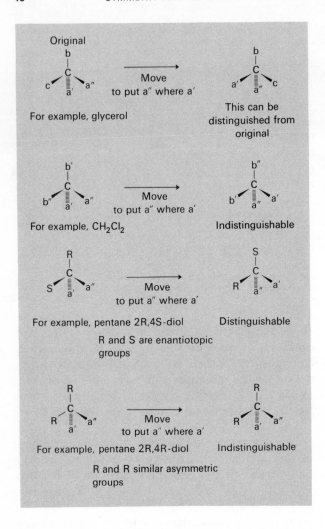

Figure 3.16

The application of Hirschmann's rule to decide which groups are distinguishable from similar groups in a molecule.

is applied by designating the identical groups as a′ and a″ and other substituents as b, c, ..., and so forth. A three-dimensional model or representation of the compound is turned so that a″ coincides with the original position of a′. If the model is now indistinguishable from the starting arrangement, it is impossible to discriminate between a′ and a″. Some examples are given in Figure 3.16. In effect this rule is a formalism of the advice given at the beginning of this chapter—MAKE A MODEL.

An important final example is that of succinic acid. The compound appears to be totally symmetric but it turns out that it is possible to distinguish between many of the pairs of hydrogens in the compound although not all.

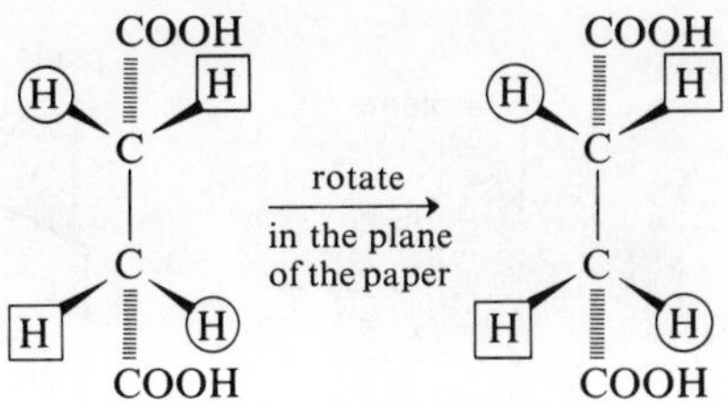

The hydrogens enclosed in circles cannot be distinguished from each other, and neither can those enclosed by squares. As shown below, however, it is possible to distinguish between certain hydrogens such as 1 and 2, 1 and 3,

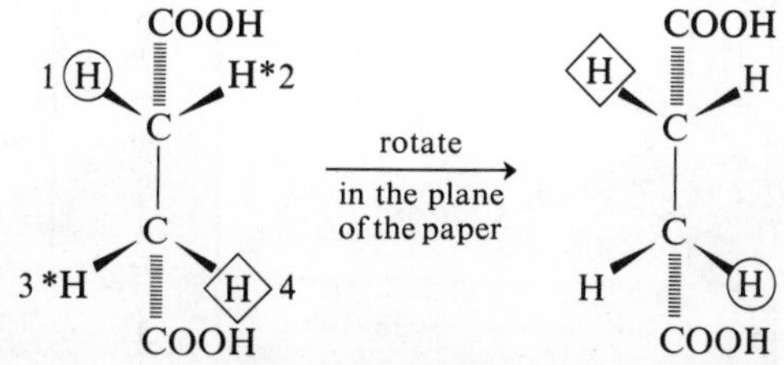

2 and 4, and 3 and 4. Neither the circled nor the squared hydrogen can be brought into either of the starred positions. This type of discrimination is exercised by the enzyme succinate dehydrogenase which catalyzes the conversion of succinate to fumarate.

3.5 SYMMETRY ELEMENTS AND POINT GROUPS

There is a completely rational and rigorous approach to the determination of whether a given molecule (or an object of any kind) is symmetric, asymmetric, or dissymmetric and whether it is possible to discriminate between like groups or atoms in it. The decision is made by performing a series of operations and, after each, concluding whether the operation has produced a molecule indistinguishable from the original, that is, by testing for symmetry. The operations involved are rotations and reflections. Objects can be classified on the basis of the kinds of rotations and reflections which must be applied to them so that they present to the observer an appearance indistinguishable from that in their original orientation. In the discussion which follows concerning molecules (which happen to be the *objects* of interest here), it must be remembered that rotations involve the whole molecule and *not* parts of it. In other words, we are not concerned with changes which occur in the structure when rotation about single bonds takes place (changes in conformation) and, in considering any molecule, we must consider it in a specific conformation. Consider the molecules (*a*) and (*b*) in Figure 3.17. They are obviously symmetrical. Molecule (*a*) is symmetrical because it is equally disposed on either

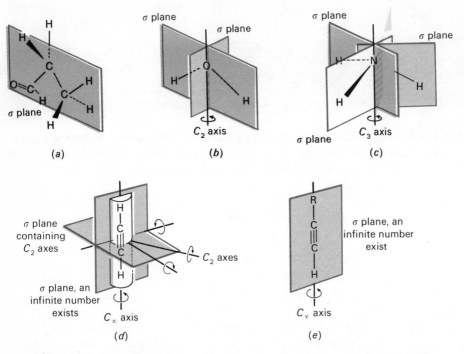

Figure 3.17   The planes and axes of symmetry in a few simple molecules. (*a*), propion-aldehyde; (*b*), water; (*c*), ammonia; (*d*), acetylene; (*e*), a substituted acetylene.

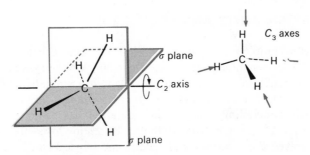

Figure 3.18   Methane, showing the planes and axes of symmetry.

side of the *plane of symmetry* (σ); (*b*) is symmetrical for the same reason, but in addition it has a second plane of symmetry normal to the first and also an axis, C, rotation about which by $2\pi/2$ rad (180 deg) produces a molecule indistinguishable from the original (*b*). This axis of symmetry is designated as a C_n axis where *n* is the number of times an indistinguishable array is produced every 2π rad. Some examples of these two types of symmetry are given in Figure 3.17. Molecule (*b*) has a C_2 axis, molecule (*c*) has three planes of symmetry and a C_3 axis of symmetry. Acetylene (d) can be rotated about its long axis an infinitesimal amount and be indistinguishable. This axis is called a *cylindrical* axis and designated C_∞ since a rotation of $2\pi/\infty$ rad produces identity. The molecule also has an infinite number of C_2 axes normal to the C_∞ axis; only three are shown. It also has an infinite number of planes of symmetry which contain (and intersect at) the C_∞ axis; one is shown in the diagram. Finally, it has a single plane of symmetry containing all of the C_2 axes. Note that in a substituted acetylene (e) the C_∞ axis and its associated planes of symmetry are present, but the C_2 axes and the plane containing them are not present. An example of a tetrahedral carbon compound is methane, shown in Figure 3.18. It has four C_3 axes, three C_2 axes, and six σ planes. The C_3 axes coincide with the C—H bonds. The C_2 axes bisect the angles between adjacent bonds and are mutually perpendicular (one is shown). Each C_2 axis lies at the intersection of two of the σ planes (two are shown). As the groups about the carbon are changed, the symmetry decreases. In Figure 3.19, a number of compounds and the symmetry elements present in them are shown. Note that the last compound contains an "asymmetric carbon atom" and, according to the definition of C axes given above, could be considered to have a C_1 axis, as could all the other compounds discussed. A C_1 axis implies that rotation through 360 deg $(2\pi/1$ rad) results in an

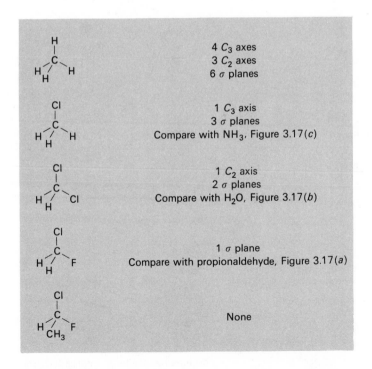

4 C_3 axes
3 C_2 axes
6 σ planes

1 C_3 axis
3 σ planes
Compare with NH_3, Figure 3.17(c)

1 C_2 axis
2 σ planes
Compare with H_2O, Figure 3.17(b)

1 σ plane
Compare with propionaldehyde, Figure 3.17(a)

None

Figure 3.19

The decrease in the number of symmetry elements with increasing differences in substitution about a single tetrahedral carbon atom.

TABLE 3.1 MOLECULAR SYMMETRY CATEGORIES

I. No Reflection Symmetry—Disymmetric Molecules

Point group designation	*Symmetry elements present*	*Examples*
C_1 (asymmetry)	None	
C_n (usually C_2)	C_n	Newman projection
D_n (dihedral symmetry n = principal axis)	C_n; one principal axis $C_n + nC_2$ axes perpendicular to it.	D_2 Newman projection

II. Reflection Symmetry—Nondisymmetric Molecules

C_s	σ planes only	Figure 3.17(a) σ = plane of paper
S_n	Rotation–reflection symmetry but no plane of symmetry	Figure 3.6 Requires rotation by $2\pi/4$ rad and reflection through a plane perpendicular to S_4

TABLE 3.1   MOLECULAR SYMMETRY CATEGORIES *(cont.)*

II. Reflection Symmetry—Nondisymmetric Molecules

Point group designation	Symmetry elements present	Examples
C_{nv}	σ planes and one C_n axis; all σ planes intersect at C_n. $C_n + n\sigma_v$ only (v = vertical)	Figure 3.17(b) = C_{2v} Figure 3.17(c) = C_{3v} Figure 3.17(e) = $C_{\infty v}$
C_{nh}	σ planes and one C_n axis. C_n not in any of σ planes; C_n perpendicular to σ planes; $C_n + \sigma_n$ (h = horizontal)	 C_2 axis σ plane bisects molecule horizontally
D_{nd}	σ planes and C_n axes; One principal C_n axis + nC_2 axes perpendicular to it; $n\sigma$ planes which contain the C_n axis and bisect the angles between C_2 axes	
D_{nh}	σ planes and C_n axes; one principal C_n axis + nC_2 axes perpendicular to it. $n\,\sigma$ planes which contain the C_n axis and bisect the angles between C_2 axes. One σ plane perpendicular to the principal C_n axis.	 + σ plane in plane of the ring
T_d (tetrahedral symmetry)	6 σ planes 3 C_2 axes 4 C_3 axes	Figure 3.18
O_h	9 σ planes 6 C_2 axes 4 C_3 axes 3 C_4 axes	 cubane

indistinguishable arrangement. This is always true—for all compounds; therefore, the C_1 axis is never considered as it cannot serve to distinguish between compounds.

There is a third operation by which the symmetry of an object can be tested; it involves rotation about an axis and reflection through a plane perpendicular to that axis. It is immaterial which process is carried out first. A molecule

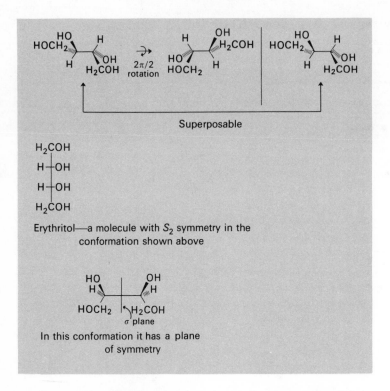

Erythritol—a molecule with S_2 symmetry in the
conformation shown above

In this conformation it has a plane
of symmetry

Figure 3.20   The effect of conformation on the symmetry of erythritol.

is said to possess reflection symmetry if, by carrying out the two operations
(rotation-reflection), a form is produced which is indistinguishable from the
original. The rotation-reflection axis required is designated S and is given
the subscript n to indicate the rotation $(2\pi/n)$ required to produce the identity.
An example is given in Figure 3.20. An S_1 axis is present where no rotation is
required, only reflection. This case is equivalent to the plane of symmetry (σ)
described above. A compound which is superimposable (superposable) on
its mirror image without rotation must possess a plane of symmetry.

On the basis of the symmetry elements, all objects can be assigned to 1 of 11
point groups as shown in Table 3.1. Molecules having no planes of symmetry
and no reflection symmetry (dissymmetric molecules) can be distinguished
from those which do (symmetric molecules), and within these major classes
distinctions can be made on the basis of whether or not they possess one or
more of the other symmetry elements.

In Figures 3.14 and 3.15, the point groups of the structures are listed. It
can be seen that apart from the totally asymmetric molecules (point group C_1),
only those in point group C_s contain similar groups that can be distinguished
from one another. Those in point groups C_1, C_n, and D_n have nonsuperimposable mirror images but similar groups in the C_n and D_n types cannot be
distinguished from one another.

REFERENCES

Allinger, N. L., and E. L. Eliel, ed., *Topics in Stereochemistry*, Vol 1, Interscience Publishers, New York, 1967. The first of a series designed to deal in detail with some of the topics summarized in standard texts and to provide an overview of recent literature. The first volume contains an article, "Stereosiomeric Relationships of Groups in Molecules," which deals with the question of what discriminations are possible between like sub-stituents.

Eliel, E. L., *Stereochemistry of Carbon Compounds*, McGraw-Hill Book Company, 1962. A more comprehensive treatment of stereochemical problems, well referenced.

Eliel, E. L., N. L. Allinger, S. J. Angyal, and G. A. Morrison, *Conformational Analysis*, Interscience Publishers, New York, 1965. This deals with conformational isomers and the stereochemical problems associated with their existence.

Hirschmann, H., "The Structural Basis for the Differentiation of Identical Groups in Asymmetric Reactions" in *Essays in Biochemistry* (S. Graff, ed.), John Wiley & Sons, Inc., New York, 1956. An excellent treatment of the title subject with special reference to biochemical problems.

Mislow, K., *Introduction to Stereochemistry*, W. A. Benjamin, Inc., New York, 1965. An excellent short treatment of the subject. A discussion of symmetry elements and point groups is presented which demonstrates the relevance of this formal approach to the analysis of stereochemical problems.

FOUR

AMINO ACIDS, PEPTIDES, AND PROTEINS

All cells and viruses contain proteins so that no matter what definition of living systems is used, proteins can be said to be *ubiquitous*. They have a variety of roles and serve as both structural and functional materials. Proteins are involved in maintaining the integrity of organelles and normal cellular function and, as catalysts, they participate in almost all reactions occurring in nature. The proteins are polymers of amino acids joined by peptide (amide) bonds. The members of the class are also referred to as *peptides* or *polypeptides*

amino acid   peptide bond

and there is no universally acceptable distinction between these terms and the term *protein*. Protein usually denotes naturally occurring polypeptides which have molecular weights of more than a few thousand daltons, but no precise molecular weight range can be associated with the term; peptide denotes

54

those with lower molecular weights. In this chapter polypeptide and protein will be used interchangeably. In general usage, peptide should be employed when referring to compounds of any degree of polymerization if the number of residues is being specified—for example, dipeptide (2 amino acids), pentapeptide (5 amino acids), and tetrahectapeptide (104 amino acids).

The most abundant of the naturally occurring amino acid polymers are the proteins. They are perhaps the most common subject for biochemical research and they are among the most complex compounds known. The amino acids from which proteins are formed are present in very small amounts in most cells and are most important to the cell as precursors of the proteins. It is most logical, however, to approach the description of protein structure and chemistry by first considering the amino acids. To a great extent the behavior of a protein is directly related to and derived from its amino acid content; the chemistry of the amino acid side chains and the peptide bond determine much of the behavior of proteins in solution. Perhaps the whole (the protein) is the sum of all its parts (the amino acids); probably it is a multiple of them.

In describing the properties of amino acids, we are concerned with the free amino acids (1) as normal constituents of living cells and (2) as building blocks for proteins. In the former case the properties are due to the presence of the α-amino carboxylic acid and the R group. In the latter the α-amino and carboxyl groups are involved in the peptide bond and the R group functions are retained.

In each section, the principles are developed first and specific examples are given by way of illustration. The goal of this chapter is to describe the chemical approaches to the elucidation of protein structure.

4.1   THE AMINO ACIDS

Eighteen α-amino acids and two imino acids are found in most proteins. All the amino acids except glycine, which is symmetrical, have the L-configuration. The structures of the R groups of the amino acids are given in Table 4.1 along with some of their physical properties and shorthand conventions for naming them. These conventions will be used later to describe the sequences of amino acids in polymers.

Many other amino acids have been isolated from natural sources, but they do not occur in proteins or do so only in specialized ones; several, with names, structures, and occurrences are given in Table 4.2. They will not be discussed further. Some of the reactions of amino acids are listed in Table 4.3; they illustrate the complexity which might be expected in the reactions of a protein formed by polymerization of these amino acids. As a reasonable first approximation, one can assume that a protein will react as would a mixture of alcohol, thiol, amine, phenol, indole, thiolether, imidazole, guanidine, carboxylic acid, and amide. Since the reactivity of each of these functional groups cannot be considered in detail, an attempt has been made to select the most important aspects of amino acid chemistry for discussion.

TABLE 4.1   AMINO ACIDS ISOLATED FROM PROTEINS

Group I—Amino acids, R—$\overset{\overset{\textstyle NH_3^+}{|}}{CH}$COOH

Name	R=	pK_a' α COOH	$pK_a'^a$ α NH_3^+	pK_a' of R^b	pI	$[\alpha]_D^{25}$ 5 N HCl	mp	Abbreviation 3 letter	Abbreviation 1 letter
Glycine	H—	2.34	9.60		5.97	0	292d	Gly	G
Alanine	CH_3—	2.35	9.69		6.02	+14.6	297d	Ala	A
Valine	CH_3\ CH— / CH_3	2.32	9.62		5.97	+28.3	315d	Val	V
Leucine	CH_3\ CH—CH_2— / CH_3	2.36	9.60		5.98	+16.0	337d	Leu	L
Isoleucine	CH_3—CH_2\ CH— / CH_3	2.36	9.68		6.02	+39.5	285d	Ile	I
Serine	$HOCH_2$—	2.21	9.15		5.68	+15.1	228d	Ser	S
Threonine	$\overset{\overset{\textstyle OH}{|}}{CH_3CH}$—	2.09	9.10		5.60	−15.0	253d	Thr	T
Cysteine	$HSCH_2$—	1.71	8.9 or 10.4	8.5 or 10.0 (3)	5.02	+6.5	—	Cys	C
Cystine	$\overset{\textstyle S—CH_2}{\underset{\textstyle S—CH_2}{\vert}}$	1.65 2.26	7.86 9.85		5.06	−232	258	Cys λ max 248 ε mol 345	
Methionine	$CH_3SCH_2CH_2$—	2.28	9.21		5.06	−22.2	283	Met	M
Aspartic acid	$HOOCCH_2$—	2.09	9.82	3.86 (100)	2.98	+25.4	269	Asp	D
Glutamic acid	$HOOCCH_2CH_2$—	2.19	9.67	4.25 (100)	3.22	+31.8	247	Glu	E
Asparagine	$\overset{\overset{\textstyle O}{\|}}{H_2N C}CH_2$—	2.02	8.8		5.41	+28.6	236	Asn	N
Glutamine	$\overset{\overset{\textstyle O}{\|}}{H_2N C}CH_2CH_2$—	2.17	9.13		5.70	+31.8	184	Gln	Q
Lysine	$H_3^\oplus N(CH_2)_3CH_2$—	2.18	8.95	10.53 (100)	9.74	+25.9	224	Lys	K
Hydroxy-lysine	$\overset{\overset{\textstyle OH}{|}}{H_3^\oplus NCH_2CH}$ CH_2CH_2—	2.13	8.62	9.67 (100)	9.15	+17.8	—	Hyl	

TABLE 4.1   AMINO ACIDS ISOLATED FROM PROTEINS *(cont.)*

Group I—Amino acids, $R-\overset{\overset{\displaystyle NH_3^+}{|}}{CH}\ COOH$

Name	R=	pK_a' α COOH	$pK_a'^a$ α NH_3^+	pK_a' of R^b	pI	$[\alpha]_D^{25}$ 5 N HCl	mp	Abbreviation 3 letter	Abbreviation 1 letter
Histidine	$-CH_2-$ HN$\overset{\oplus}{\underset{C}{}}$NH	1.82	9.17	6.0 (9)	7.59	+11.8	227	His	H
Arginine	$H_2\overset{\oplus}{N}-\overset{\overset{\displaystyle HNH}{\|\|}}{C}-NH(CH_2)_3-$	2.17	9.04	12.48 (100)	10.76	+27.6	230–44d	Arg	R
Phenyl-alanine	$-CH_2-$	1.83	9.13		5.48	−4.47	283	Phe λ max = 258 ε = 195	F
Tyrosine	HO$-$$-CH_2-$	2.20	9.11	10.07 (0)	5.67	−10.0	342	Tyr λ max = 275 ε_{275}^M = 1,340	Y
Tryptophan	$-CH_2--$ N	2.38	9.39		5.88	+2.8	283	Trp λ max 278 ε_{280}^M = 5,550	W

Group II Imino acids

Name	R=	pK_a' α COOH	$pK_a'^a$ α NH_3^+	pK_a' of R^b	pI	$[\alpha]_D^{25}$ 5 N HCl	mp	Abbreviation 3 letter	Abbreviation 1 letter
	COOH H N	1.99	10.60		6.30	−60.4	220	Pro	P
	H								
Hydroxy-proline^c	HO COOH H N H	1.92	9.73		6.33	−50.5	270	Hyp	

^{a} The α-amino and α-carboxyl group ionizations are not present in the polymers except at the beginning and end of the chain.

^{b} Figures in parentheses indicate percentage of ionized group bearing a charge at pH 7.0.

^{c} Found only in collagen.

TABLE 4.2   NATURALLY OCCURRING AMINO ACIDS NOT FOUND IN PROTEINS[a]

β-Alanine	$H_3N^{\oplus}CH_2CH_2COO^{\ominus}$	A constituent of anserine (β-alanyl-1-methyl-L-histidine) and carnosine (β-alanyl-L-histidine), which are present in muscle, and of pantothenic acid (a vitamin) and its derivative, coenzyme A. It is present also in the prosthetic group of acyl carrier protein.
γ-Aminobutyric acid	$H_3^{\oplus}NCH_2CH_2CH_2COO^-$	Of wide occurrence, it is formed from L-glutamic acid and is converted to succinate. It is present in brain tissue.
Betaine (*N,N,N*-trimethylglycine)	$(CH_3)_3^{\oplus}N-CH_2COO^-$	Found in all tissues—it is a biological methylating agent in the formation of methionine from homocysteine.
Homocysteine	$\overset{\overset{\displaystyle NH_3^{\oplus}}{\mid}}{HSCH_2CH_2CHCOO^-}$	This is a precursor of methionine in microbes.
Homoserine	$\overset{\overset{\displaystyle NH_3^{\oplus}}{\mid}}{HO-CH_2CH_2-CHCOO^-}$	This is an intermediate in the synthesis of threonine and homocysteine from aspartate.
Ornithine	$\overset{\overset{\displaystyle NH_3^{\oplus}}{\mid}}{H_3^+N\,CH_2CH_2CH_2CHCOO^-}$	This is an intermediate in the formation of urea.
Citrulline	$\overset{\overset{\displaystyle O}{\parallel}}{H_2N-C}-NH-CH_2CH_2CH_2\overset{\overset{\displaystyle NH_3^{\oplus}}{\mid}}{CHCOO^-}$	This is formed from citrulline in the urea cycle and is the precursor of arginine.

[a] H. A. Sober, Ed., *Handbook of Biochemistry Selected Data for Molecular Biology*, p. B 20. The Chemical Rubber Co., Cleveland, Ohio (1968).

TABLE 4.3 REACTIONS OF AMINO ACID R GROUPS AND REAGENTS FOR MODIFYING R GROUPS IN PROTEINS[a]

Lysine
R group $-CH_2-CH_2-CH_2-CH_2-NH_3^{\oplus}$

Reagent	Product	Comment	Other reactive groups
CH_3C (=O) $-O-$ CH_3C (=O) **Acetic anhydride**	$CH_3\overset{O}{\overset{\|}{C}}-\overset{H}{N}-$ **Acetate**	Decrease in $\oplus$ charge, stable to $OH^{\ominus}$, hydrolyzed in acid.	ser, thr, tyr, cys
H_2C-C(=O) $-O-$ H_2C-C(=O) **Succinic anhydride**	$^{\ominus}O-\overset{O}{\overset{\|}{C}}-CH_2-CH_2-\overset{O}{\overset{\|}{C}}-\overset{H}{N}-$ **Succinate**	$\oplus$ charge converted to $\ominus$; hydrolyzed in acid.	ser, thr,
(tetramethylglutaric anhydride structure) **Tetramethylglutaric anhydride**	(tetramethylglutarate structure) $C-COO^{\ominus}$ ... $C-\overset{H}{N}-$ **Tetramethylglutarate**	Converted to acid-stable cyclic imide by acid treatment; allows analysis of lysine modification by amino acid analysis.	
$HC-C$(=O) $-O-$ $HC-C$(=O) **Maleic anhydride**	$HC-COO^{\ominus}$ ‖ $HC-\overset{O}{\overset{\|}{C}}-\overset{H}{N}-$ **Maleate**	Unstable below pH 6.0 due to participation by neighbouring $-COOH$; stable above pH 6.0.	
$F_3C-\overset{O}{\overset{\|}{C}}-SCH_2CH_3$ **Ethylthiotrifluoroacetate**	$F_3C-\overset{O}{\overset{\|}{C}}-\overset{H}{N}-$ **Trifluoroacetate**	Can be removed with piperidine.	
(acetyl imidazole structure) $O=\overset{}{C}-CH_3$ **Acetyl imidazole**	$CH_3\overset{O}{\overset{\|}{C}}-\overset{H}{N}-$ **Acetate**	Reagent has preference for phenolic OH.	

TABLE 4.3   **REACTIONS OF AMINO ACID R GROUPS AND REAGENTS FOR MODIFYING R GROUPS IN PROTEINS**[a] *(cont.)*

Lysine (cont.)

R group $-CH_2-CH_2-CH_2-CH_2-NH_3^{\oplus}$

Reagent	Product	Comment	Other reactive groups
CS_2 Carbon disulfide	 Dithiocarbamic acid	Alkali-stable, acid-labile; some selectivity for α-amino groups at pH 6.9.	
$N{\equiv}C-O^{\ominus}$ Cyanate	 Carbamate	Stable in alkali; other reactive groups are unstable in alkali	
  O-Methylisourea	 Guanidinium derivative	No loss of charge; slow reaction; pH 10–11 for 2–5 days, specific for lysine.	
  Imido ester	 Amidine	No loss of charge; reacts at pH 8.5–9.5.	
  Fluorodinitrobenzene (FDNB)	 Dinitrophenyl (DNP)	Stable; chloro derivative is less reactive; reagent enters hydrophobic regions and reacts with nearby residues.	cys, his, tyr
  Trinitrobenzenesulfonic acid	 Trinitrophenyl	Fairly high specificity.	
HNO_2 Nitrous acid	$-CH_2OH$	Loss of $\oplus$ charge.	
  2-Methoxy-5-nitrotropone	 2-Amino-5-nitrotropone	Absorbs at 420 nm; can be removed with hydrazine (N_2H_4).	

TABLE 4.3   REACTIONS OF AMINO ACID R GROUPS AND REAGENTS FOR MODIFYING R GROUPS IN PROTEINS[a] *(cont.)*

Lysine (cont.)
R group $-CH_2-CH_2-CH_2-CH_2-NH_3^\oplus$

Reagent	Product	Comment	Other reactive groups
$H_2C=O$ Formaldehyde	$-N\overset{CH_2OH}{\underset{CH_2OH}{<}}$ Dimethylol derivative	pK_a' of amino group is lowered, allowing it to be titrated (formol titration).	
$\overset{H}{RC=O}$ or $RR'C=O$ Aldehyde or ketone	$-N=C\overset{/}{\underset{\backslash}{}} \xrightarrow{NaBH_4} \overset{H}{-N}-\overset{H}{C}\overset{/}{\underset{\backslash}{}}$ Schiff base Amine	Schiff base is unstable; fixed by reduction.	

Cysteine
R group $-CH_2SH$

Reagent	Product	Comment	Other reactive groups
$CH_3C\overset{O}{\underset{O}{<}}$ $CH_3C\overset{O}{\underset{O}{<}}$ Acetic anhydride	$CH_3\overset{O}{\overset{\|}{C}}-S-CH_2-$ Thiolacetate	Removed with dilute $OH^\ominus$ or NH_2OH.	
$O_2N-\langle\bigcirc\rangle-F$ $\underset{NO_2}{}$ Fluorodinitrobenzene	$O_2N-\langle\bigcirc\rangle-S-CH_2-$ $\underset{NO_2}{}$ S-Dinitrophenyl	Not very specific, reacts near or in hydrophobic regions; can be removed with mercapto-ethanol. Chloro derivative; more selective for $-SH$ groups.	lys, his, tyr
$ICH_2C\overset{O}{\underset{O^\ominus}{<}}$ Iodoacetate	$-CH_2-S-CH_2C\overset{O}{\underset{O^\ominus}{<}}$ S-Carboxymethyl	Not very selective; very variable reactivity.	met, his, lys asp, glu
$ICH_2C\overset{O}{\underset{NH_2}{<}}$ Iodoacetamide	$-CH_2-S-CH_2C\overset{O}{\underset{NH_2}{<}}$ S-Carbaminomethyl		
RCOOOH Peracids	$-CH_2SO_3^\ominus$ Cysteic acid	Used as a pretreatment before acid hydrolysis.	cys, try

TABLE 4.3   REACTIONS OF AMINO ACID R GROUPS AND REAGENTS FOR MODIFYING R GROUPS IN PROTEINS[a] *(cont.)*

Cysteine (cont.)
R group —CH_2SH

Reagent	Product	Comment	Other reactive groups
N-Ethylmaleimide (NEM)	*N*-Ethylsuccinimide	Quite specific.	lys (slight)
N-(4-Dimethylamino-3,5-dinitro-phenyl maleimide)		Allows colorimetric estimation of substitution; pK_a of dimethyl ammonium group depends on its environment used as a probe of —SH environment.	
$H_2C{=}\overset{H}{C}{-}CN$ Acrylonitrile	$-CH_2{-}S{-}CH_2CH_2CN$	Very selective at pH 8.0; other groups react at higher pH.	lys, arg, his
$HSCH_2COO^{\ominus}$ Thioglycolate	$-CH_2{-}S{-}S{-}CH_2COO^{\ominus}$ Disulfide	Readily reversible with thiols.	
$ClHg{-}\langle\bigcirc\rangle{-}COO^{\ominus}$ *p*-Chloromercuribenzoate	$-CH_2S{-}Hg{-}\langle\bigcirc\rangle{-}COO^{\ominus}$	Extent of reaction can be estimated spectrophotometrically.	
$(CH_3)_2\overset{\oplus}{N}{-}\langle\bigcirc\rangle{-}N{=}N{-}\langle\bigcirc\rangle{-}HgOAc$ Azobenzene mercurial	$-CH_2S{-}Hg{-}\langle\bigcirc\rangle{-}N{=}N{-}\langle\bigcirc\rangle{-}\overset{\oplus}{N}(CH_3)_2$	The pK'_a of the dimethyl ammonium ion depends on its local environment.	

Cystine
R group —$CH_2{-}S{-}S{-}CH_2^-$

$HSCH_2CH_2OH$ Mercaptoethanol	$2{-}CH_2{-}S{-}SCH_2CH_2OH$ Mixed disulfide	A readily reversible reaction; all RSH compounds can react similarly.	
RCOOOH Peracids	$2{-}CH_2SO_3H$ Cysteic acid		
$Na_2S_2O_3$ Sulfide	$-CH_2S^{\ominus}Na^{\oplus}$ $-CH_2SSO_3^{\ominus}Na^{\oplus}$		

TABLE 4.3   REACTIONS OF AMINO ACID R GROUPS AND REAGENTS FOR MODIFYING R GROUPS IN PROTEINS[a] *(cont.)*

Methionine
R group $-CH_2-CH_2-S-CH_3$

Reagent	Product	Comment	Other reactive groups
$ICH_2COO^{\ominus}$ Iodoacetate	$-CH_2-\overset{\oplus}{S}-CH_3$ $\qquad\quad\; \underset{\displaystyle CH_2COO^{\ominus}}{\mid}$ Carboxymethyl	Not very selective.	cys, his, lys, asp, glu
H_2O_2 HCOOOH Peroxides	$\underset{\displaystyle CH_2}{\overset{\displaystyle CH_3}{\mid}} S{=}O \;\rightarrow\; O{=}\underset{\displaystyle CH_2}{\overset{\displaystyle CH_3}{\mid}}S{=}O$ Sulfoxide Sulfone	Fairly selective at pH 2–3; can be reduced with thiols.	cys, try
H_2: Pt or Pd Catalytic reduction	$-CH_3 + HSCH_3$ α-amino butyric acid + methanethiol		
 N(Bromoacetyl)-3-nitro-4-hydroxy aniline		In chymotrypsin; only methionine reacts.	

Serine and threonine
R group $-CH_2OH$ and $-CHOH-CH_3$ (less reactive)

 5-Dimethylaminonaphthyl sulfonyl chloride (DANS)	 $-CH_2OSO_2$ Dansyl	Only activated serines can react; derivative is fluorescent and fluoroescence depends on environment.	lys
 Diisopropyl fluoro phosphate	$-CH_2O\overset{\displaystyle O}{\overset{\displaystyle \|}{P}}-(OCH(CH_3)_2)_2$	Reagent is very toxic; only activated serines react.	

TABLE 4.3 REACTIONS OF AMINO ACID R GROUPS AND REAGENTS FOR MODIFYING R GROUPS IN PROTEINS[a] *(cont.)*

Serine and threonine (cont.)
R group —CH_2OH and —$CHOH$—CH_3 (less reactive)

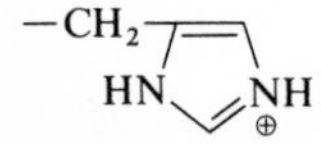

Reagent	Product	Comment	Other reactive groups
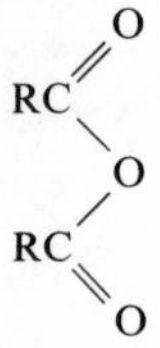 Acid anhydrides	Esters	Derivatives unstable in alkali.	lys

Histidine
R group —CH_2—

$ICH_2COO^\ominus$

Iodoacetate

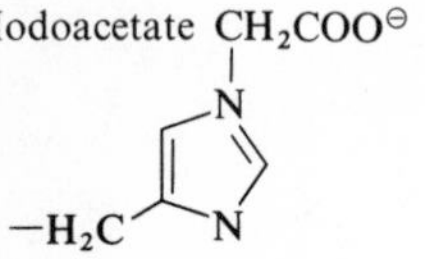

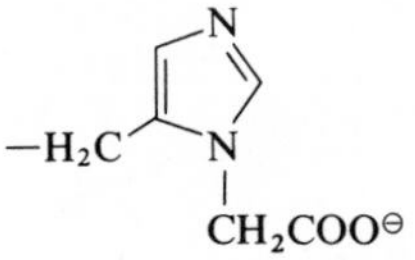

		Not very selective.	met, cys, lys asp, glu

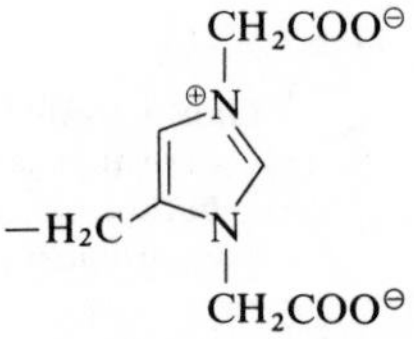

Carboxymethyl derivatives

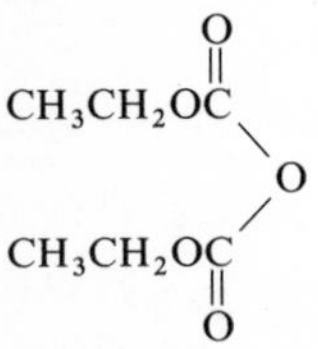

Diethyl pyrocarbonate

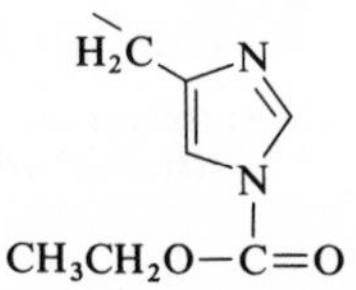

Ethyl carboxamide

TABLE 4.3   REACTIONS OF AMINO ACID R GROUPS AND REAGENTS FOR MODIFYING R GROUPS IN PROTEINS[a] (cont.)

Tryptophan
R group   —CH₂—

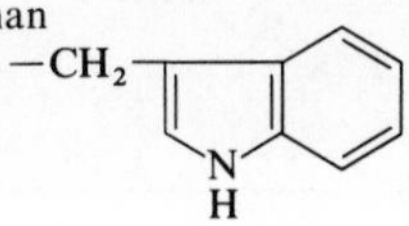

Reagent	Product	Comment	Other reactive groups
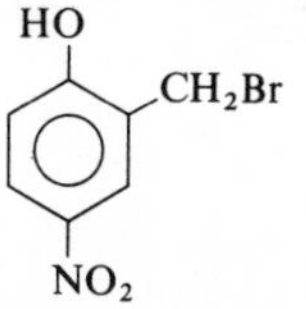 2-Hydroxy-5-nitrobenzyl bromide	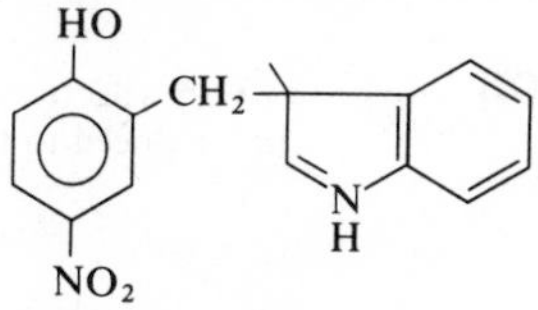	Reacts at low pH; derivative absorbs at 420 nm in OH$^{\ominus}$.	cys, met, tyr (in OH$^{\ominus}$), H_2O
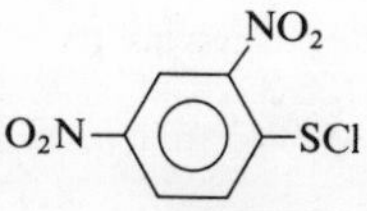 2,4-Dinitrophenylsulfenyl chloride	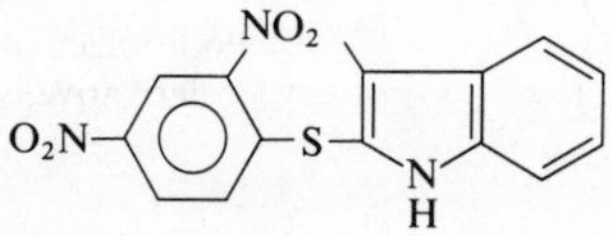	cys derivative can be decomposed with thiols.	cys
H_2O_2 Hydrogen peroxide	?		cys, met
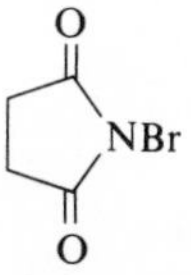 N-Bromosuccinimide	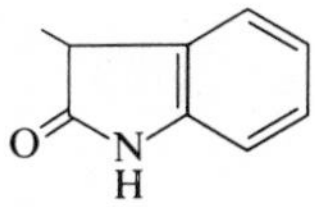	Can be used to determine tryptophan content of polypeptides (see text).	his, cys, arg, lys, tyr

Tyrosine
R group —CH₂——OH

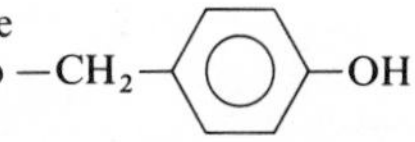

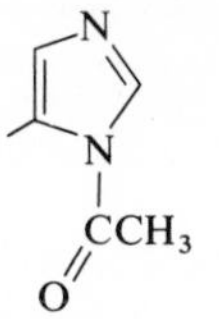 Acetyl imidazole	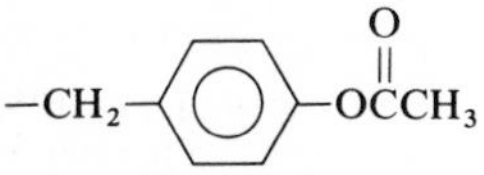 Acetate	Readily removable with NH₂OH; fairly specific; no reaction with cys.	lys
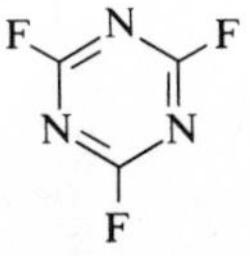 Cyanuric fluoride (cyF)	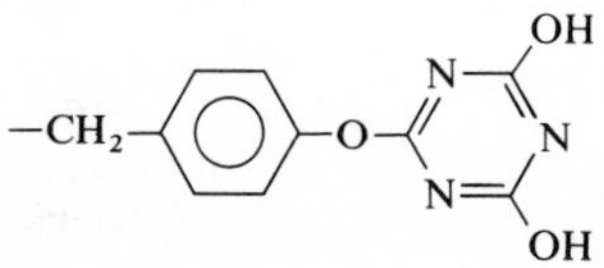	Relatively nonspecific reaction; tyrosyl derivative can be estimated spectrophotometrically.	lys, try, his cys

TABLE 4.3   REACTIONS OF AMINO ACID R GROUPS AND REAGENTS FOR MODIFYING R GROUPS IN PROTEINS[a] (cont.)

Tyrosine (cont.)
R group $-CH_2-$⟨benzene ring⟩$-OH$

Reagent	Product	Comment	Other reactive groups
$O_2N-\overset{\displaystyle NO_2}{\underset{\displaystyle NO_2}{C}}-NO_2$ Tetranitromethane	$-CH_2-$⟨benzene ring with NO_2⟩$-OH$ 3-Nitrotyrosine	Very selective at pH 8.0; can be reduced to the amino derivative.	
$I_3^{\ominus}$ Iodine	$-CH_2-$⟨benzene ring with I, I⟩$-OH$ 3,5-Diiodotyrosine	Poor selectivity; mono iodo derivative is also formed.	cys, his, try
⟨Diazonium-1H-tetrazole structure⟩ Diazonium-1H-tetrazole	$-CH_2$⟨benzene ring, $N=N-$tetrazole⟩$-OH$ Tyrosine azo-1H-tetrazole	Tyrosyl derivative absorbs at 550 nm; histidyl derivative at 480 nm.	his

Arginine
R group $-CH_2CH_2CH_2N-\overset{\displaystyle NH_2^{\oplus}}{\underset{}{C}}-NH_2$ (with H)

Reagent	Product	Comment	Other reactive groups
$\overset{\displaystyle HC=O}{\underset{\displaystyle HC=O}{CH_2}}$ Malonaldehyde	$-CH_2-N-C\overset{\displaystyle N=CH}{\underset{\displaystyle N-CH}{\diagup\diagdown}}CH$ (with H) A pyrimidine	Reaction only proceeds in strong acid (6–12 N HCl).	
⟨Cyclohexane-1,2-dione structure⟩ Cyclohexane-1,2-dione	$-CH_2-N=C$⟨cyclopentane ring fused with N, N, O⟩	Requires 0.05–0.2 N OH$^{\ominus}$; a benzilic acid rearrangement occurs.	
$\overset{\displaystyle HC=O}{\underset{\displaystyle H_2COH}{}}$ Glyoxal	?	pH 8.6–9.2; can follow reaction of arginine colorimetrically.	

TABLE 4.3   REACTIONS OF AMINO ACID R GROUPS AND REAGENTS FOR MODIFYING R GROUPS IN PROTEINS[a] (cont.)

Glutamate aspartate
R group $-CH_2-CH_2COO^{\ominus}$ and $-CH_2COO^{\ominus}$

Reagent	Product	Comment	Other reactive groups
ROH—HCl	$-CH_2C\overset{\displaystyle O}{\underset{\displaystyle OR}{\big\langle}}$	acid catalysis disrupts polypeptide structure.	
Alcohol	Ester		
CH_2N_2	$-CH_2C\overset{\displaystyle O}{\underset{\displaystyle OCH_3}{\big\langle}}$	Difficult to achieve full substitution.	
Diazomethane	Methyl ester		
① CH_3OH—HCl or CH_2N_2	$-CH_2CH_2OH$	Vigorous conditions cannot be used for conformation studies; C-terminal residues give α-amino alcohols.	
② $LiAlH_4$ lithium borohydride in tetrahydrofuran at reflux	Alcohol		

[a] See References 1 and 4 at the end of this chapter.

[b] Cohen, L. A., "Group Specific Reagents in Protein Chemistry," *Annual Reviews of Biochemistry*, **37**, 695 (1968).

A.   Ionic Properties of Amino Acids

The ionic properties of the amino acid R groups are among the important determinants of their reactivity in proteins, certainly in aqueous solution. For example, most reactions of amino groups involve the base (RNH_2) rather than the acid ($RNH_3^{\oplus}$) and the equilibrium between these two forms determines the rates of these reactions. For this reason, it is necessary to consider the pKa' values of the ionizable groups and factors which alter them.

Briefly, all ionizable groups can be treated as acids and all ionizations spoken of as acid dissociations. There is no need to consider a pOH scale and a pH scale since for every sample considered pH + pOH = 14. For the weak acid HA which dissociates

$$HA \; \underset{\uparrow}{\rightleftharpoons} \; H^{\oplus} + \underset{\uparrow}{A^{\ominus}}$$

$$\text{acid} \qquad \text{conjugate base}$$

so that in dilute solution only a small proportion is ionized, an apparent dissociation constant can be written.

$$K'_a = \frac{[H^{\oplus}][A^{\ominus}]}{[HA]}$$

Rearranging gives

$$[H^{\oplus}] = \frac{K'_a[HA]}{[A^{\ominus}]}$$

Since pH $= \log(1/[H^{\oplus}])$, we can write $\log(1/[H^{\oplus}]) = \log(1/K_a) + \log([A^{\ominus}]/[HA])$ or pH $= pK'_a + \log[A^{\ominus}]$. The term pK'_a bears the same relation to K'_a as pH bears to $[H^{\oplus}]$. Note that the primed term, K'_a, indicates an apparent dissociation constant—one that is measured under specified conditions of ionic strength, and so forth. The intrinsic dissociation constant K_a is independent of experimental variables and is much more difficult to evaluate. However, both pK_a and pK'_a vary with temperature. The equation developed above is known as the Henderson–Hasselbach equation. It predicts certain important relationships.

First, if one adds to a weak acid its salt ($A^{\ominus}$) so that the concentration of $A^{\ominus}$ is made equal to HA (the undissociated acid), the pH will be equal to pK'_a. An easy way to measure the latter!

Second, the lower the pK'_a, the stronger the acid.

Third, it predicts what the pH will be, relative to pK'_a, as the ratio of $A^{\ominus}$ to HA is altered. In effect, it predicts that *all* weak acids will show the same dependence of pH on the ratio $[A^{\ominus}]/[HA]$. Now the ratio of $[A^{\ominus}]/[HA]$ is what varies as a weak acid is titrated so that the titration curves for weak acids should all have the same shape and differ only by their position on the pH scale (Figure 4.1), being symmetrical about their respective pK'_a values. It is a useful approximation to consider that the ionizing species is completely in the *acid* form (HA) 2 pH units below pK'_a and completely in the *base* form

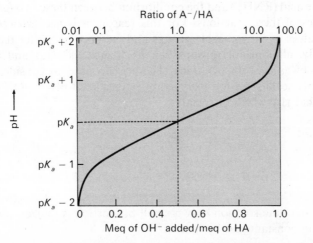

Figure 4.1   The titration curve of a weak acid, HA, showing the relationship of the ratio $A^{\ominus}/HA$ to pH. All weak acids have titration curves of this general shape.

$(A^{\ominus})$ 2 pH units above pK'_a. It must be kept in mind, however, that there is some of each form at all pH values.

Fourth, it predicts that if a small proportion of a strong acid (one that is completely ionized) is added to a solution containing a weak acid and its salt, the pH change produced would be much less than that produced by its addition to a similar volume of water.

1. Consider 100 ml of solution containing 10 mmoles of a weak acid HA and 20 mmoles of $A^{\ominus}$ with HA having $pK'_a = 7.0$.

2. The pH of the solution is $7.0 + \log \frac{20}{10} = 7.30$.

3. To this is added 1 mmole of hydrochloric acid in 1 ml of water (1 ml of $1NH^{\oplus}$).

4. The reaction $H^{\oplus} + A^{\ominus} \rightleftharpoons HA$ takes place. HA is only slightly dissociated so we can consider this process to go to completion.

5. The ratio of $[A^{\ominus}]/[HA]$ has thus become $\frac{19}{11}$ and pH $= 7.0 + \log \frac{19}{11} = 7.24$, a change of 0.06 *pH units*.

6. If we add 1 ml of 1 N hydrochloric acid to 100 ml of water at pH 7.0, the pH becomes 2.0, *a change of* 5.0 *pH units*.

A similar result would be observed if a highly dissociated base were added. This type of resistance to change in pH is termed *buffering* and is very important to the maintenance of the biological status quo.

For all amino acids, the important pK'_a values are for the α-carboxyl group (approximately 2.5) and for the α-amino group (approximately 9.6). At pH 6.0, the carboxyl group will be completely ionized and the amino group will still be in the acid form. This form is referred to as the *zwitterion*. At pH 12.0,

$$
\begin{array}{ccc}
\overset{\oplus}{N}H_3 & \overset{\oplus}{N}H_3 & NH_2 \\
| & | & | \\
R-C-COOH & \longrightarrow \quad R-C-COO^{\ominus} & \longrightarrow \quad R-C-COO^{\ominus} \\
| & | & | \\
H & H & H \\
\\
pH = 1.0 & pH = 6.05 & pH = 12.0 \\
\text{(net charge } +1) & \text{(net charge 0)} & \text{(net charge } -1)
\end{array}
$$

both acid groups will be titrated and the molecule will have a net charge of -1. The responses of amino acids in an electric field (electrophoresis) toward ion-exchange resins and toward chemical reagents such as 2,4-dinitrofluorobenzene are all affected by pH. A reference point of use in discussing these effects is the pH value at which the molecule has a net charge equal to 0 – the isoelectric point (pI). In the simple case being discussed, the isoelectric point must be exactly halfway between the pK'_a values of the amine and carboxyl groups. If the values were 2.50 and 9.60, respectively, then the pI would be at pH 6.05. In practice, the amino acid would carry only a very small charge in the pH range from 4.50 to 7.60. The proportion of charge would increase above and below these values. In fact, a titration curve can be obtained by measuring electrophoretic mobility as a function of pH.

The titration curve of a simple amino acid is shown in Figure 4.2. For those amino acids that have other ionizable groups, more complex titration curves

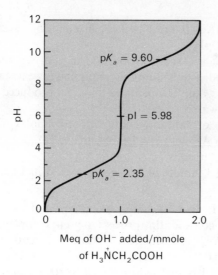

Figure 4.2

The titration curve of a simple amino acid such as glycine.

are obtained. The pI values for such amino acids are also the pH value at which the net charge is zero, and they depend on the character and pK_a' of the additional ionizing group. If ionization of the third group gives rise to a negatively charged base and if it is titrated in the same pH range as the α-carboxyl group, then the pI will be reached when one negative charge per molecule is produced, that is, between the pK_a values for the two groups giving negative bases.

$$9.82 \ ^{\oplus}H_3N-\underset{\underset{\underset{COOH \ 3.86}{|}}{\overset{\overset{COOH \ 2.09}{|}}{C}}}{C}-H \quad + \ 1OH^- \longrightarrow \ ^{\oplus}H_3N-\underset{\underset{\underset{COOH}{|}}{\overset{\overset{COO^-}{|}}{C}}}{C}-H \quad + \ ^{\oplus}H_3N-\underset{\underset{\underset{COO^-}{|}}{\overset{\overset{COOH}{|}}{C}}}{C}-H$$

$$\text{aspartic acid} \qquad\qquad\qquad\qquad 88\% \qquad\qquad 12\%$$
$$\text{pH} = 0 \qquad\qquad\qquad\qquad\qquad \text{pH} = 2.98$$

For lysine, however, pI = 9.74, which is halfway between the pK_a' for the two amino groups.

$$8.95 \ ^{\oplus}H_3N-\underset{\underset{\underset{CH_2NH_3^{\oplus} \ 10.53}{|}}{\overset{\overset{COO^-}{|}}{C}}}{C}-H \quad + \ 1OH^- \longrightarrow \ H_2N-\underset{\underset{\underset{CH_2NH_3^{\oplus}}{|}}{\overset{\overset{COO^{\ominus}}{|}}{C}}}{C}-H \quad + \ ^{\oplus}H_3N-\underset{\underset{\underset{CH_2NH_2}{|}}{\overset{\overset{COO^{\ominus}}{|}}{C}}}{C}-H$$

$$\text{pH} = 6.0 \qquad\qquad\qquad\qquad\qquad \text{pH} = 9.74$$

Although simple formulas can be written for calculating pI values, it is best to approach the problem from the point of view of the groups involved and the change which takes place as they are titrated. The proportions of various forms present can be calculated using the Henderson–Hasselbach equation.

B. The Influence of Neighboring Groups on Ionization

The pK'_a for the carboxyl group of glycine is 2.3; that of acetic acid is 4.8. The increased acid strength of glycine is due to the presence of the strong electron-withdrawing ammonium group and to the spatial proximity of its positive charge (*field effect*); both of these effects increase the ease with which the proton ($H^{\oplus}$) can leave and thus lower pK'_a.

$$\overset{\overset{\displaystyle NH_3^{\oplus}}{|}}{H_2C-COOH} \qquad pK'_a = 2.3$$

glycine (acid form)

$$\overset{\overset{\displaystyle NH_3^{\oplus}}{|}}{H_2C-CH_3} \qquad pK'_a = 10.7$$

$$\overset{\overset{\displaystyle NH_3^{\oplus}}{|}}{H_2C-COO^{\ominus}} \qquad pK'_a = 9.6$$

$$H_3C-COOH \qquad pK'_a = 4.8 \qquad \text{ethylammonium ion} \qquad \text{glycine (zwitterion)}$$

acetic acid

The pK'_a of the ammonium group of glycine is more difficult to explain. Glycine zwitterion would be expected to be a weaker acid than ethyl ammonium ion due to the spatial proximity of the negative charge on the carboxylate anion, making removal of $H^{\oplus}$ more difficult. The opposite result is observed. Perhaps ethylammonium ion is not an appropriate standard; consider instead ammonium ion ($NH_4^{\oplus}$), which has $pK'_a = 9.3$. Relative to this standard, the glycine zwitterion is a weaker acid ($pK'_a = 9.6$), indicating the possibility of some stabilization by the neighboring, diffuse (spread over two oxygen atoms) negative charge. Since the inductive effect of the carboxylate anion is small (that is, it neither donates nor withdraws electrons), the pK'_a value of ethylamine might be increased because of electron donation by the ethyl group. In all probability neither ammonium nor ethylammonium ion is completely satisfactory as a basis for comparison, neither are explanations based on charge separation or inductive effects since the whole system including the solvent is influenced by neighboring groups. There is some value in continuing the evaluation, however, since certain general effects are apparent. Consider glycylglycine (pK'_a of $-COOH = 3.1$, pK'_a of $NH_3^+ = 8.2$). The carboxyl group is a much weaker acid than that of glycine but is still stronger than that of acetic acid.

inductive electron withdrawal

The change may be due to the increased distance between the carboxyl and ammonium groups (relative to glycine) and to electron withdrawal by the neighboring amide bond (relative to acetic acid). On the other hand, in the zwitterion, the ammonium group is an appreciably stronger acid than that of glycine since the negative carboxylate anion is significantly farther away and the peptide bond withdraws electrons. Although a quantitative analysis of structural effects on ionization constants is not warranted, there can be no doubt of the importance of field and inductive effects on the ionization of amino acids and the R groups.

In general, neighboring electron-withdrawing groups such as amides, hydroxyls, and thiols will increase the acid strength (decrease pK_a') of an ionizing group. The influence of neighboring charged groups may not be so great but will cause significant shifts in pK_a'. Since these latter effects are not directional (they are transmitted in all directions), they are probably important in determining pK_a' values of groups on proteins. In addition, the extent to which charge effects are transmitted is influenced by the dielectric constant of the medium, which may be quite different on the surface of the protein from that in the bulk solvent.

To illustrate the effect of changes in charge structure on the ionization of other groups, consider a solution containing one equivalent of cysteine at about pH 6.0. We can titrate two groups over the pH range of 6.3 to 12.8 which appear to have pK_a' values of 8.3 and 10.8. Which one is due to —SH and which one to —NH$_3^\oplus$? A study of this titration in which the concentration of —S$^\ominus$ (absorption at 240 nm) as well as change in pH was monitored indicated that four dissociations are involved which have the pK_a values shown. Loss of the first proton from either —SH or —NH$_3^\oplus$ increases the pK_a for the remaining group by 1.5 pH units.

Whereas the titration of an amino acid usually gives a titration curve with a number of distinct steps, the titration of a protein gives a curve which rises fairly gradually. The large number of ionizing functions involved along with the fact that each function is influenced by its environment causes this effect (Figure 4.3). An additional related effect can be seen in this figure, namely,

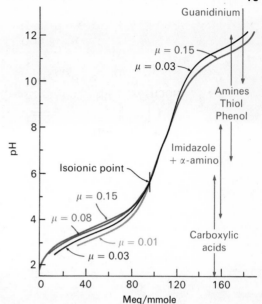

Figure 4.3

The titration curve of bovine serum albumin showing the effect of ionic strength on the ionization of functional groups. On the right the ranges in which the various kinds of groups are titrated are indicated. Redrawn from C. Tanford, S. A. Swanson, and W. S. Shore, *J. Am. Chem. Soc.*, **77**, 6,414 (1955). Copyright © 1955 by The American Chemical Society. Reprinted by permission of the copyright owner.

the influence of the ionic strength of the medium. This is discussed in some detail by Van Holde in this series, (Chapter 3), but it is pertinent to comment here that the ionic strength of the medium has a significant effect on ionizing groups. In general, as the ionic strength of the medium is increased, the pK_a' values of ionizing groups decrease. It must be stressed that the intrinsic pK_a is not affected by changes in the medium.

C. General Reactions of Amino Acids

The reactions of biochemical importance for all amino acids are deaminations, decarboxylations, and cleavage of the α–β carbon bonds.

Deamination is usually accomplished in chemical systems by treatment with nitrous acid (HONO). The product, an α-hydroxy acid, retains the

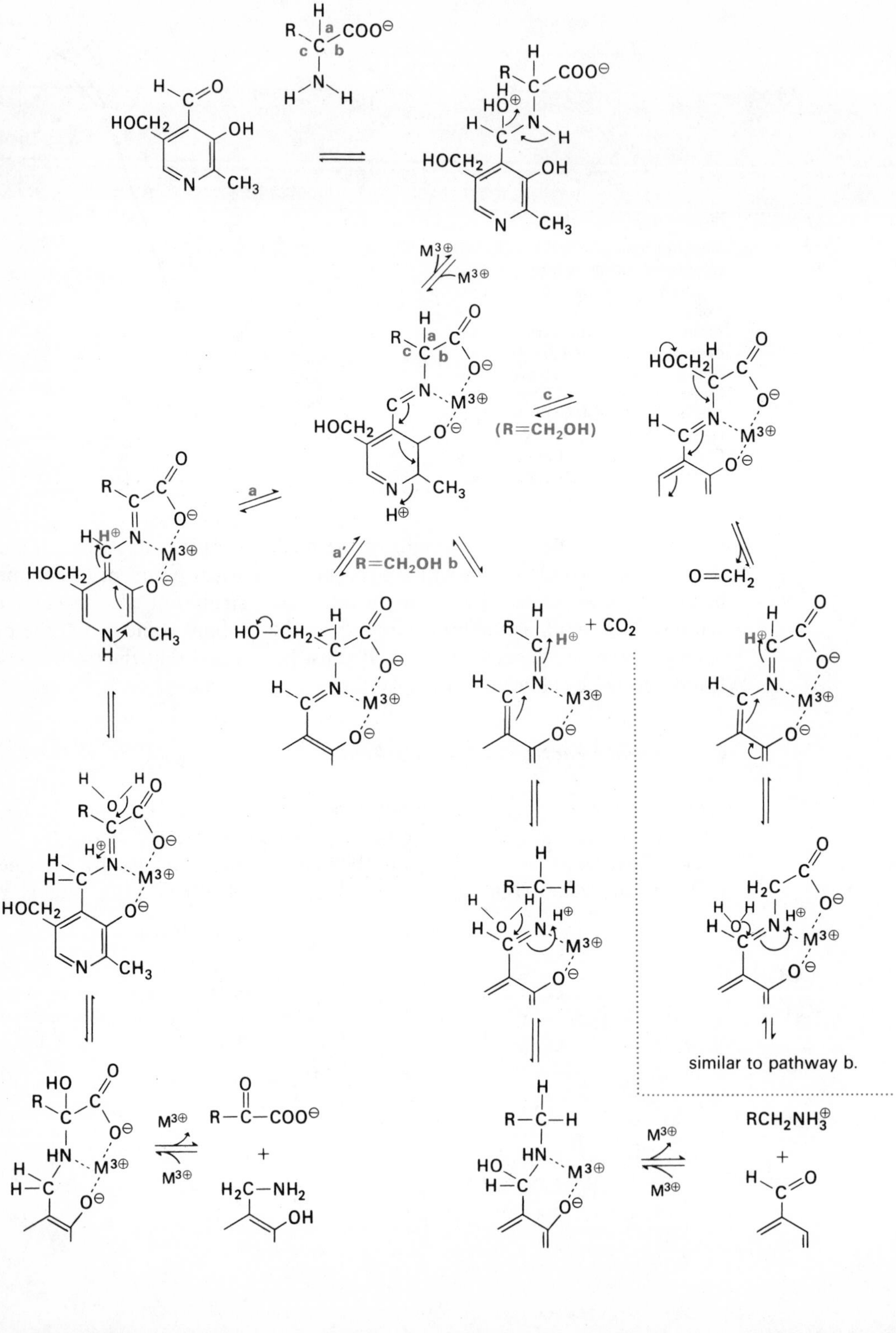

HOCH2
OH
CH3
COO⊖
M3⊕
c
a
b
(R=CH2OH)
a′
R=CH2OH b
+ CO2
O=CH2
similar to pathway b.
R—C—COO⊖
H2C—NH2
OH
RCH2NH3⊕

configuration of the starting amine so that the L-amino acids are convenient starting materials for the preparation of the corresponding alcohols. Retention of configuration is considered to involve participation of the carboxyl group and the intermediate formation of an α lactone. Alternatively, the diazonium ion may be so tightly solvated that water is concentrated on the side of the leaving group and the carbonium ion is always attacked from that side.

Deaminations can also be effected by formation of Schiff base intermediates of pyridoxal which, when chelated with a suitable metal, can undergo facile reversible reactions involving any of the bonds around carbon 2 as shown in Figure 4.4. This rather complex system is exceedingly important since it serves as a very good model for the biological systems in which pyridoxal phosphate and specific enzymes catalyze similar conversions. The C—H bond (a) can be broken to give a keto acid and pyridoxamine (pathway a). Since all the steps are reversible, exchange of keto and amino functions between compounds having different R groups can occur. An alternate possibility (pathway a') involving bond a is the formation of an α–β unsaturated amino acid. The first step is illustrated for the dehydration of serine (R = —CH$_2$OH) and a similar intermediate can be proposed for the desulfuration of cysteine (R = —CH$_2$SH). The pyridoxal complex of the unsaturated intermediate from serine is susceptible to attack by indole, with the formation of tryptophan, and may be involved in the biosynthesis of that compound.

The decarboxylation of an amino acid can be accomplished by migration of electrons from bond b to form the C=N derivative (pathway b). Reversal of the steps then produces the amine and regenerates pyridoxal.

Loss of the R group by migration of the electrons from the C—C bond of the amino acid to the C—N position can only occur when a group such as an —OH is present on the β carbon (pathway c). The reaction with serine produces formaldehyde and glycine.

In all the reactions catalyzed by pyridoxal involving Schiff base formation, reversible protonation of the ring nitrogen provides for the flow of electrons out of, and into, the bonds being altered. The metal probably serves to maintain a planar arrangement which facilitates the formation of conjugated systems in the reaction site.

In model systems such as the one described, although it is possible to achieve some control over the proportions of products formed by varying pH and metal ion, the reaction invariably gives a mixture of products. In biological systems in which this type of reaction occurs on the surface of an enzyme, there is no requirement for metal ion, and the pyridoxal phosphate derivative is utilized. It should be noted that the enzyme-catalyzed reactions are highly specific and each enzyme catalyzes a single pathway. There is good evidence, however, that the same types of intermediates are involved as those described for the model system.

Figure 4.4

Reactions of α-amino acids catalyzed by pyridoxal. Reaction can take place at all the bonds to the α carbon: a = reaction of the C—H bond; b = reaction at the C—COO$^-$ bond; c = reaction at the R—C bond (see text).

The specific reactions of individual amino acids will not be considered here in any detail. Some of them are listed in Tables 4.3, 4.4, and 4.5. These will be used to describe the approaches to the elucidation of protein structure and the synthesis of peptides. Ionization constants, isoionic points, specific rotations, extinction coefficients, absorption maxima, and melting points are listed in Table 4.1. The chemistry of the individual amino acids is quite predictable from consideration of the chemistry of the α-amino and carboxyl groups and the particular R group involved. The R groups are studied most often in proteins where the α-amino and carboxyl groups are masked in the peptide bond, and it is in that context that the chemistry of the R group will be considered.

4.2 PEPTIDE SYNTHESIS

The peptide bond is the amide bond between two amino acids. It can be represented formally as the bond formed by splitting out a molecule of water between the carboxyl and amino groups of the contributing amino acids.

$$
\begin{matrix} NH_2 \\ | \\ R-C-COOH \\ | \\ H \end{matrix}
+
\begin{matrix} NH_2 \\ | \\ R'-C-COOH \\ | \\ H \end{matrix}
\longrightarrow
\begin{matrix} NH_2 & O \\ | & \diagup\!\!\diagup \\ R-C-C & \quad COOH \\ | & \diagdown \quad | \\ H & N-CH \\ & | \quad | \\ & H \quad R' \end{matrix}
+ H_2O
$$

Much of our knowledge of protein structure and function comes from studies of synthetic peptides. In addition, within the last few years it has been demonstrated that peptides having large molecular weights and biological activity can be synthesized, making precise correlations between structure and function possible.

To synthesize a peptide, it is necessary to protect those groups on the amino acids which are capable of reaction but which are to be free in the product by using blocking agents; these agents must be selected so that they can later be removed without disturbing the peptide bond(s) formed. A selection of suitable blocking groups for the thiol, amine, alcohol, and carboxylate functions is given in Table 4.4. The choice of blocking agent depends not only on the nature of the group to be protected but also on the nature of other reactants to be used later in the synthesis. Blocking groups are chosen such that their chemical stability is as low as possible compared to the peptide bond.

A number of important methods of forming peptide bonds have been developed (Table 4.5). All involve activation of the carboxyl or amine groups to be joined by reagents which do not disturb the blocking groups used. It is usual to activate the carboxyl group by replacing —OH with a better leaving group. The earlier methods utilized the acid chloride or azide, but more recently other derivatives have been used. Most of them are anhydrides of oxyacids, but others are esters of acidic alcohols or phenols. Although activation of

TABLE 4.4 BLOCKING GROUPS FOR PEPTIDE SYNTHESIS[a]

Group	Reagent	Product	Abbreviation	Removal
Amine $R-NH_2$	Carbobenzoxychloride	Benzyloxycarbonyl	Z	Na + liquid NH_3, HBr + glacial HAc, or H_2 + Pd
	Triphenylmethylchloride (trityl chloride)	Trityl	Trt	H_2—Pd
	Phthalic anhydride	Phthalyl	Pht	Hydrazine (H_2N-NH_2)
	Trifluoroacetic anhydride	Trifluoroacetyl	Tfa	Dilute alkali
	$(CH_3)_3C-O-\overset{\overset{\displaystyle O}{\|}}{C}Cl$ t-Butyloxycarbonyl chloride	$(CH_3)_3C-O-\overset{\overset{\displaystyle O}{\|}}{C}-\overset{H}{N}R$ t-Butyloxycarbonyl	Boc	HCl—HAc
Carboxyl				
RCOOH	CH_3CH_2OH CH_3OH R'OH Alcohols	Benzyloxy (benzy ester) Ethoxy (ethyl ester) Alkoxy	O Bzl O Et	Dilute alkali. H_2—Pd in the case of benzyl esters
	Thiophenol	Phenylthio	S Ph	$OH^{\ominus}$

TABLE 4.4 BLOCKING GROUPS FOR PEPTIDE SYNTHESIS[a] (cont.)

Group	Reagent	Product	Abbreviation	Removal
Guanyl NH_2 $|$ $C=NH$ $|$ NH $|$ R	$H+$	NH_2 $||$ $R-NH-C-NH_2^+$ Guanidinium	$H^\oplus$	
	⟨benzene⟩$-CH_2O\overset{O}{\overset{||}{C}}C2$ Carbobenzoxychloride	NH $||$ O $R-N-C-NH-\overset{||}{C}-OCH_2$⟨benzene⟩ $|$ $C-OCH_2$⟨benzene⟩ $||$ O Benzyloxycarbonyl	Z	$Na-NH_3$ or $HBr-HAc$ or H_2-Pd
Sulfhydryl RSH	⟨benzene⟩CH_2Cl Benzylchloride	$R-S-CH_2$⟨benzene⟩ Benzyl thioether	S Bzl	H_2-Pd, $Na-NH_3$
Alcohol or Phenol ROH				
$-$⟨benzene⟩$-OH$	Carbobenzoxychloride	O $||$ $RO-COCH_2$⟨benzene⟩ Benzyloxycarbonyl	Z	H_2-Pd
	Acetic anhydride	O $||$ $ROCCH_3$ Acetate	Ac	$Na-NH_3$, OH^-
	ClO_2S-⟨benzene⟩$-CH_3$ Tosyl chloride	O $||$ $ROS-$⟨benzene⟩$-CH_3$ $||$ O Tosylester	Tos	$Na-NH_3$
Imidazole ⟨imidazole ring⟩	None required Benzyl chloride	⟨imidazole ring⟩ CH_2⟨benzene⟩ Benzyl	Bzl	$Na-NH_3$

[a] T. Wieland and H. Determann, "The Chemistry of Peptides and Proteins," *Annual Review of Biochemistry*, **35**, 651 (1966).

TABLE 4.5   METHODS FOR FORMING PEPTIDE BONDS[a]

I. Carboxyl group activation

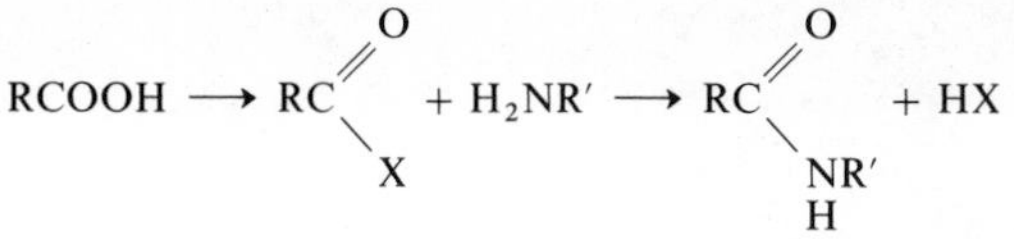

$$RCOOH \longrightarrow RC \underset{X}{\overset{O}{\diagdown}} + H_2NR' \longrightarrow RC \underset{\underset{H}{NR'}}{\overset{O}{\diagdown}} + HX$$

$$-X = -C_2, -N_3, -O_2CR'', -OP(OR'')_2, -OPO(OR'')_2,$$
$$-OCH_2CN, -OC_6H_4NO_2, -SC_6H_4NO_2$$

II. Amino group activation

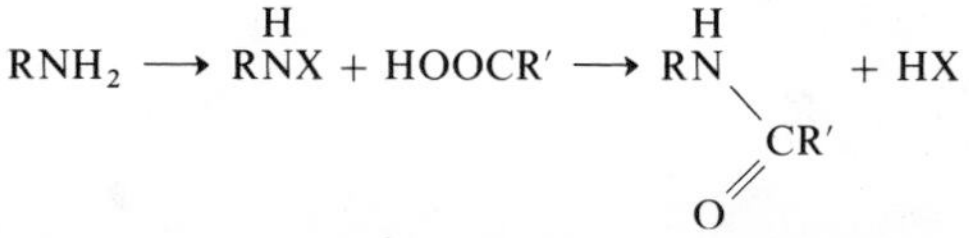

$$RNH_2 \longrightarrow \overset{H}{RNX} + HOOCR' \longrightarrow RN \underset{\underset{O}{\overset{\diagup}{CR'}}}{\overset{H}{\diagdown}} + HX$$

or

$$RNH_2 \longrightarrow RN{=}X + HOOCR' \longrightarrow RN \underset{\underset{O}{\overset{\diagup}{CR'}}}{\overset{H}{\diagdown}}$$

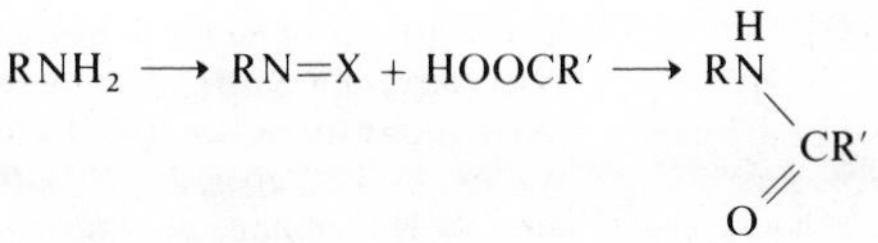

$$-X = -P(OR'')_2$$
$$=X = C{=}O, =P{=}NR$$

[a] T. Wieland and H. Determann, "The Chemistry of Peptides and Proteins," *Annual Review of Biochemistry*, **35**, 651 (1966).

the amino group is less frequently used, it can be accomplished by formation of amides of phosphorous or other acids which cause the nitrogen to be susceptible to nucleophilic attack by the carboxyl group.

A useful special case are the *N*-carboxy-α-amino acid anhydrides (NCA derivatives) which can condense with free amino acids at pH 10.2 to form peptides (Figure 4.5). This method was applied to the formation of about 40 percent of the peptide bonds in the synthesis of a tetrahectapeptide (104 amino acids) (ribonuclease *S*-protein) which had biological activity.

In addition to methods which involve the activation and isolation of "active" amino acid derivatives, methods have been devised which allow the activation and condensation to be performed in a single operation. As seen below, these methods have been utilized to make peptide synthesis much less tedious and significantly simpler. The methods involve use of "dehydrating" agents such as dicyclohexylcarbodimide (DCC), 3-isoxazolium salts, or tetraethylpyrophosphite for the condensations. The first two react to activate the carboxyl function and the last mentioned reacts to activate the amino group (Figure 4.6).

Perhaps the most significant advance in peptide synthesis is the solid-phase method (the process is illustrated in Figure 4.7). The desired carboxyl terminal *N*-*t*-butyloxycarbonyl amino acid is attached to a resin support (partially methylated polystyrene, 1 to 2 percent cross-linked) which can be isolated by

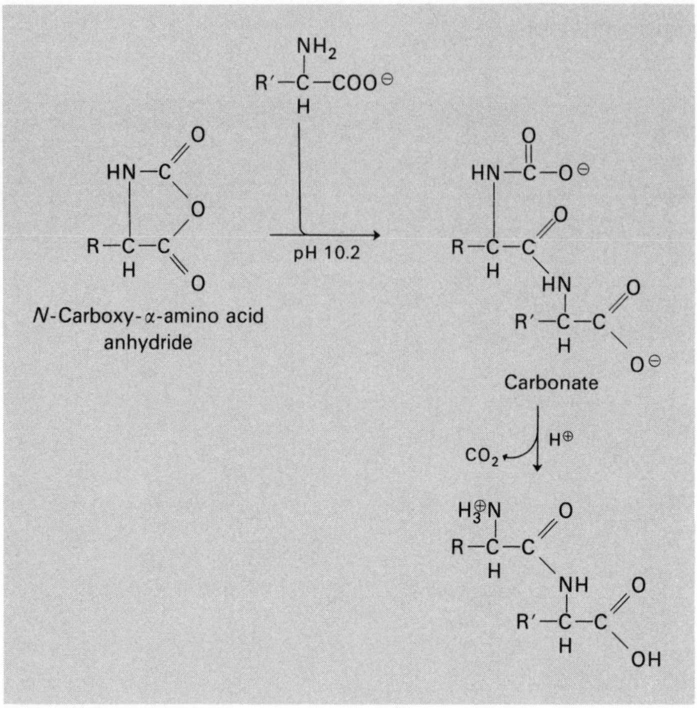

NH_2

$R'-\overset{|}{\underset{H}{C}}-COO^{\ominus}$

$HN-C\overset{O}{\underset{\diagdown}{\parallel}}$

$R-\overset{|}{\underset{H}{C}}-C\overset{O}{\underset{O}{\diagup}}$

N-Carboxy-α-amino acid
anhydride

$\xrightarrow{\text{pH } 10.2}$

$HN-\overset{O}{\overset{\parallel}{C}}-O^{\ominus}$

$R-\overset{|}{\underset{H}{C}}-C\overset{O}{\underset{HN}{\diagup}}$

$R'-\overset{|}{\underset{H}{C}}-C\overset{O}{\underset{O^{\ominus}}{\diagup}}$

Carbonate

$CO_2 \diagdown \overset{H^{\oplus}}{\diagdown}$

$H_3^{\oplus}N$

$R-\overset{|}{\underset{H}{C}}-C\overset{O}{\underset{NH}{\diagup}}$

$R'-\overset{|}{\underset{H}{C}}-C\overset{O}{\underset{OH}{\diagup}}$

Figure 4.5

The synthesis of peptides using the N-carboxy-α-amino acid anhydride (NCA) derivatives. See R. G. Denkewalter, D. F. Veber, F. W. Holly, and R. Hirschman, *J. Am. Chem. Soc.*, **91:2,** 502 (1969) and the four communications following.

filtration and purified by washing. The N blocking group is removed with trifluoroacetic acid in methylene chloride and, after isolation of the product by filtration, the next amino acid is added using the carbodimide procedure. The advantage of this method lies in the fact that purification is achieved by filtration which allows excess reagents to be used and undesirable products to be removed by simple washing, provided they are not joined to the polymer. Suitably blocked amino acids are now commercially available as in the chloromethylated polymer. The process has also been automated. The advantages of this technique are clearly demonstrated by the fact that bradykinin (a nonapeptide) was synthesized in 68 percent overall yield in 8 days. Each cycle, from cleavage of N-t-butyloxycarbonyl to isolation of the elongated N-t-butyloxycarbonyl peptide, requires 4 hr. The technique has also been used to synthesize ribonuclease—an enzyme containing 124 amino acids —which catalyzes the hydrolysis of phosphodiester bonds of ribonucleic acid. The synthetic product possessed the biological activity and physical characteristics of the natural material. The synthesis involved 369 chemical reactions and 11,931 automated steps without any intermediate isolation. The amino acid composition of ribonuclease is given in Table 4.6. Opposite each amino acid with a potentially reactive R group is listed the blocking agent used to protect it during the synthesis. These were selected so that they could withstand treatment with trifluoroacetic acid in methylene chloride to remove the N-t-butyloxycarbonyl groups and yet be removed by treatment with a

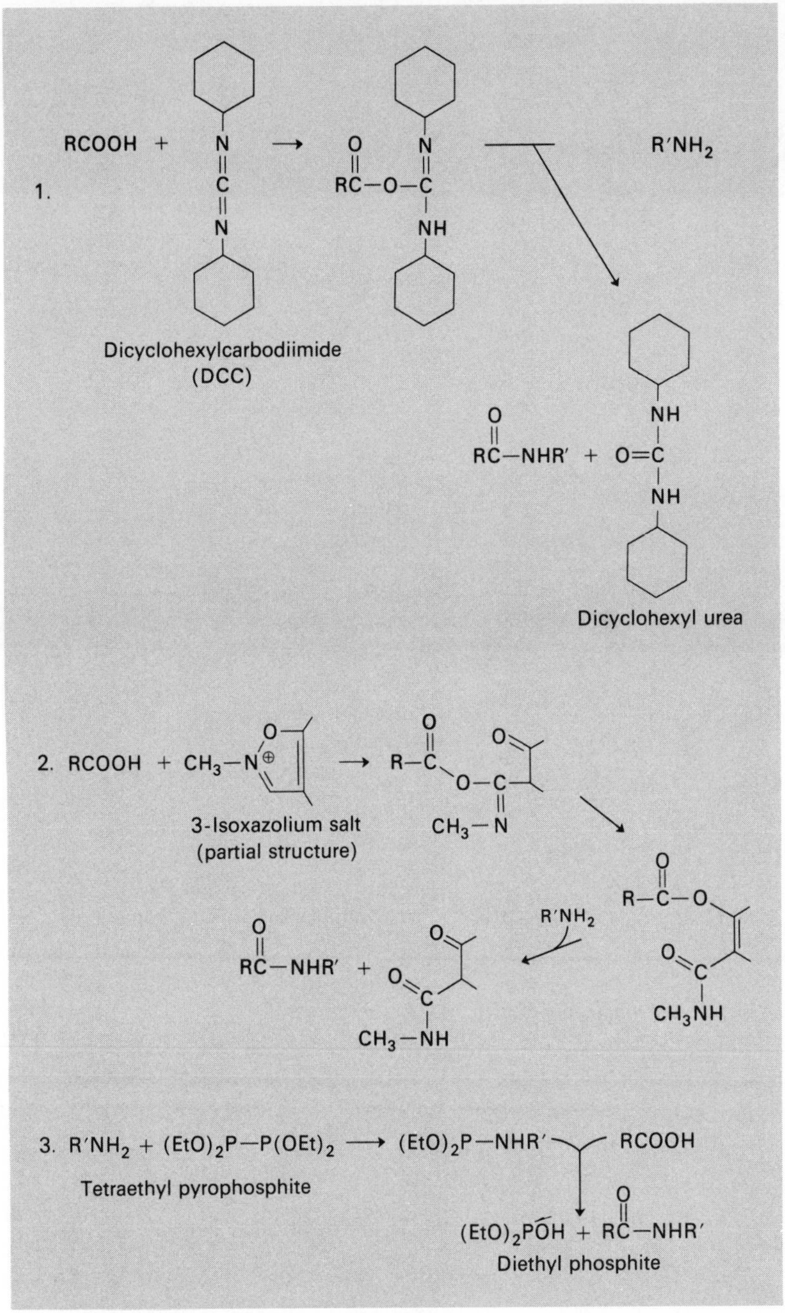

Figure 4.6   Reagents for peptide synthesis which allow activation and condensation to be carried out in a single step.

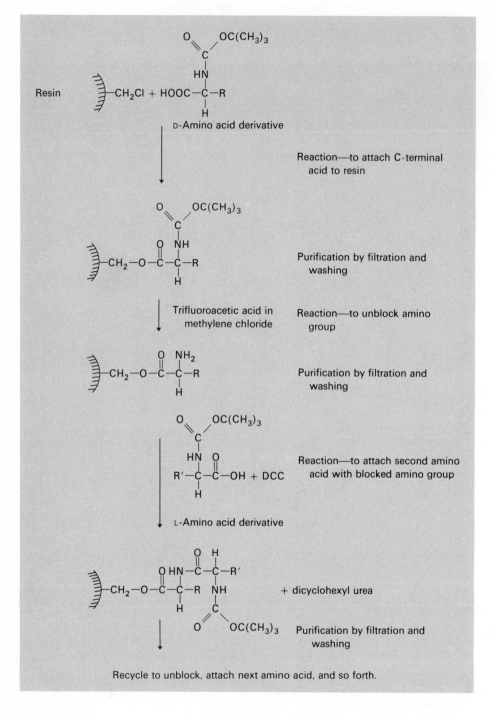

Figure 4.7   The Merrifield method for peptide synthesis. See B. Gutte and R. B. Merrifield, *J. Am. Chem. Soc.*, **91:2,** 501 (1969) for the application of this method to the synthesis of an enzyme with ribonuclease *A* activity.

TABLE 4.6   THE COMPOSITION OF RNase AND THE BLOCKING GROUPS USED IN ITS SYNTHESIS[a]

Amino acid	Moles/Mole of RNase	Blocking Groups
lys	10	$Z = -\overset{\displaystyle O}{\overset{\|}{C}}-O-CH_2-\langle\!\!\rangle$
his	4	None
arg	4	$Z(NO)_2 = -\overset{\displaystyle O}{\overset{\|}{C}}-O-CH_2-\langle\!\!\rangle NO_2$
asp	15	$O\ Bzl = -OCH_2-\langle\!\!\rangle$
thr	10	$Bzl = -CH_2-\langle\!\!\rangle$
ser	15	$Bzl = -CH_2-\langle\!\!\rangle$
glu	12	$O\ Bzl = -OCH_2-\langle\!\!\rangle$
pro	4	None
gly	3	None
ala	12	None
cys	8	$Bzl = -CH_2-\langle\!\!\rangle$
val	9	None
met	4	$O = =O, \text{sulfoxide } \overset{\displaystyle CH_3}{\underset{\|}{S}}=O$
ile	3	None
lev	2	None
tyr	6	None
phe	3	None

[a] B. Gutte and R. B. Merrifield, *J. Am. Chem. Soc.*, **91**:2, 501 (1969).

mixture of hydrofluoric acid, trifluoroacetic acid and anisole at 0–15°C. This latter treatment is sufficiently mild to preserve the bond between the peptide and the resin support. The only blocking group remaining was the sulfoxide derivative of methionine, which was reduced with mercaptoethanol to give the methionine side chain. The solid-phase method of peptide synthesis bears an interesting similarity to the biological process of protein synthesis in which carboxyl activated amino acids (attached to transfer ribonucleic acid *t*RNA) are added to the N terminus of a growing peptide chain which is bound to a solid support (the polysome). The polysome, in addition to providing a surface for reaction, provides the information directing the sequence of amino acids in the growing protein.

4.3   PEPTIDE STRUCTURE DETERMINATION

The determination of the structure of a polypeptide is a formidable task. It can be undertaken at several levels—amino acid composition, amino acid sequence (that is, the covalent structure), conformation of the polypeptide chain, or the location of R groups. In many naturally occurring polypeptides (proteins), multiple polypeptide chains and/or nonprotein components are present. These complications raise questions concerning numbers of chains, the manner in which they interact, the way in which nonprotein components are involved in the structure, and so forth. In this section the methods which can be used to examine polypeptide structure at increasingly complex levels are described.

A. Amino Acid Composition

The peptide bond can be cleaved in aqueous acid or alkali with the release intact of the amino acid residues. The usual conditions employed are $6\,N$ hydrochloric acid at 110°C under nitrogen for 20 hr. All amino acids except tryptophan, serine, threonine, and cysteine (also cystine) are stable under these conditions. Tryptophan is completely destroyed but the destruction of serine, threonine, and cysteine is incomplete. The presence of traces of oxygen decreases the recovery of these amino acids and of methionine, which is partially oxidized to the sulfoxide or sulfone; cysteine is converted in part to cysteic acid. To obtain relatively accurate values for the partially destroyed amino acids, hydrolyses are performed for varying lengths of time—24, 48, and 72 hr,

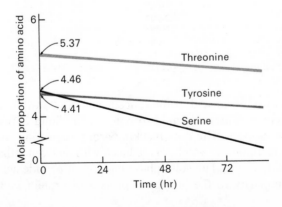

Figure 4.8

The evaluation of the serine, tyrosine, and threonine content of a polypeptide. Values obtained by amino acid analysis after 24, 48, and 72 hr of hydrolysis in $6\,N$ HCl at 110°C are plotted versus time and the "true" value is obtained by extrapolation to zero time. Note that the values so obtained are not integers, an indication of the inaccuracy of the method which has a maximum precision of $\pm 2\%$.

(a) Analysis by ninhydrin

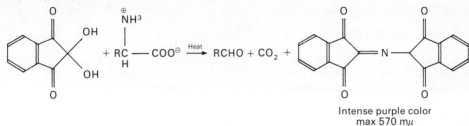

Intense purple color
max 570 mμ

(b)

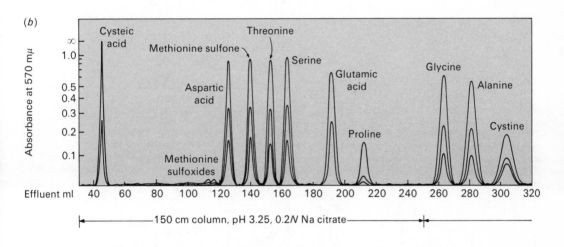

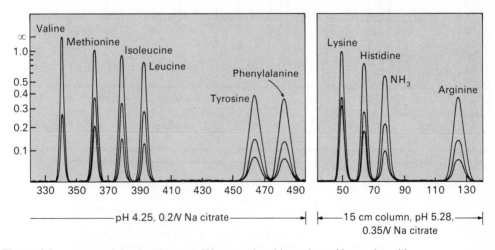

Figure 4.9   Record of the absorbance at 570 nm produced by amino acids reacting with ninhydrin following their separation on an automatic amino acid analyzer. (a) The ninhydrin color reaction. Proline and hydroxyproline give a yellow pigment. The color yield varies with amino acid structure but is highly reproducible. (b) Record showing buffers used to elute the amino acids from sulfonated polystyrene. D. H. Spackman, W. H. Stein, and S. Moore, *Anal. Chem.*, **30**, 1,190 (1958). Copyright © 1958 by the American Chemical Society. Reprinted with permission of the copyright owner.

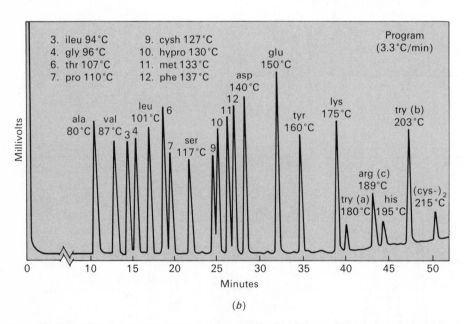

(a)

(b)

Figure 4.10   Gas-liquid chromatographic analysis of a mixture of amino acids as *n*-butyl-
N^α-trifluoroacetyl derivatives. (*a*) The preparation of derivatives. (*b*) Record
of the separation on a column containing diethyleneglycol succinate/ethylene-
glycol succinate methyl silicone polymer (0.75/0.25%) on chromosorb W
with N_2 carrier. (1) diacyl derivative, (2) monoacyl derivative, (3) acylated for
18 hr and added to the prior mixture. [W. Gehrke and F. Shahrokhi. *Anal.
Biochem.*, **15**, 97 (1966)]. Reprinted with permission of Academic Press, Inc.

for example—and the molar proportion of amino acid versus time plotted.
Extrapolation to zero time gives a fairly accurate value (Figure 4.8).

The quantitative analysis of the amino acid mixture obtained by acid hydrolysis is accomplished either by ion-exchange chromatography using sulfonated

polystyrene resins or by gas-liquid chromatography. The ion-exchange procedure is fully automated and a typical separation is shown in Figure 4.9. The relative concentrations of amino acids are determined by their reaction with ninhydrin. The gas-liquid chromatographic procedures are not yet as well standardized but offer possibilities for more rapid analysis. One of the more promising methods is illustrated in Figure 4.10. This procedure requires the reproducible, quantitative formation of volatile derivatives of all the amino acids. To be practical, the same reagents must be used to derivatize all the amino acids in the mixture. Although this is difficult to accomplish, recently reported procedures are promising.

Tryptophan can be determined following hydrolysis of the polypeptide with $2N$ alkali at 100°C. This treatment results in the destruction of arginine, cystine, serine, and threonine and racemizes (produces a DL mixture) the other amino acids. Tryptophan can also be determined colorimetrically by several methods. For example, the decrease in absorbance in the protein at 280 nm following treatment with N-bromosuccinimide can be used. The indole residue is converted to an oxindole derivative which has its absorption maximum at 250 nm (Table 4.3).

B. Amino Acid Sequence-Covalent Structure

Perhaps this section should be titled *the determination of polypeptide covalent structure*. The covalent structure consists of three parts: the amino acid structure, the peptide bonds between specific amino acids forming a unique sequence, and the disulfide bonds which may be present and which link different regions of the polypeptide chain together. The disulfide covalent bonds are unusual. They can be formed from suitably situated thiol groups under mild oxidizing conditions such as molecular oxygen at pH 7.0, and they can be cleaved by exchange with thiol groups under equally mild conditions. The covalent structure of the amino acids is taken as established and is not discussed further. The methods used to establish amino acid sequence are described first and then the location of disulfide bridges.

Nomenclature. By convention, peptide sequences are written with the N terminus to the left, using the abbreviations given in Table 4.1. For example, a dipeptide formed from glycine and alanine with glycine at the amino end is termed "gly-ala" or "G-A." With the exception of a few cyclic polypeptides, polypeptide chains have amino and carboxyl termini which are usually present as the $-NH_3^\oplus$ and $-COO^\ominus$ forms. Thus, only one amino acid in the chain has a free α-amino group and only one has a free α-carboxyl group. Sequencing procedures depend on the identification of these groups. A procedure which allows only their identification is useful in establishing the number of terminals and therefore the number of polypeptide chains. However, a reagent or procedure which allows both the identification of the terminal group and the recovery of the remaining polypeptide can be applied repeatedly and a sequence established. Both types of reagents are listed in Table 4.7.

TABLE 4.7　REAGENTS USED TO IDENTIFY TERMINAL AMINO ACIDS IN PEPTIDES[a]

1. Amino terminal

Reagent	Reacts with	Product

Fluorodinitrobenzene
　(FDNB)

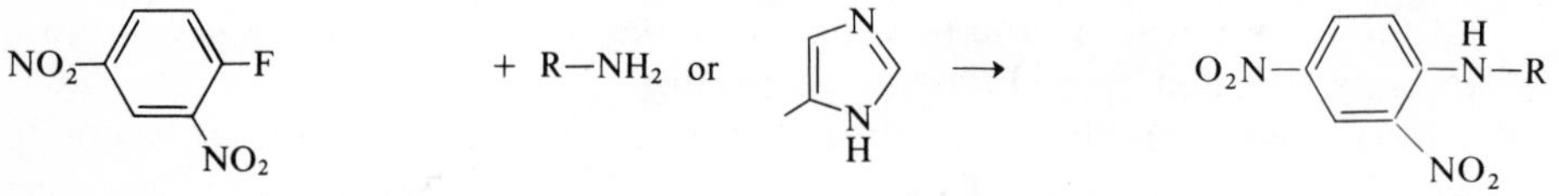

In dilute aqueous $NaHCO_3$
　at room temperature

Characteristic yellow compound stable
to acid hydrolysis

Amino terminal residue can be separated after hydrolysis of the modified peptide and identified by comparison to standards.

Dansyl chloride (DNS—Cl)

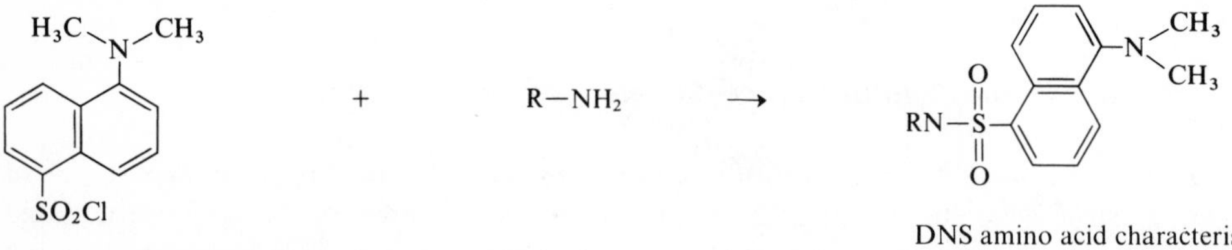

DNS amino acid characteristic fluorescence stable to acid hydrolysis

These derivatives can be detected in smaller amounts than the DNB derivatives and are more stable toward hydrolysis. The products, which include those from the side chains of lys and tyr, can be separated by electrophoresis and compared with standards.

Phenylisothiocyanate

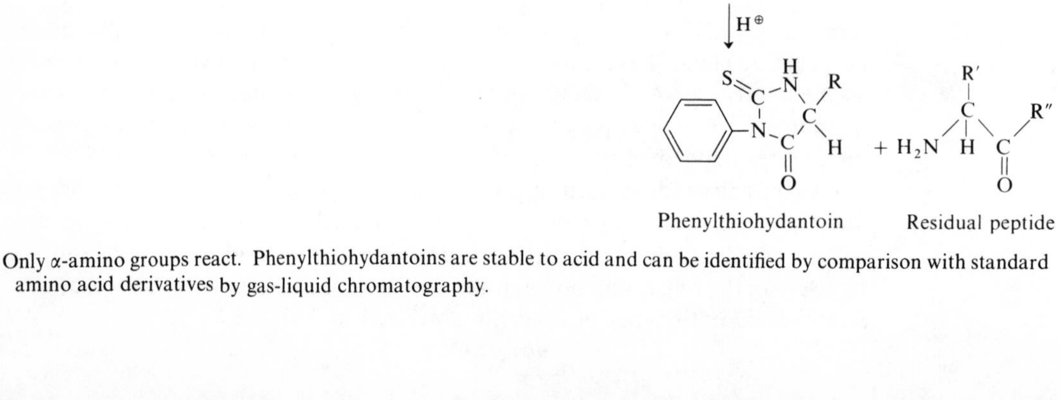

Phenylthiocarbonate

Phenylthiohydantoin　　　　Residual peptide

Only α-amino groups react. Phenylthiohydantoins are stable to acid and can be identified by comparison with standard amino acid derivatives by gas-liquid chromatography.

TABLE 4.7   REAGENTS USED TO IDENTIFY TERMINAL AMINO ACIDS IN PEPTIDES[a] (cont.)

1. Amino terminal

Reagent	Reacts with	Product

Potassium cyanate

$K^{\oplus \ominus}NCO$ + [amino acid structure] $\xrightarrow{pH\ 8}$ [carbamoyl derivative structure] $\xrightarrow{H^\oplus}$ [hydantoin structure] $+ H_2NR'$

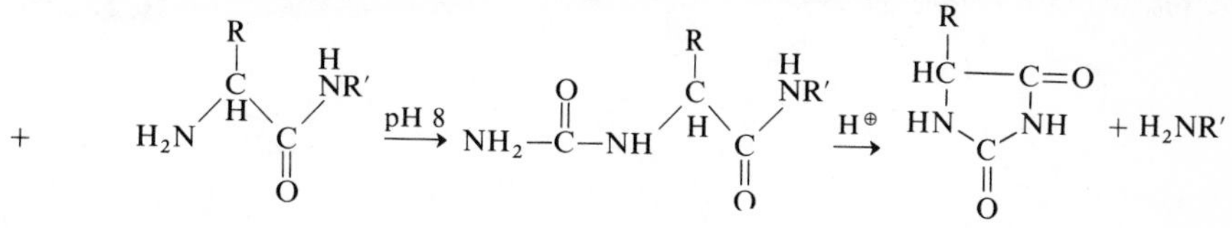

Reaction very similar to that for phenylthiohydantoin.   Carbamoyl derivative   Hydantoin

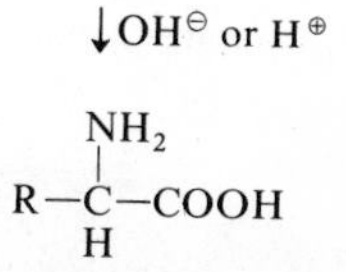

$\downarrow OH^\ominus$ or $H^\oplus$

$$\underset{H}{\overset{NH_2}{R-C-COOH}}$$

N-terminal amino acid

Leucine amino peptidase, an enzyme from swine kidney + [peptide structure] $\xrightarrow{H_2O}$ [N-terminal amino acid] + [residual peptide]

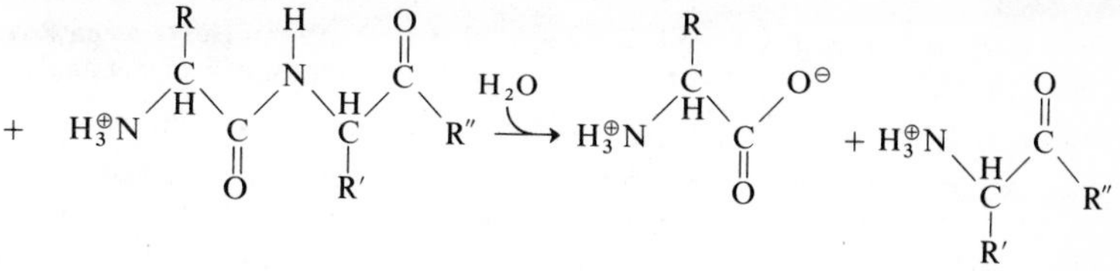

N-terminal amino acid   Residual peptide

The enzyme is specific for L-amino acids. It acts most rapidly when R is alkyl or aryl but will attack bonds involving other amino acids. It will not cleave the amide bond of pro. Since the residual peptide is also susceptible to attack by the enzyme, amino acids are released in series.

2. Carboxyl terminal

Hydrazine

N_2H_4 + [peptide structure] $\rightarrow$ [hydrazide structure]

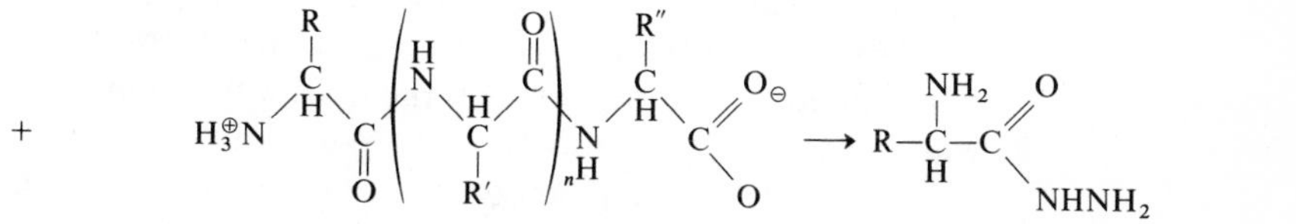

Anhydrous, 90°C, 20–100 hr, with resin catalyst. The free amino acid from the C-terminal end can be isolated and identified.

Hydrazides of all amino acids except

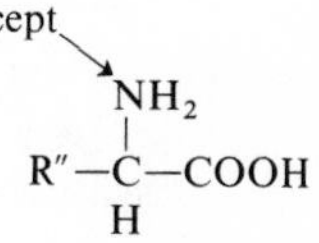

$$\underset{H}{\overset{NH_2}{R''-C-COOH}}$$

TABLE 4.7   REAGENTS USED TO IDENTIFY TERMINAL AMINO ACIDS IN PEPTIDES[a] (cont.)

2. Carboxyl terminal

Reagent	Reacts with	Product
Methanolic-hydrochloric acid		

(a) $CH_3OH-HCl$

$$RC\overset{O}{\underset{OH}{\diagdown}} \xrightarrow{(a)} RC\overset{O}{\underset{OCH_3}{\diagdown}} \xrightarrow{(b)} RCH_2OH$$

Lithium borohydride

Alcohol from C-terminal amino acid

(b) $LiBH_4$

C-terminal amino acids yield amino alcohols which can be identified by comparison with standards.

Carboxypeptidases A and B + from swine pancreas

Residual peptide   C-terminal amino acid

The A enzyme cleaves best when R is an alkyl or aryl. It will not cleave when R or R′ is pro but will remove gly and acidic amino acids slowly. Lys and arg are also cleaved very slowly or not at all.

The B enzyme removes lys and arg rapidly and all others very slowly, pro not at all.

Since the residual peptide may be a substrate, the amino acids may be released in sequence. Careful kinetic analysis of amino acid release can give the sequence of up to 10 residues under ideal conditions.

[a] *Methods in Enzymology*, XI (C. H. W. Hirs, ed.) Section II, Academic Press, New York, 1967, pp. 125–168, articles by G. R. Stark, W. R. Gray, H. Fraenkel-Conrat, and Chun Ming Tsung; and R. P. Ambler.

By far the most useful of the sequencing procedures is the phenylisothiocyanate method for removal and identification of the N-terminal amino acid. This method has recently been automated and has been used to determine the sequences in large polypeptides. The utility of the method lies in the fact that the reaction is very nearly quantitative. If it were not, then the residual peptide would be impure and the second application of the reaction would produce two phenylthiohydantoins; the third, three; and so forth. It can be appreciated that at the nth step the question of which of the various products was due to the nth amino acid would be difficult to answer.

The two enzymatic methods listed in the table can be applied in some cases. The enzymes have fairly broad specificity—that is, they can remove terminal amino acids of various kinds from peptides of a range of sizes; however, neither enzyme attacks dipeptides. The application of these enzymes to sequence determinations is complicated by the fact that the first reaction produces a peptide with one less amino acid which can also be cleaved by the enzyme.

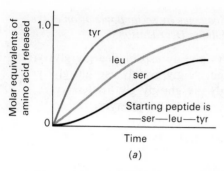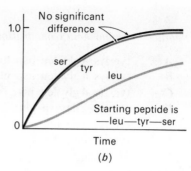

Figure 4.11   The release of carboxypeptidase of amino acids from peptides having the terminal sequences shown. (*a*) All bonds cleaved at the same rate. (*b*) Ser removed slowly; tyr cleaved rapidly; leu cleaved at same rate as tyr.

If the enzyme acts at the same rate on both the original and product peptides, then the release of amino acids one, two, three, and so forth should follow the time course shown in Figure 4.11(*a*). If the second amino acid is released much faster than the first and the third at an intermediate rate, however, the events shown in Figure 4.11(*b*) would be observed. In this case, it is very difficult to decide which of the amino acids (ser or tyr) is on the terminus except when relative rates of release are known from study of model compounds.

Most sequence determinations are limited to small peptides; therefore, larger polypeptides need to be cleaved into smaller pieces for sequence determination. The pieces are then separated, purified, and sequenced. It is usually necessary to use two different cleaving reagents to produce two different sets of peptides from the starting polypeptide so that a unique sequence can be worked out. A hypothetical example is given below.

A small protein, MW ~ 6,000, has N-terminal leu, C-terminal tyr, contains no tryptophan, and after hydrolysis in 6 *N* HCl for 12 hr at 100°C gives 6 tyr, 6 lys, 2 leu, 12 gly, 7 ile, 4 asp, 3 his, and 3 glu.

We can write:

leu————————————————————————————tyr

Experiment 1:

Treatment with carboxypeptidase which removes amino acids serially from the carboxyl terminal end: tyr, gly, leu, his (trace), asp (trace) are released.

Experiment 2:

Treatment with chymotrypsin cleaves on the carboxyl side of phenylalanine and tyrosine and gives six peptides.

Peptide	Amino Acid Composition	Sequence by serial application of the Edman Degradation
A	tyr·lys·4 gly·leu·asp·his·	lys·gly·gly·asp·his·leu·gly·gly·tyr·
B	tyr·lys·4 gly·leu·asp·glu·	leu·lys·gly·gly·gly·asp·glu·gly·tyr·
C	tyr·lys·4 gly·leu·his	gly·gly·glu·gly·gly·his·lys·tyr·
D	tyr·lys·2 ile·asp·glu·	ile·ile·asp·lys·glu·tyr·
E	tyr·lys·2 ile·his·	ile·his·lys·ile·tyr·
F	tyr·lys·ile·asp·	asp·ile·lys·tyr·

Deduction:

1. Peptide *B* must be N-terminal in the polypeptide since it is the only one with an N-terminal leucine.
2. Only peptides *A* and *B* contain leucine and *B* is the N-terminal peptide; therefore, peptide *A* must be the C-terminal peptide (leucine was released by carboxypeptidase).

We can write:

$$
B \overline{\quad\quad} \left| \begin{array}{c} CDEF \\ \text{order} \\ \text{unknown} \end{array} \right| \overline{\quad\quad} A
$$

N-terminal   C-terminal

Experiment 3:

The polypeptide is treated with trypsin, which cleaves on the carboxyl side of basic amino acids. Seven peptides are released, separated, and their composition (not sequence) determined. These are designated *M, N, O, P, Q, R, S.*

1. Peptide *M* is a dipeptide containing only tyrosine and lysine. Now the chymotrypsin peptides each contain 1 molecule of lysine, and they must fit together in an arrangement which can be diagrammed (↓ indicates the position of peptide cleavage by trypsin):

Chymotrypsin peptides

leu · lys—tyr—lys—tyr—lys—tyr—lys—tyr—lys—tyr   lys—tyr
　　　　B　　　　　　　　　　　　　　　　　　　　　　　*A*

tyr · lys
M

Peptide *M* could only arise from the sequence lys · tyr · lys which would only exist if peptide *C* preceded peptide *A*. We can now write:

B (DEF) C · A

2. Peptide *N* contains 4 gly · asp · his · tyr; it is equal to peptide *A* without its N-terminal lysine. This does not help to elucidate the structure but confirms the assignment above.

3. Peptide *O* contains 4 gly · glu · his · tyr · ile · lys. It has the same composition as peptide *C* except that it contains ile. Now the C-terminal tyr of peptide *C* has already been acounted for (see item 1 of Experiment 3) and the tyrosine in this peptide (*O*) must have come from the C terminus of another of the chymotrypsin peptides—and this peptide must have had the C-terminal sequence lys · ile · tyr. This sequence is present in peptide *E*.

We can now write:

$$B \quad (DF) \quad E.C.A.$$

4. Peptide *P* contains 4 gly · 2 asp · glu · tyr · ile · lys. It must correspond to peptide *B* plus a portion of the peptide on its C terminus. This portion must contain asp · ile · lys ·, which is found as a sequence in peptide *F*.

We can now write:

$$B.F.D.E.C.A. \qquad \text{as the total sequence.}$$

No sequence studies were necessary on the trypsin peptides.

Some of the reagents which can be used to cleave specific bonds are listed in Table 4.8. Note that a number of enzymes are listed. These natural catalysts are themselves polypeptides; however, since they are catalysts, only small amounts are used and they offer the advantage of increased specificity over

TABLE 4.8   THE CLEAVAGE OF PEPTIDE BONDS[a]

I. Chemical Methods

Cyanogen bromide (CNBr), cleaves at methionyl residues

Peptidyl homoserine   Peptide
lactone

Polypeptide must be unfolded to expose methionyl residues. This is the only highly specific chemical cleavage which gives a high yield.

TABLE 4.8   THE CLEAVAGE OF PEPTIDE BONDS^a (cont.)

I. Chemical Methods

Proton ($H^\oplus$), cleaves at aspartyl residues

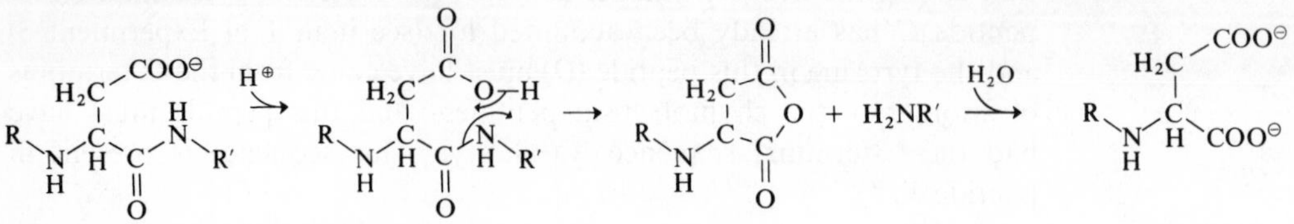

Acyl rearrangement (N $\longrightarrow$ 0), cleaves at seryl and threonyl residues

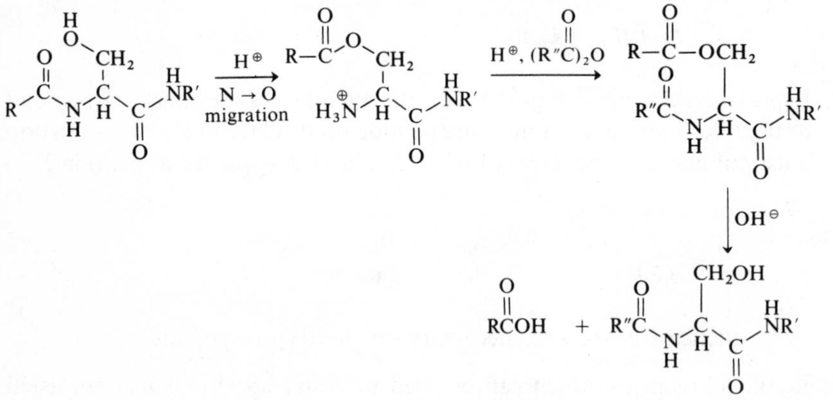

Peptide or   Acyl peptide
amino acid

N-Bromosuccinimide

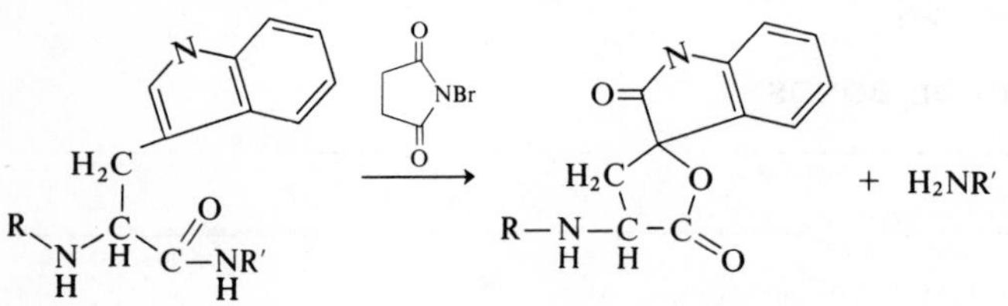

Tryptophanyl peptide   Peptidyl lactone   Peptide

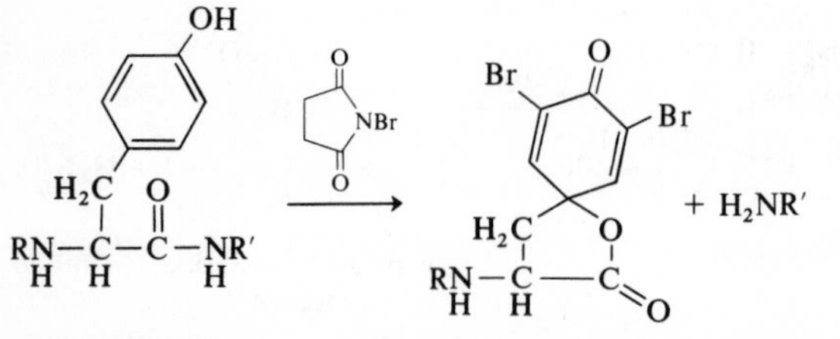

Tyrosyl peptide   Peptidyl lactone   Peptide

Histidyl peptides are cleaved much more slowly.

TABLE 4.8   THE CLEAVAGE OF PEPTIDE BONDS^a (cont.)

I. Chemical Methods

Electrolytic oxidation

$$\text{RN}-\underset{\text{H}}{\overset{\text{H}}{\text{C}}}-\overset{\text{O}}{\underset{}{\text{C}}}-\underset{\text{H}}{\text{NR}'} \longrightarrow \text{RN}-\underset{\text{H}}{\overset{\text{H}}{\text{C}}}-\overset{}{\underset{\text{H}}{\text{C}}}=\overset{\oplus}{\text{NR}'} \xrightarrow[\text{H}_2\text{O}]{\text{pH 3.0}} \text{RN}-\underset{\text{H}}{\overset{\text{H}}{\text{C}}}-\overset{}{\text{C}}\diagdown_{\text{O}} + \text{R}'\text{NH}_2$$

Tyrosyl peptide   Peptidyl lactone   Peptide

S-aminoethylation + trypsin

$$\text{Br}-\text{CH}_2\text{CH}_2-\text{NH}_2 \longrightarrow \underset{\overset{\oplus}{\text{H}}_2\text{Br}^{\ominus}}{\overset{\text{H}_2\text{C}-\text{CH}_2}{\text{N}}} + \text{R}-\underset{\text{H}}{\overset{\text{SH}}{\text{N}}}-\underset{\text{H}}{\overset{\text{CH}_2}{\text{C}}}-\underset{\text{O}}{\overset{}{\text{C}}}-\underset{\text{H}}{\text{NR}'} \longrightarrow \text{RN}-\text{C}-\text{C}-\text{NR}'$$

Cysteinyl peptide   S-Aminoethylated peptide

↓ H$_2$O-trypsin

S-Aminoethyl derivative is an analog of lysine and trypsin and can act on the adjacent peptide bond.

Peptide

the more vigorous chemical methods. It is usually necessary to treat a poly-peptide to unfold it before enzymatic hydrolysis can be accomplished. Un-folding can be accomplished by treatment with acid or with agents such as urea or guanidinium chloride. This pretreatment may also utilize reagents to cleave disulfide bonds such as mercaptoethanol, performic acid, or iodo-acetate since the folding they produce may inhibit enzyme action. The location

TABLE 4.8   **THE CLEAVAGE OF PEPTIDE BONDS**[a] *(cont.)*

II. Enzymatic Methods

	Cleavage Point	Identity of R group	Source, MW and pH optimum
Trypsin	2	Lysine or arginine (modified cysteine)	Pancreas 24,000 8–9
Chymotrypsin	2	Rapidly—tyrosine, tryptophan, phenylalanine Slowly—asparagine, glutamine, histidine, leucine, lysine, methionine, serine, threonine	Pancreas 24,300 7–8
Pepsin	1 or 2	Rapidly—tyrosine, tryptophan, phenylalanine, leucine Slowly—glutamic, cysteine, cystine, alanine	Gastric juice 32,700 1.5–2.0
Papain	2	Most rapidly—arginine, lysine Slowly—most amino acids	Papaya 21,000 Broad pH range
Subtilisin	—	Non specific	B. Subtilis 27,500
Pronase	—	Non specific	S. Griseus

[a] *Methods in Enzymology*, XI, C. H. W. Hirs, ed., Section V, Academic Press, New York, 1967, pp. 211–324, many contributing authors.

of disulfide bridges of cystine between parts of a peptide chain requires special techniques. An effective approach is illustrated in Figure 4.12(*b*).

In a similar way the location of a modification in a specific R group can be established. A polypeptide is subjected to a chemical or enzymatic degradation procedure and the resulting peptides are separated on a large sheet of filter paper by chromatography in one direction followed by electrophoresis at right angles to the chromatographic development. In this way, the peptides are distributed over the paper in an array which depends on the polypeptide structure, the digestion, and the separation procedures. If the separation procedures are held constant, then changes in the structure will show up by the disappearance of some spots and the appearance of others (Figure 4.13).

C. Factors Determining Peptide Conformation

On the basis of the foregoing discussion, a polypeptide may appear to be a stringlike molecule—a linear polymer. In many instances, this is found not to be the case due to four characteristics of polypeptides. First, because they

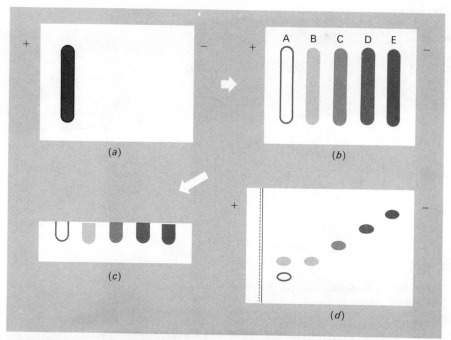

Figure 4.12   Diagonal electrophoresis—a method for establishing the location of disulfide bonds. (a) Mixture of peptides from enzymic digestion of polypeptide streaked on paper and subjected to electrophoresis. Disulfide bonds are intact. (b) Peptides separated and guide strip cut off. Peptides are not stained and separation is not visible. (c) Guide strip exposed to formic acid vapor to oxidize —S—S— linkages to —SO₃H   HO₃S—, thus cleaving disulfides. (d) Guide strip is sewn to a second sheet, subjected to electrophoresis under the same conditions, and then stained. Peptides not on a diagonal contain cysteic acid, therefore peptide B has disulfide bridge and can be located on and removed from the first electrophoretogram.

contain a variety of different functional groups, there are possibilities for inter-actions between groups which will stabilize or destabilize a specific confor-mation. These interactions range from very weak ones due to van der Waals forces and hydrogen bonds to strong ones such as disulfide bonds† between cysteine residues in different parts of the chain. Second, certain conformations of the peptide chain are forbidden by steric repulsions between bulky groups. Third, the solvent in biological systems—water—plays a role either by solvation of hydrophilic groups or by its inability to solvate hydrophobic groups, requiring that they be sequestered in cavities in the solvent (see Section 2.4). Fourth, the peptide bond itself has a planar structure and is always *trans*.

---

† Disulfide bonds should not be considered as contributing to conformational stability since they are covalent. It is convenient to include them in this discussion, however, because they result in folding of the polypeptide chain in much the same way as the noncovalent bonds do.

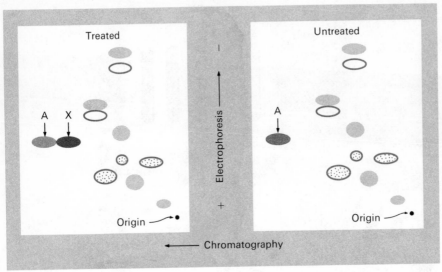

Figure 4.13   A diagrammatic representation of comparative peptide maps to show the effect of carboxypeptidase treatment on a polypeptide. Treatment with carboxy-peptidase (in 4 *M* urea to cause unfolding) released leucine. The protein was then dialyzed to remove urea and the released leucine and then treated with trypsin. A sample of polypeptide subjected to urea treatment but not to carboxypeptidase treatment was also dialyzed and digested. The treated and untreated samples were subjected to chromatography and then electrophoresis on separate sheets of paper and the sheets stained with ninhydrin. The two maps are identical except for the decreased intensity of spot A in the treated map and the appearance in it of a new spot X. Spot A must be due to the carboxyl terminal peptide and can now be isolated and its sequence determined.

The peptide bond is shown in Figure 4.14 with the bond distances and angles estimated by X-ray crystallography. The resonance contributor shown in Figure 4.14(*b*) is sufficiently important to require that the six atoms enclosed by the line in the figure remain in a plane. In a polypeptide, all the atoms in the backbone lie in these planar regions and the only flexible elements are the bond between the α and the carbonyl carbon (C^α—C') and the bond between the nitrogen and the α carbon (N—C^α). Thus the chain can assume only those conformations which result from rotations about these two bonds. In Figure 4.14(*c*) a pair of adjacent peptide bonds is shown arranged so that the angles of rotation about both the C^α—C' bond (ψ) and the N—C^α bond (ϕ) are zero. This is the conformation of the fully extended chain. When the chain is viewed from the N terminus and clockwise rotations are performed, the rotations are considered to be positive.

It should be clear that even the restriction in the number of conformers produced by the planarity of the peptide bond allows an enormous number of conformers to exist. As the chain length is increased, however, the possibility also increases that parts of the structure, well separated in terms of their position in the backbone, will be brought into position to interact. Many rotational

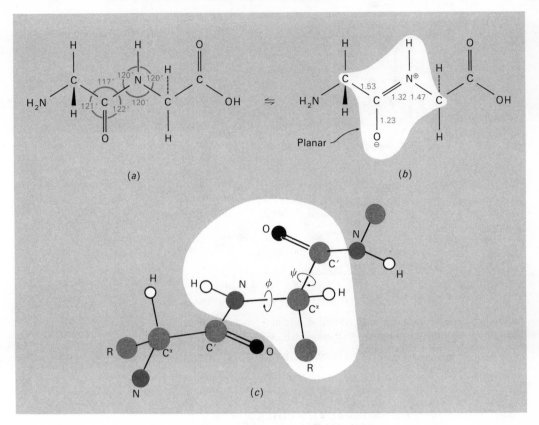

Figure 4.14   The peptide bond. (*a*) Bond distances and angles established by X-ray crystallographic analysis. (*b*) The contributing resonance form which results in the planar configuration of the bond. (*c*) A stereoscopic diagram of a pair of peptide bonds in the fully extended configuration. ϕ and ψ refer to the angles within a single amino acid residue—enclosed by the line in the diagram. Both ϕ and ψ equal 180° in this projection and adjacent peptide bonds are coplanar.

conformers are impossible because they would bring such groups into over-lapping positions. (No space can be occupied by more than one atom!) It is possible to construct a map showing those combinations of ψ and ϕ which are possible for a simple polypeptide. Such a map is shown in Figure 4.15. The areas included within the dashed lines encompass the allowed rotational angles for poly-L-alanine, assuming that the contact distances (that is, the sum of the van der Waals radii of the atoms) are smaller than usually allowed (Table 4.9).

As pointed out above, there are four factors which tend to stabilize conformations in a peptide chain. One of them—the *trans*-planar arrangement of the peptide bond together with the requirement that no two atoms come within less than the sums of their van der Waals radii—has been shown to limit the conformations available. If the further provision is made that conformations will be stabilized by the formation of weak bonds between suitable

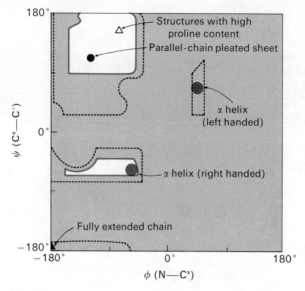

Figure 4.15   Conformational map (exclusion diagram) for polypeptides. Conformations allowed by the van der Waals contact distances shown in Table 4.9. – – – – outer limit distances; ——— normally allowed distances. (Following the recommendations of the IUPAC–IUB Commission on Biochemical Nomenclature.)

TABLE 4.9   NORMAL AND OUTER LIMIT CONTACT DISTANCES BETWEEN ATOMS IN POLYPEPTIDES[a]

Contact[c]	Normally allowed (Å)	Outer[b] limit (Å)
C...C	3.20	3.00
C...O	2.80	2.70
C...N	2.90	2.80
C...H	2.40	2.20
O...O	2.80	2.70
O...N	2.70	2.60
O...H	2.40	2.20
N...N	2.70	2.60
N...H	2.40	2.20
H...H	2.00	1.90

[a] G. N. Ramachandran, C. Ramakrishnan, and V. Sasisekharan, *J. Mol. Biol.*, 7, 95 (1963).

[b] Outer limit distances were obtained by analysis of structural data on a variety of amino acids and peptides.

[c] Contact distance equals the sum of the van der Waals radii of the atoms.

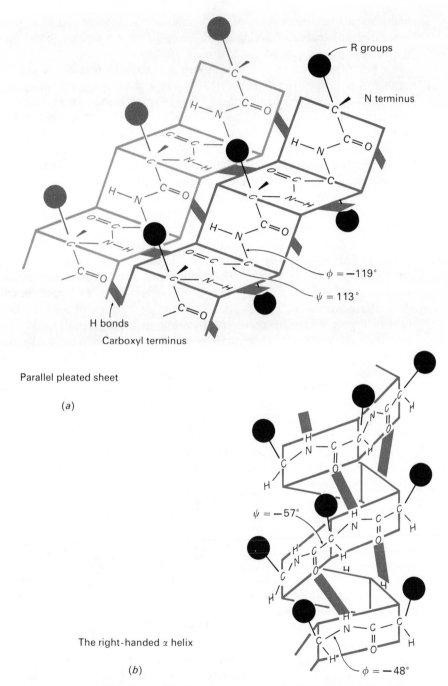

Parallel pleated sheet

(a)

The right-handed α helix

(b)

Figure 4.16   Hydrogen-bond stabilized structures for polypeptide chains. (a) Parallel pleated sheet; hydrogen bonds between adjacent chains are shown in color. The planar regions of the peptide bonds are represented by the planes. (b) The α helix showing planar peptide bonds and hydrogen bonds between every fourth bond.

neighbors, further restrictions of conformation can be expected. The peptide bond itself is capable of acting as an acceptor in a hydrogen bond (at the oxygen) and as a donor (at the hydrogen attached to nitrogen).

If a polypeptide chain is folded in regular ways to produce a maximum number of hydrogen bonds between —N—H and O=C—, a number of distinct conformations can be produced. Two of these are shown in Figure 4.16; in each the planar regions of the peptide bond are shown as plates and the angles between adjacent plates are determined by ϕ and ψ. The first (a) is a pleated sheet or β structure in which polypeptide chains are arranged in parallel with interchain hydrogen bonds stabilizing the structure. The R groups of adjacent amino acid residues are above and below the plane of the sheet. The values of ϕ and ψ are given in the figure. A similar structure is possible with the adjacent chains *anti*-parallel (that is, with the amino terminus of one chain adjacent to the carboxyl terminus of its neighbors).

The second conformation is a right-handed helix which has 3.6 residues per turn. The structure is stabilized by intrachain hydrogen bonds. This structure is termed the α *helix*. There are other stable regular arrangements possible in which all peptide linkages are involved in hydrogen bonding. All of those which are important fall within (or very close to) the allowed areas of Figure 4.15. The presence of proline in a polypeptide produces an important exception to the generalities discussed above. This cyclic imino acid and its hydroxy analog do not have free rotation about the N—C$^\alpha$ bond and cannot fit into a regular α helix or pleated sheet. Their presence in a sequence will tend to disrupt regular structures. If prolines are present in large proportion, however, other types of regular structures can be formed which fall within the allowed areas (Figure 4.15).

Other types of interactions which might be important in stabilizing polypeptide conformations are those which involve the R groups (R—R' interactions and R—peptide bond interactions). These interactions are not as likely to produce regular structures such as the hydrogen bonds discussed above, but they are equally important in determining stable conformations which are not required to be regular. Some of the types of interaction are diagrammed in Figure 4.17 where an indication is also given of the strengths of the interactions —that is, the stability they can confer on a particular conformation.

It is currently thought that hydrophobic interactions are probably the most important in determining the way in which a polypeptide chain folds. The argument is as follows: A polypeptide having 300 amino acid residues would, if completely arranged in an α helix, be about 300 Å long. If it contained proline, it would resemble a coil spring with hinges in it at each proline residue and would easily fold upon itself to form a globular structure. This folding would tend to bring polar groups which can solvate readily to the outside and nonpolar groups to the inside to form hydrophobic interactions. The polypeptide would thus tend toward a spherical shape with a polar and charged surface and a lipid-like interior. The driving force for the formation of hydrophobic bonds appears to be the increase in entropy which occurs when they form

		Approximate Stabilization Energy kcal mole^{-1}
 $>C=O \cdots H-N<$	Hydrogen bond between peptides	2–5
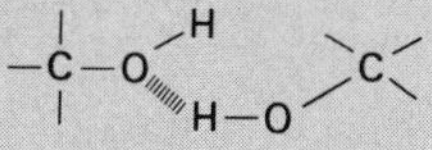 $-C-O \cdots H-O-C<$ (with H above O)	Hydrogen bond between neutral groups	2–5
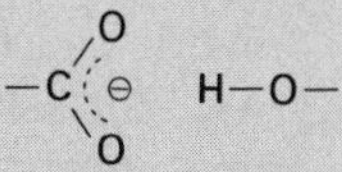 $-C(O)(O)^{\ominus} \quad H-O-$	Hydrogen bond between neutral and charged groups	2–5
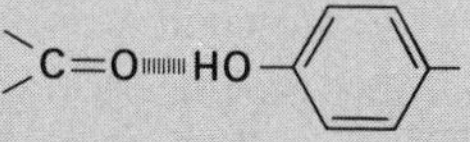 $>C=O \cdots HO-\text{(ring)}-$	Hydrogen bond between peptide and R group	2–5
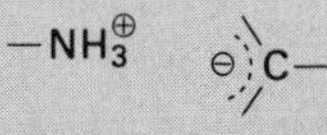 $-NH_3^{\oplus} \quad {}^{\ominus}:\!\!\!\backslash C-$	Hydrogen bond or ionic bond between charged groups strongly dependent on distance	<10
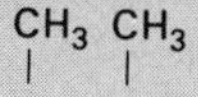 $CH_3 \; CH_3$	Hydrophobic interaction	0.3
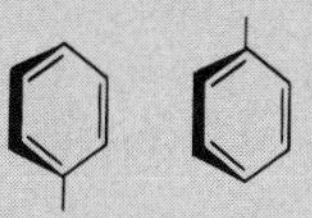 (two aromatic rings)	Hydrophobic interaction —stacking of aromatic rings	1.5
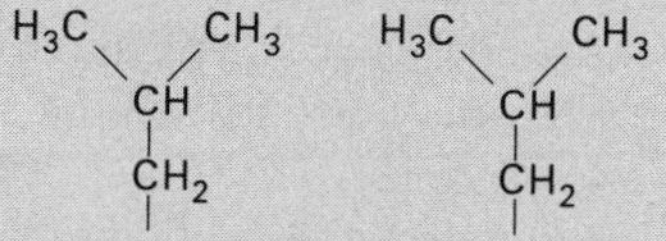 $H_3C \backslash \; CH_3 \; / \; CH \; / \; CH_2$ (two groups)	Hydrophobic interaction	1.5
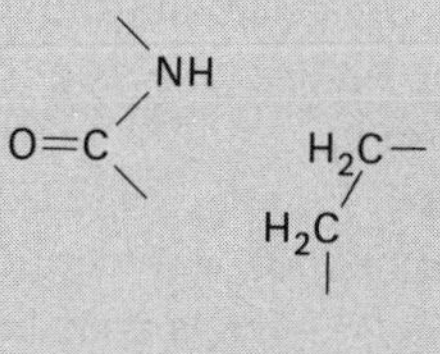 $O=C \backslash NH / \qquad H_2C- \; H_2C$	Hydrophobic interaction between R group (or part of R group, as in lysine) and the peptide bond. This interaction usually involves the hydrophobic R groups of the amino acid contributing the carboxyl group to the bond.	0.3 per CH_2
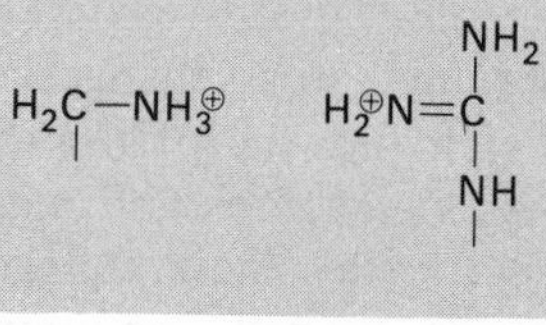 $H_2C-NH_3^{\oplus} \quad H_2^{\oplus}N=C(NH_2)(NH)$	Repulsive interactions between similarly charged groups, strongly dependent on distance	< −5

Figure 4.17   Noncovalent bonds and interactions in polypeptides.

rather than the release of heat. This is shown in a model system in which the transfer of a hydrocarbon from an organic (nonpolar) phase to water is examined.

$$T = 291°C$$

$$\Delta S^0 = -14 \text{ entropy units}$$

$$\text{Benzene}_{\text{liquid}} \longrightarrow \text{Benzene}_{\text{in H}_2\text{O}} \qquad \Delta H^0 = 0$$

$$\Delta G^0 = +4{,}070 \text{ cal/mole}$$

Clearly, the reverse reaction has $\Delta G^0 = -4{,}070$ cal/mole, which is entirely due to the $-T\Delta S^0$ term ($-291 \times +14 = -4{,}070$). Although ΔH^0 for some similar processes is not zero, in all cases the entropy term determines the sign of ΔG^0. The increase in entropy when the organic molecule enters the organic phase may be due to the fact that in the aqueous phase the molecule was surrounded by highly organized clathratelike water structures (see Section 2.4) which were released when the transfer occurred. The increased randomness of the solvent would thus provide the energy for "hydrophobic bonding."

The noncovalent bonds described above are all individually weak. It is easy to break one or several and, because of this, it is usually relatively easy to cause a significant conformational change in a polypeptide chain. In addition, there will be many different conformers possible which differ in shape much more than they do in stability. Several conformers may be in dynamic equilibrium with each other. Factors which alter the hydrogen-bonding capabilities of water, the ionic forms of R groups, and the affinities of hydrophobic residues for each other will all tend to produce conformational changes in polypeptides.

Although stabilizing interactions have been focused on above, it should be realized that some groups within a polypeptide may repel each other (similarly charged or polar groups) and the repulsions between these groups will be an important force in establishing polypeptide conformations.

The disulfide bond is the only covalent bond, other than the peptide bond, commonly contributing to polypeptide structure. The presence of one or more disulfide bonds greatly restricts the conformations available to a peptide chain since they constrain what might be widely separated residues in the linear sequence within a few angstroms of each other.

This discussion of the factors influencing conformations of polypeptide chains can be summarized as follows: The stable conformer under a specified set of conditions will usually be little different in stability from a group of other conformers with which it may be in equilibrium. Gross changes in conformations should be produced by heating, by changing the pH or the dielectric constant of the medium, by adding reagents which complex strongly with the polypeptide backbone of R groups, and so forth. These changes may result in conformers which are quite stable under the new conditions and which may not revert to the original conformation on returning to the original conditions. Conformations stabilized by the covalent disulfide bonds

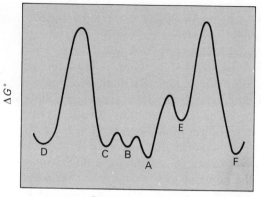

$\Delta G°$

Reaction coordinate

Figure 4.18

Diagram of the relative stabilities of a number of conformers of the same polypeptide. A is the most stable conformation. B and C will be present in small amounts because of their small differences in stability and the low energy barriers to their formation. D is almost as stable as A but will not normally be present because of a high energy barrier. It may represent a denatured form or one in which disulfide bonds have been made or broken. E is a relatively unstable conformer having fewer weak interactions. F is stable but a different conformation from A and D.

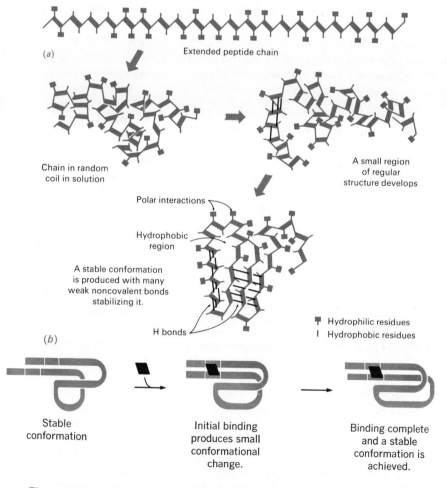

(a)

Extended peptide chain

Chain in random coil in solution

A small region of regular structure develops

Polar interactions

Hydrophobic region

A stable conformation is produced with many weak noncovalent bonds stabilizing it.

H bonds

⊤ Hydrophilic residues
Ι Hydrophobic residues

(b)

Stable conformation

Initial binding produces small conformational change.

Binding complete and a stable conformation is achieved.

Figure 4.19   The development of stable conformations in polypeptides. (a) The role of weak bonds in the development of stable structures. (b) The occurrence of a conformational change during the binding of a small molecule to a protein.

would be expected to be more resistant to changes in the environment than those which are not. These features of polypeptide conformational stability are represented by the energy diagram of Figure 4.18.

The stabilization of a structure by many weak bonds allows cooperative changes to occur within the structure. For example, if a single weak interaction, such as a hydrogen bond, is formed in a randomly arranged polypeptide which is undergoing rapid conformational change in solution, it will increase the likelihood that other weak bonds will form in the same region. If a sufficient number of weak bonds form, a stable conformation will result [Figure 4.19(a)]. The process is similar to zippering a zipper. In addition, if a stable conformation of a polypeptide is capable of binding an ion or small molecule, it may do so by forming fairly strong interactions such as coordinate covalent bonds or multiple ionic bonds. The formation of these bonds can weaken existing noncovalent interactions, resulting in a cascade of events within the polypeptide structure with the eventual production of a new, stable conformation containing the small molecule [Figure 4.19(b)]. Such cooperative interactions are very important in the functioning of natural polypeptides and hormones.

The next section deals with methods that can be used to study the conformations of polypeptide chains.

D. The Determination of Polypeptide Conformation

It is always more difficult to determine the conformation of a compound than it is to determine its covalent structure. Because of the weak and readily interconvertible bonds involved in stabilizing different conformations, the system must be studied under conditions which will not disrupt weak bonds, and only physical methods meet this criterion. For example, spectroscopic methods (ultraviolet, infrared, nuclear magnetic resonance) can be used to examine hydrogen bonding or changes in the environment of R groups in a polypeptide without disrupting the structure being observed. It is beyond the scope of this book to discuss the application of spectroscopic techniques to conformational analyses of polypeptides in detail; only an indication of the kinds of information that can be obtained is attempted.

X-Ray Diffractometry. The ultimate goal of a conformational analysis of a polypeptide is to determine the location of every one of its atoms in space. This is not possible since some groups will usually be in constant rapid motion. For example, the R group of lysine may be relatively easily positioned at the β carbon but not at the $-NH_3^+$ group since this group is sticking out into the solvent and will have a finite probability of being anywhere within a volume determined by the length of the $-CH_2-CH_2-CH_2-CH_2-$ side chain and the presence of other space-filling groups. It is possible in principle, however, to locate all the atoms involved in stable bonds under conditions where conformational motion is reduced to a minimum—that is, the solid state. This has been accomplished by X-ray diffraction analysis of crystalline proteins.

```
                                        10                                                   20
α, val ——— leu -ser -pro -ala -asp -lys -thr -asn -val -lys -ala -ala -try -gly -lys -val -gly -ala -his -ala -gly -glu -tyr-
β, val -his -leu -thr -pro -glu -glu -lys -ser -ala -val -thr -ala -leu -try -gly -lys -val -asg ——————— val -asp -glu -val-

                            30                                       40
α, gly -ala -glu -ala -leu -glu -arg -met -phe -leu -ser -phe -pro -thr -thr -lys -thr -tyr -phe -pro -his -phe ——— asp -leu-
β, gly -gly -glu -ala -leu -gly -arg -leu -leu -val -val -tyr -pro -try -thr -gln -arg -phe -phe -glu -ser -phe -gly -asp -leu-

                                                            Attachment to heme
                                                                    ↓
   50                                                                    60
α, ser -his -gly -ser ——————————————— ala -gln -val -lys -gly -his -gly -lys -lys -val -ala -asp -ala -leu -thr-
β, ser -thr -pro -asp -ala -val -met -gly -asg -pro -lys -val -lys -ala -his -gly -lys -lys -val -leu -gly -ala -phe -ser-
                                                                    ↑

                                                                              Attachment to heme
                                                                                      ↓
              70                                  80                                                  90
α, asn -ala -val -ala -his -val -asp -asp -met -pro -asn -ala -leu -ser -ala -leu -ser -asp -leu -his -ala -his -lys -leu-
β, asp -gly -leu -ala -his -leu -asp -asn -leu -lys -gly -thr -phe -ala -thr -leu -ser -glu -leu -his -cys -asp -lys -leu-
                                                                                      ↑

                                      100                                           110
α, arg -val -asp -pro -val -asn -phe -lys -leu -leu -ser -his -cys -leu -leu -val -thr -leu -ala -ala -his -leu -pro -ala-
β, his -val -asp -pro -glu -asn -phe -arg -leu -leu -gly -asn -val -leu -val -cys -val -leu -ala -his -his -phe -gly -lys-

                          120                                     130                                            140  141
α, glu -phe -thr -pro -ala -val -his -ala -ser -leu -asp -lys -phe -leu -ala -ser -val -ser -thr -val -leu -thr -ser -lys -tyr -arg.
                                                                                                                      146
β, glu -phe -thr -pro -pro -val -gln -ala -ala -tyr -gln -lys -val -val -ala -gly -val -ala -asn -ala -leu -ala -his -lys -tyr -his.
```

Figure 4.20 The amino acid sequence of the α and β chains of human hemoglobin arranged to show similarities in amino acid sequences. Long bonds do not indicate breaks in the chains; they are used to obtain maximum matching.

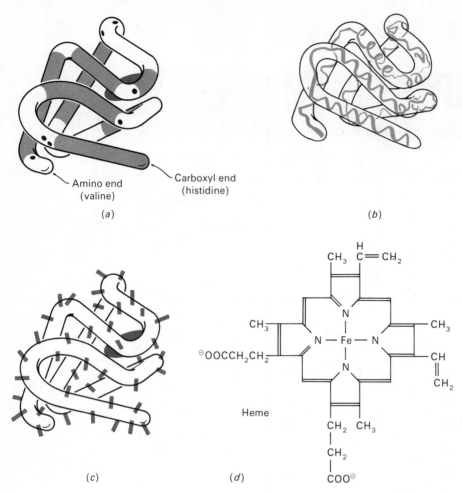

Figure 4.21   (a) β Chain of human hemoglobin showing helical regions in color and heme group as a dark disk to the rear. Proline residues are indicated by dots. (b) β Chain of human hemoglobin showing folding of the peptide backbone. Wider lines are closer to the viewer. (c) β Chain of human hemoglobin showing the approximate distribution of hydrophilic groups on the surface of the protein. (d) Heme—the prosthetic group of hemoglobin.

Provided suitable crystals are available, it is possible to utilize this technique to establish both the covalent and conformational structure of the polypeptide backbone of a protein in the solid state. The X-ray method has already been applied to several proteins and the results of the analysis of one of them, hemoglobin, are presented diagrammatically in Figure 4.21.

Hemoglobin is a globular protein; it is the protein of the red blood cell responsible for oxygen transport. Each molecule of hemoglobin from the normal adult human contains four polypeptide chains; two α chains, each with 141 amino acid residues, and two β chains, each with 146 amino acid residues. The amino acid sequences of the α and β chains as determined by procedures similar to those given above are shown in Figure 4.20. Each chain is folded in a highly convoluted fashion. The folding of the β chain as determined by X-ray diffraction is shown, somewhat diagrammatically, in Figure 4.21. Three representations are used to show (a) the disposition of the folded chains, (b) the polypeptide backbone, and (c) the location of the hydrophilic groups. Note that proline residues fall in nonhelical regions.

The helix content of hemoglobin is unusually high. In most proteins, only a few percent of the amino acid residues are engaged in helical regions or other regions of regular structure (such as pleated sheet). Each polypeptide chain of hemoglobin contains a molecule of heme [Figure 4.21(d)] which is shown as a dark disk in Figure 4.21(a), (b), and (c). The iron atom at the center of the heme molecule forms coordinate covalent linkages with the nitrogens of the protoporphyrin and with the imidazole group of histidine at position 87 in the α chain and at position 92 in the β chain. The sixth position is occupied by a water molecule which may act as a bridge to histidines 58 and 63 of the α and β chains, respectively. The heme fits into a cleft in the hemoglobin molecule and a number of aromatic side chains are oriented parallel to the prophyrin ring. The phenylalanines at 43 and 46 on the α chain and at 42 and 45 on the β chain are involved. When the molecule is oxygenated, significant changes in conformation occur which bring the β chains about 7 Å closer to one another. The change in conformation of the individual chains is slight. The oxygen displaces the water from the sixth coordination position and forms a strong interaction with the iron. This causes a marked change in the pK'_a of the imidazole group (from 7.71 to 7.16) so that oxygenation of hemoglobin is associated with the release of protons. In addition, the conformational change produced by the addition of one molecule of oxygen is transmitted through the protein and results in an increase in the affinity of the other hemes for oxygen.

The preceding brief description of the structure of hemoglobin as ascertained by X-ray diffractometry, and knowledge of the amino acid sequence clearly illustrates the power of the method. It is capable of providing answers to all questions of structure except those concerning the conformation of the molecule in solution and the position of groups having rotational freedom in the crystal. In addition, the technique can only be applied to polypeptides which can be crystallized with heavy atom substituents to serve as markers. It is expensive and time-consuming. Although X-ray diffractometry will have to be used on all proteins about which detailed knowledge of the structure is required, it needs to be complemented by other approaches which can simplify the crystallographic problem. In addition, the other approaches can be applied to solutions and thus provide information on the situation which is not accessible to X-ray analysis.

Polypeptides in Solution. Both spectroscopic and chemical methods can be used to examine the conformations of proteins in solution. They have a common shortcoming when applied to the problem, and it is worth keeping it in mind. Both approaches are used in an attempt to determine the positions of R groups within the molecule. This is usually done by considering the molecule to have groups in a small number of different environments, ascribing properties to the group in each of those environments and then applying an analysis to decide how many groups are in each environment. The simplest model utilizes the generality that a globular protein has a nonpolar inside and a polar outside and considers groups to lie inside or outside (buried or exposed). Obviously, buried groups will have different physical properties and chemical reactivities than exposed ones. Equally as obvious, the model is too simple. Groups on a protein surface may be in more or less polar regions of the surface, may be partly inside the protein, may be influenced by neighboring charged groups, may be hydrogen bonded, and so forth. There will be similar variations in the environments of groups inside the protein and all of these factors will influence activity and reactivity. All that can be observed is a physical property or a chemical reactivity of an R group, and the interpretation of the property or reactivity in terms of position of the R group must be made with caution. In addition, spectroscopic methods provide average values. For example, a similar spectrum would be expected from two phenolic residues, both situated half in and half out of a protein, as from one com-

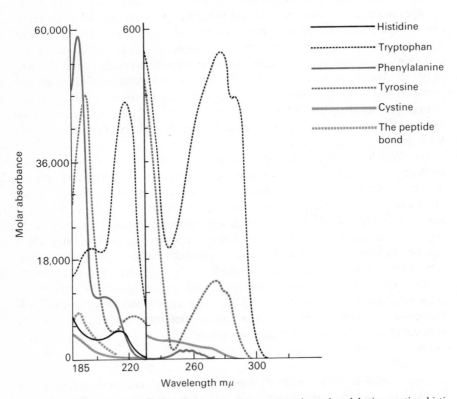

Figure 4.22   The absorption spectra of tyrosine, tryptophan, phenylalanine, cystine, histidine, and the peptide bond.

pletely in and one completely out. The lesson to be learned is that position can only be *inferred* from the property being evaluated; positions are not being measured directly.

Even though interpretations must be made with caution, the spectroscopic and chemical approaches to the determination of the positions of R groups are exceedingly valuable. The positions of R groups and thus the conformation of a polypeptide can be determined in terms of (1) the amino acid sequence, (2) the positions of the R groups (buried or exposed), (3) whether the groups are involved in specific interactions with other R groups, (4) whether the groups are engaged in interactions between the polypeptide and compounds which can bind to it, and (5) whether given pairs of groups are close together.

The conformation can also be thought of in terms of the general shape and size of the molecule—aspects which are studied by procedures such as viscometry, sedimentation, ultracentrifugation, osometry, and light scattering. These methods are not discussed in this book (however, see Van Holde in this series).

Ultraviolet Spectroscopy. There are three strongly absorbing R groups present in polypeptides. The phenolic group of tyrosine, the indolic group of tryptophan, and the phenyl group of phenylalanine absorb in the near ultraviolet ($\sim$280 nm) (Figure 4.22). In addition, cystine and the ionized thiol groups of cysteine absorb very weakly. Imidazole absorbs weakly at 211 nm and peptide bonds below 200 nm. In most naturally occurring polypeptides, phenylalanine does not make a major contribution to the spectrum so that tyrosine and tryptophan are the principal absorbing species. As pointed out above, two regions can be expected in most polypeptides—a polar, exterior, and a nonpolar, interior. Since the absorbing groups of polypeptides have different absorption characteristics in aqueous and nonaqueous media, those on the interior of a polypeptide would be expected to absorb differently from those on the exterior or from those half in and half out. The effect of organic solvents on the absorbance of tryptophan and tyrosine are shown in Figure 4.23(*a*) and (*b*). Note that the spectra shown are difference spectra obtained by comparison of a solution of the chromophore in water with an identical concentration of the material in a different solvent. This method is termed *difference spectroscopy* and allows rather small differences in absorbance to be observed with precision.

Polypeptide chains having a unique amino acid composition and a unique conformation will have a characteristic ultraviolet absorption spectrum. For most, the absorption will have a maximum at approximately 280 nm, and the molar absorptivity at that wavelength can conveniently be used to calculate the concentrations of polypeptide solutions. For many naturally occurring polypeptides (proteins), absorptivity values of approximately 1.0 are found for 1.0 mg of protein/ml using a 1-cm light-path cuvette. This is due to proteins of large molecular weights having fairly similar amino acid compositions.

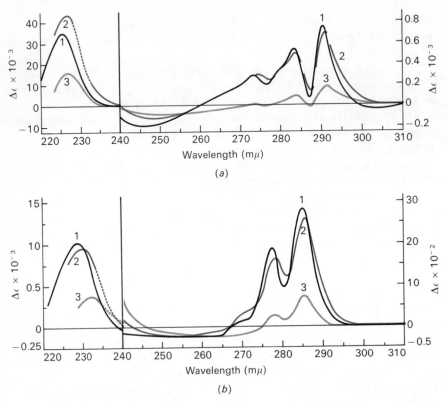

Figure 4.23 (a) Solvent-induced difference spectra of tryptophan methyl ester (1) 40%
ethanol, (2) 28% dioxan, (3) 3.2 *M* urea; all against aqueous solution. (b) Solvent-induced difference spectra of tyrosine methyl ester. (1) 40% ethanol,
(2) 28% dioxan, (3) 4 *M* urea; all against aqueous solution. From J. E. Bailey,
G. H. Beavan, D. A. Chignell, and W. B. Grutzer, *European J. Biochem.*, 7,
5 (1968), with permission of the Federation of European Biomedical Societies.

The effect of various agents on the spectrum of a polypeptide can be used
to determine whether tyrosine or tryptophan groups are "inside" or "outside."
For example, if the spectrum of a polypeptide in water is subtracted from the
spectrum in 40 percent ethanol, the difference spectrum should be interpretable
in terms of the number of tyrosine and tryptophan side chains that are exposed
to the solvent—that is, on the outside (Figure 4.24). It is assumed that no conformational change is occurring with the change in solvent. Whether conformational changes are occurring can be checked by other methods, but they
can also be checked by examining the effect of smaller proportions of ethyl
alcohol on the difference spectrum. If there are no conformational changes,
the relationship of change in absorbance to proportion of ethyl alcohol should
be linear. A change in conformation would be expected to produce curvature
or a discontinuity in the plot. The technique of locating chromophores in a
polypeptide by observing changes in absorbance on the addition of solutes

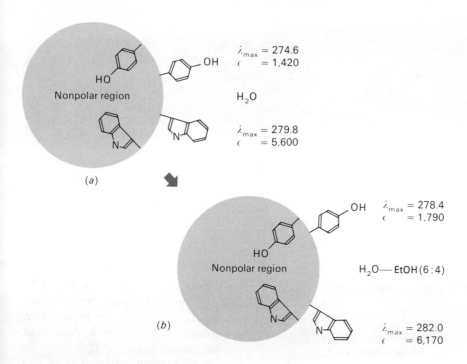

$\lambda_{max} = 274.6$
$\epsilon = 1,420$

H_2O

$\lambda_{max} = 279.8$
$\epsilon = 5,600$

(a)

$\lambda_{max} = 278.4$
$\epsilon = 1,790$

H_2O—EtOH(6:4)

(b)

$\lambda_{max} = 282.0$
$\epsilon = 6,170$

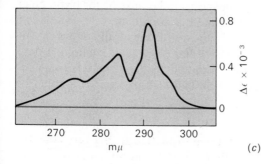

(c)

Figure 4.24   The determination of the number of chromophore groups exposed to solvent by solvent perturbation using difference spectroscopy. (a) Polypeptide in water. (b) Polypeptide in 40% aqueous ethyl alcohol. (c) Spectrum generated with (a) in the reference beam and (b) in the sample beam of a double beam spectrophotometer.

to the solvent is termed *solvent perturbation*. Clearly, perturbants other than ethyl alcohol can be used, and if a series of perturbants having similar chemical properties of increasing size is used, deductions can be made regarding the position of the perturbed groups on the surface of the polypeptide. If a chromophore is in a crevice, small perturbants may reach it, whereas large ones will not (Figure 4.25).

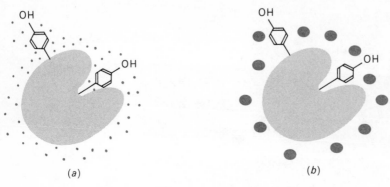

Figure 4.25   The location of tyrosine residues by solvent perturbation. (*a*) Tyrosine residues are both perturbed by a small molecule in the solvent. (*b*) Large perturbants can only affect the exposed tyrosine. The tyrosine in the cleft is perturbed by the small molecule but not the large one.

If a change in conformation occurs in a polypeptide which brings groups out of or into the interior of the molecule, a similar change in the spectrum results. If one tryptophan residue moves to the interior of the polypeptide, a difference spectrum like that shown in Figure 4.23(*a*1) will result. A tyrosine molecule moving in would give the spectrum shown in Figure 4.23(*b*1). Note that if one of each moves, the spectrum will be a combination of these two but will strongly resemble the tryptophan spectrum because $\Delta\varepsilon$ for tryptophan is approximately 800, whereas $\Delta\varepsilon$ for tyrosine is approximately 30.

An obvious experiment, which should be carried out in all studies of this kind, is to obtain the spectrum of the polypeptide in a totally unfolded state so that all the chromophore groups are exposed to the same environment. In addition, the total number of tryptophan and tyrosine residues should be established.

Infrared Spectroscopy. The examination of polypeptides by infrared spectro-scopy is difficult for three reasons. First, the polypeptide gives a very complex spectrum because it appears as a mixture of 20 different amino acids would. However, each amino acid residue in a polypeptide has a common feature— the amide bond

$$-\overset{\displaystyle O}{\underset{\displaystyle N-}{\overset{\displaystyle \|}{C}}}{}_{\displaystyle H}$$

The absorptions of this bond are the amide I band at 1,600 to 1,700 cm^{-1} due to C=O stretching and the amide II band at 1,500 to 1,550 cm^{-1} due to N—H in plane bending and C—N stretching. The second difficulty arises from the fact that water absorbs strongly at 1,650 cm^{-1} directly in the amide I region, and studies in solution must be carried out in D_2O. Since amide hydrogens can exchange with deuterium, band shifts are produced. The third difficulty is the requirement that for precise analysis the polypeptide is best observed in the solid phase as a cast film. Not all polypeptides can be examined this way.

If these difficulties can be overcome, the amide I band can be used to evaluate the amount and kind of regular structure in which amide groups are involved. Bands at 1,632 and 1,645 cm^{-1} have been shown to reflect the presence of pleated sheet structure with parallel polypeptide chains [Figure 4.16(a)]. A band at 1,658 cm^{-1} reflects a disordered polypeptide backbone and bands at 1,650 and 1,652 cm^{-1} reflect the presence of α helix [Figure 4.16(b)].

Optical Rotatory Dispersion and Circular Dichroism. The optical activity of a polypeptide is the sum of the contributions of the amino acid residues plus the contributions which arise from the folding of the polypeptide chain in a specific conformation. The α helix is an obvious source of optical activity since it is totally asymmetric (Figure 3.8). It has been shown with model polypeptides such as a copolymer of L-tyrosine (5 percent) and L-glutamic acid (95 percent) that the change in optical rotation with wavelength can be fitted to the two-term Drude equation developed by Moffitt and Yang

$$[R]_\lambda = a_0 \frac{\lambda_0^2}{\lambda^2 - \lambda_0^2} + b_0 \frac{\lambda_0^4}{(\lambda^2 - \lambda_0^2)^2}$$

where

$$[R]_\lambda = \frac{3}{n^2 + 2} \cdot \frac{\text{MRW}}{100} [\alpha]_\lambda$$

n = index of refraction at wavelength λ

MRW = mean residue weight

$[\alpha]_\lambda$ = specific rotation at wavelength $\lambda = \dfrac{\text{observed rotation} \cdot 100}{\text{conc. in g/100 ml} \cdot \text{length of light path in decimeters}}$

The values of a_0 and b_0 are obtained by plotting $[R]_\lambda(\lambda^2 - \lambda_0^2)/\lambda_0^2$ against $\lambda_0^2/(\lambda^2 - \lambda_0^2)$ where λ_0 is selected to give a straight line; λ_0 is usually 212 nm, as shown in Figure 4.26. The L-tyrosine-L-glutamic acid polymer exists in solution as a helix at pH 4.0 and in a random conformation at pH 7.0. The value of a_0 is found to vary significantly with solvent and has a large positive value (between 200 and 700). The value of b_0 is fairly independent of solvent and is typically near -630 for right-handed α helical structures and is 0 for antiparallel pleated sheets and random structures. The analysis of the ultraviolet ORD spectra of natural polypeptides gives b_0 values which have been interpreted in terms of percent of α helix in the structure. This interpretation should be accepted with caution since the contributions of stable nonhelical conformations to b_0 cannot be evaluated.

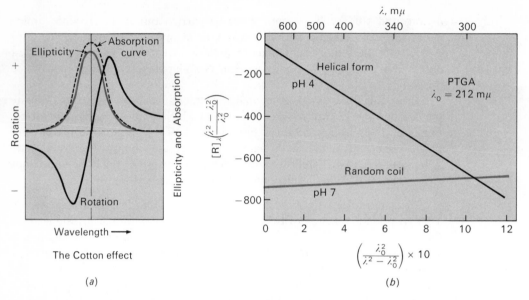

Figure 4.26 (a) An optical rotatory dispersion curve showing the relationship of an absorption band to the Cotton effect. (b) Graphical determination of a_0 and b_0 for a copolymer of tyrosine and glutamic acid: Slope of the line equals b_0 and intercept equals a_0. [P. Urnes and P. Doty, *Advan. Protein Chem.*, **16**:401 (1961), with permission of Academic Press, Inc.]

When an optically active substance absorbs light, an anamolous change in rotation with wavelength is observed within the band which is termed the *Cotton effect* [Figure 4.26(*a/b*)]. This effect refers to the increase in rotation to a maximum, followed by a decrease through zero to a minimum. The Cotton effect is due to the phenomenon that right and left circularly polarized light is absorbed to a different extent by optically active substances (circular dichroism). Since the plane-polarized light used to observe optical activity is composed of right and left circularly polarized components, it is converted, within an absorption band, to elliptically polarized light. The ellipticity $[\theta'] = \pi/\lambda(\varepsilon_l - \varepsilon_r)$, where ε_l and ε_r are the molar absorbances of the material for left and right circularly polarized light, respectively.

The circular dichroism method gives bands which are relatively narrow compared to the ORD bands even though the two phenomena share a common base. Using model compounds, some correlations between ellipticity and peptide chain conformations have been made. Some values are given in Table 4.10.

PMR Spectroscopy. The proton magnetic resonance spectrum of a polypeptide is very complex since there are many different kinds of hydrogen atoms having different chemical shifts. With the development of more powerful spectrometers and use of multiple scanning techniques, however, it is possible to identify and observe changes in specific functional groups. For example,

TABLE 4.10   CIRCULAR DICHROMISM PARAMETERS FOR POLYPEPTIDE CONFIGURATIONS[a]

Conformation	Band position	Ellipticity	Rotational strength
		$[\theta'] \times 10^{-3}$	$R_k \times 10^{-40}$
α helix	221–2	−30.4	−(17–22)
	207	−(28–29)	−(12–14)
	190–2	52–55	30–40
Random	235–8	~0.2	−(0.05–0.15)
	217	~2	1–2
	196–7	−(25–35)	−(14–16)
β structure	217	−(14–17)	−11
antiparallel pleated sheet	195	21	14

[a] S. N. Timasheff and M. J. Gorbunoff, *Ann. Rev. of Biochem.*, **36**, 13 (1967).

the hydrogens on the imidazole ring of histidine can be identified. In cases where there are a relatively small number of histidine residues and where they reside in sufficiently different environments, they can be observed and their perturbation (by titration) examined (Figure 4.27).

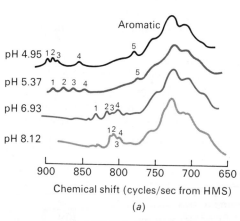

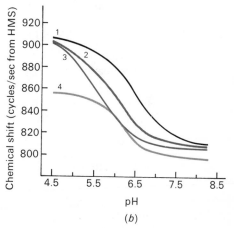

(a)   (b)

Figure 4.27   The proton magnetic resonance spectrum of ribonuclease. (a) 100 Mc pmr spectra of the aromatic region of 0.012 M RNase A in deuteroacetate buffer (pH 4.95, 100 CAT scans; pH 5.37, 14 CAT scans; pH 6.93, 51 CAT scans; pH 8.12, 57 CAT scans). Peaks 1 to 4 are C-2 imidazole peaks of the four histidine residues. Peak 5 is a C-4 imidazole resonance. The envelope labeled "aromatic" includes three other C-4 imidazole peaks as well as peaks from six tyrosine and three phenylalanine residues. (b) Titration curves of C-2 peaks of histidine residues of RNase A. Curves 1–4 correspond to peaks 1 to 4 of (a). Approximate pK's are curve 1, 6.3; curve 2, 5.9; curve 3, 5.6; curve 4, 6.1. [D. H. Meadows, J. L. Markley, J. S. Cohen, and O. Jardetzky, *Proc. Nat. Acad. Sci.*, **58**, 1,307 (1967), with permission of the National Academy of Sciences.]

Chemical Methods. If a polypeptide is treated with a reagent which reacts specifically with one of the types of R groups present in the molecule, the number of groups reacting might be expected to depend on their position in the molecule and to be interpretable in terms of the conformation of the molecule. In some cases, groups on the inside might be less accessible than those on the outside—particularly if the reagent is water-soluble. However, each of the susceptible groups on the surface should possess a different reactivity which reflects its particular environment. In addition, the conditions under which the reaction is attempted will affect both determinants of reactivity and so will the extent to which reaction has occurred. The result is that the chemical modification of a protein really gives information only on the relative reactivities of R groups under highly specified conditions. Whether the reactivity observed is determined by the position of the group and can be interpreted in conformational terms is often open to question. In other words, rapid reaction does not equal "outside."

Although chemical modification is usually attempted under mild conditions which, it is hoped, do not cause conformational changes in the polypeptide, the lack of conformational change cannot be assumed; it must be established.

Of the 18 amino acid residues, 7 are chemically unreactive and are not susceptible to modification under mild conditions. They are the R groups of glycine, alanine, valine, leucine, isoleucine, phenylalanine, and proline. The others can all be caused to react with some reagents.

In Table 4.3 some of the reagents which have been used to modify R groups in polypeptides are listed. Most of the reactions entail attack by the nucleophilic groups of the polypeptide ($-SH, OH, NH_2$) on the reagent. There are a few examples of electrophilic substitutions such as nitration of tyrosine, the coupling of histidine and tyrosine with diazonium-1H-tetrazole, the reaction of tyrosine with iodine, and the deamination of lysine. In general, the unprotonated forms of the nucleophilic groups are most reactive and the selectivity and extent of reaction are very pH dependent. Since the guanidinium group of arginine has a high pK_a, it is very unreactive compared to other groups and its substitution has not been extensively studied.

The introduction into a polypeptide of a substituent which possesses a characteristic property such as pK_a', ultraviolet spectrum, or fluoresence can be used to explore the environment near the reaction site since these properties are sensitive to changes in the polarity of the medium. For example, 5-dimethylaminonaphthylsulfonyl derivatives show changes in pK_a' and fluorescence which can be used in this way. Substituents so used have been referred to as *reporter groups.*

As stated above, each polypeptide presents a unique situation in that the reactivities of functional groups are dependent on many factors besides their location inside or outside the polymer. Hydrogen bonding, field effects, polarity of the local environment, and the steric requirements of the reaction will also influence the ease with which modification can be achieved. In some cases, "outside" groups may be prevented from reaction; in others, some groups may be hyper-reactive. In addition to determining whether a group is inside or outside, specific modifications can be related to amino acid sequence so that the position of the peptide backbone can be inferred.

The use of hydrogen ion to determine the position of groups in a polypeptide has been very common. Those which can be examined are aspartic and glutamic acids, histidine, cysteine, tyrosine, lysine, and arginine and, as shown in Figure 4.3, these groups are titrated in different pH ranges. In the case of tyrosine, conversion to the phenolate ion by removal of a proton produces a large change in ultraviolet absorbance and the titration of tyrosyl residues in a polypeptide with alkali can be followed both electrometrically and spectrophotometrically by the increase in absorbance at 243 or 295 nm. The increases in molar absorbance on deprotonation at these wavelengths are 11,000 and 2,330, respectively, allowing the determination to be made with small amounts of material. Only the tyrosyl residues on the exterior of the polypeptide are titrated under ordinary conditions of low ionic strength. The number observed can then be compared to the number observed when the polypeptide is completely unfolded in 6 M guanidinium chloride. The type of data obtained is shown in Figure 4.28, where the protein ribonuclease has a total of six tyrosines, three of which are available for titration in the natural conformation.

The titration of other residues in polypeptides usually can be observed only by changes in pH (electrometrically). To make decisions concerning the availability of groups to the solvent requires interpretation of differences between the titration in dilute salt and in a medium (such as 6 M guanidinium chloride) which exposes all the groups for titration. Such determinations are usually less precise than those involving tyrosyl residues.

The usual way in which a group specific reagent is used in examining a polypeptide structure is very similar to that described above for the titration of tyrosyl residues. The polypeptide is brought into reaction with the reagent

Figure 4.28

The spectrophotometric titration of residues in ribonuclease in the native (1 M KCl) and denatured state (6 M guanidinium chloride). [Y. Nozaki and C. Tanford, *J. Am. Chem. Soc.*, **89**, 763 (1967). Copyright © 1967 by the American Chemical Society. Reprinted by permission of the copyright owner.]

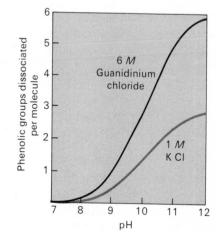

under carefully controlled conditions of pH and temperature. The modification is allowed to proceed and at intervals the amounts and kinds of residues reacting are estimated. In some cases, estimates are based on color changes; in others, on differences observed in amino acid composition: in others, by using radioactive labeling in the reagent. In most cases, the number of groups reacting in the completely unfolded polypeptide should also be examined. Depending on the kind of information sought from the study, the location of substituted groups in the polypeptide may be determined by sequencing the polypeptide.

An example of the application of this technique to the question of whether there is a change in conformation on dissolving a crystalline protein in water is provided by studies on myoglobin. Myoglobin is a protein from muscle which is structurally very similar to the β chain of hemoglobin. The X-ray crystallographic analysis of myoglobin has been achieved so that the conformation of the crystal is known. The protein contains 12 histidine residues. Only 6 of these residues show normal reactivity toward bromoacetate and hydrogen ion (see below). The other 6 are "buried," in the sense that they do not react. If the protein is unfolded, all the histidyl (imidazole) residues are available for reaction. The crystallographic analysis also indicates that half of the histidyl residues will be protected. This work has been extended to show that chemical reactivity in the crystal is similar but not identical to that in solution. The reagent, being quite small, can diffuse into the crystal and react. Further, the residues reacting have been located by sequence analysis following cleavage of the protein with trypsin, and the reacted histidyl residues were shown to be those located on the exterior by the crystallographic analysis. This work shows that in this case the conformation deduced from crystallographic analysis persists in solution and that the chemical modification technique gives valid information on the "location" of residues. It was observed that, following prolonged treatment with bromoacetate, more than 6 histidyl residues had been derivatized indicating that the protein undergoes conformational changes which bring masked residues into reactive locations. Whether these conformational changes are induced by the substitution cannot be deduced.

Information on the relative position of R groups in a polypeptide in solution can be obtained using difunctional reagents. These molecules have two reactive groups capable of forming bonds with two R groups (Table 4.11). When these reagents react, they fix the distance between the two groups. By degradation of the polypeptide and identification of the derivatized peptides using techniques such as fingerprinting by electrophoresis, the position of the groups in the peptide sequence can be established. The R groups which are found to have been linked by the reagent must have come as close to each other as the length of the reagent after the first R group had reacted. Thus, it can be inferred that they lie close together in a stable conformation or that conformational changes occur in solution which bring them close together.

TABLE 4.11   BIFUNCTIONAL REAGENTS[a]

Reagent	Groups Reacting Preferentially		Comment
Azophenyldimaleimide	$-CH_2SH$	pH 9.0	Can cleave azo linkage with dithionite
$\alpha_2\alpha'$-Dibromo(ordiiodo)-p-xylenesulfonic acid	$-CH_2SH$ $-CH_2SH$ $-CH_2NH_2$	pH 7.0 pH 9.0	Imidazole groups can react
p,p'-Difluoro-m,m'-dinitrodiphenyl sulfone	$-CH_2NH_2$ ⟨◯⟩—OH	pH 9.0	Sulfhydryl and imidazole can also react; products are yellow
2,2'-Dicarboxy-4,4'-azophenyldiisocyanate	$-CH_2NH_2$		The urea derivatives produced are unstable in acid; mild cleavage reactions are required
Dimethylimidoester (n can have any value)	$-CH_2NH_2$		Product amidine is stable to acid $\overset{\oplus}{\underset{\parallel}{N}}H_2$ $-\overset{\parallel}{C}-NH-CH_2-$ Charge on lysine is preserved
"Glutaraldehyde"	Fairly nonspecific $-CH_2NH_2$		Glutaraldehyde itself is not the active agent. A trimer with α–β unsaturated aldehyde groups is involved. Used to fix proteins in histological techniques.

[a] Finn Wold, "Bifunctional Reagents," *Methods in Enzymology* (C. W. H. Hirs, ed.), Academic Press, New York, p. 617, 1967.

The chemical methods described so far give information on the distribution and relative reactivities of reactive R groups in the protein structure. A chemical method is available for examining the extent to which peptide bonds are involved in relatively stable hydrogen-bonded structures, such as the α helix or pleated sheet. This method is based on the fact that many hydrogens in a polypeptide can exchange with those in the solvent. Protons on oxygen, sulfur, and nitrogen can all exchange readily, including those on the amide nitrogen of the peptide bond. Those amide hydrogens engaged in hydrogen bonds should exchange less readily than those not so protected. A protein is placed in D_2O or in H_2O containing tritium, and the incorporation of hydrogen isotopes into the protein is allowed to proceed to completion. The solvent is removed and the labeled protein redissolved in H_2O; then the release of isotope into the solvent is measured. The position of a break in the isotope release versus time plot reflects the proportion of the exchangeable hydrogen not readily available to the solvent and is usually interpreted as equal to the amount of hydrogen-bond stabilized structure. Data obtained on the release of deuterium from denatured insulin are shown in Figure 4.29. The curves can be fitted if it is assumed that there are 4 kinds of hydrogens exchanging; 60 with $k = \infty$, 6 with $k = 35$, 15 with $k = 20$, and 8 with $k = 0.1$.

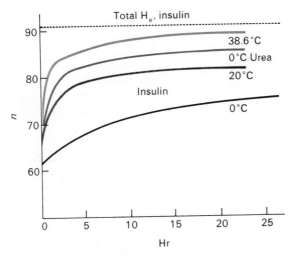

Figure 4.29   Hydrogen-deuterium exchange on insulin showing the effect of temperature and urea. [A. Hvidt and K. Linderstrøm-Lang, *Compt. Rend. Trav. Lab. Carlsberg,* **29**, 385 (1955), with permission of Danish Science Press.]

Summary. In this section, it has been shown that polypeptides can be synthesized and degraded. Structures can be established with regard to amino acid sequence, disulfide bridges, and the conformation of the peptide backbone—the last with less precision and more difficulty. The occurrence of interactions between many R groups can be inferred and some kinds of R groups can be located inside or outside the protein. All aspects of the conformation of polypeptides in solution are difficult to establish, however, and the generality that polar groups will be on the surface and nonpolar groups on the interior should be accepted with the constant reservation that this will not be true in many specific cases.

Nomenclature. The discussion of protein (polypeptide) conformation has been presented without attempting to label the various levels of structural organization. This has been done because it is very difficult to define terms with precision and have them apply to most situations. However, there are terms defining aspects of polypeptide structure in common use with which the student should be familiar. They are

Primary structure: This refers to the unique amino acid sequence of the polypeptide chain as determined by the peptide bonds (not the disulfide bonds).

Secondary structure: This refers to those parts of the polypeptide chain stabilized in formal arrangements by hydrogen bonding.

Tertiary structure: This refers to the folding of the polypeptide chain into its unique three-dimensional arrangement which is stabilized by covalent (disulfide) and noncovalent forces (hydrophobic interactions, hydrogen bonds, ionic interactions, and so forth.

Quaternary structure: This refers to the structure involved in fitting polypeptide chains together to form multichain protein complexes.

Many of the same kinds of forces are involved in stabilizing secondary, tertiary, and quaternary structures. The disulfide bonds are part of the covalent structure and contribute to the structure in much the same way as peptide bonds do. It is difficult to decide in many cases whether an event which changes protein conformation is attributable to its effect on secondary, tertiary, or quaternary structure. For these and other reasons, the use of the terms becomes imprecise and may tend to obscure rather than illuminate.

Denaturation. For most naturally occurring polypeptides, a number of conformations are possible. At one extreme is the conformation in which all noncovalent interactions and disulfide bonds are broken and the polymer exists in a flexible linear arrangement or random coil. At the other extreme, we can place that conformation which has the most "structure"—that is, the maximum of nonbonded interactions and disulfide bonds. In their natural environment (in the cell), proteins exist in the state in which they can play their biological role—that is, their *native state*. In general, but not always, this state will be similar to that in which the maximum of nonbonded interactions and disulfide bonds are present.

It is impossible to isolate a protein and at the same time maintain it in the natural environment. Therefore, it is probable that all isolated proteins are structurally slightly different from the same protein in the cell. To provide a suitable reference state for study, however, it is usual to define the native state as that state of the isolated protein in which it can demonstrate its maximum biological activity.

Almost all studies on proteins are aimed at understanding the biological role of the protein in relationship to its function and usually require that the effect of various changes on biological function be ascertained. In some cases, the change in activity produced by a change in structure can be reversed when the agent producing the structural change is removed; in other cases, this cannot be accomplished. The term *denaturation* has been applied to changes in structure which cause changes in function.

Denaturation can be reversible or irreversible. The term is not usually applied to changes in the primary structure—that is, cleavage of the peptide chain.

The random-coil arrangement represents the completely *denatured* form for most proteins, and agents which can produce this state are very useful. Perhaps the best of these is 5 to 6 M aqueous guanidinium chloride (guanidine hydrochloride),

$$\left[\begin{array}{c} NH_2^{\oplus} \\ \| \\ H_2N-C-NH_2 \end{array} \right] Cl^-$$

containing 0.1 M mercaptoethanol ($HOCH_2CH_2SH$). The guanidinium chloride is capable of disrupting all the noncovalent interactions in a polypeptide by direct interactions with various parts of the chain and/or by altering the "structure" of the water to decrease its role in hydrophobic interactions and hydrogen bonding. The mercaptoethanol cleaves all disulfide bonds by forming mixed disulfides.

$$Pr_1-S-S-Pr_2 + 2HOCH_2CH_2SH \rightleftharpoons Pr_1S-S \cdot CH_2CH_2OH + Pr_2S-S-CH_2CH_2OH$$

High concentrations (6 to 8 M) of urea are also capable of causing complete unfolding of most polypeptides; so are high concentrations of hydrogen ions, acetic acid, and so forth. It should be clear that any agent which can disrupt hydrogen bonds or can alter water structure will alter protein structures.

In some cases, native proteins can be caused to show a greater amount of structure—that is, appear to tighten up when solutes are added which increase the structure of water. For example, sulfates have been found to increase the tendency of polypeptides to form regular structures.

E. The Determination of Numbers of Polypeptide Chains

In proteins having more than one polypeptide chain, the least sophisticated question that can be asked about the structure is "How many chains are there?" The answer can be obtained by determining the molecular weight of the native protein and the average molecular weight of the totally denatured protein—that is, in 6 M guanidinium chloride containing mercaptoethanol. The ratio of the molecular weights serves to establish the number of polypeptide chains (see Van Holde, in this series).

Chemical determination of numbers of polypeptide chains depends on knowledge of the molecular weight of the native protein and the fact that the ends of each polypeptide chain are unique in that they have free α-amino groups or α-carbonyl groups belonging to the terminal amino acids. If these can be derivatized, released, and quantified, the number released per protein is equal to the number of chains present. Reagents which have been used to carry out this type of analysis are shown in Table 4.7. In some proteins, the terminal amino acids are protected in the native conformation, and it is necessary to denature before treatment. Some proteins appear to have the α-amino group of the N-terminal amino acid present as an *N*-acetyl derivative and a micromethod has been developed to estimate acetate after total hydrolysis of such proteins.

4.4 PROTEINS

The proteins are *naturally occurring* polypeptides of high molecular weight. The methods of establishing their structures and conformations are those discussed in the preceding sections. In this section, some of the complications found in proteins but not in polypeptides will be considered. The principal complication is the presence of non-amino acid components in many proteins. To offset the complications is the fact that many proteins have activities which can be used as indicators of the changes wrought in their structures by chemical or physical means.

TABLE 4.12 SIMPLE PROTEINS CLASSIFIED BY SOLUBILITY

Proteins	Solubility	Occurrence
Albumins	Soluble in water and dilute solutions	Almost all systems
Globulins		
(a) Euglobulins	Insoluble in water. Soluble in dilute salt solutions. Insoluble at 50% saturation with $(NH_4)_2SO_4$	Almost all systems
(b) Pseudoglobulins	Sparingly soluble in water. Soluble in dilute salt. Insoluble at 50% saturation with $(NH_4)_2SO_4$	Almost all systems
Prolamins	Insoluble in water. Soluble in 50–90% aqueous ethanol	Plants
Glutelins	Insoluble in water, dilute salt, and aqueous ethanol Soluble in dilute acids or alkali	Plants
Scleroproteins	Insoluble in most solvents	
(a) Collagens	Soluble in dilute acid. Insoluble at neutral pH. Boiling in water converts to the water-soluble form gelatin (contain large amounts of glycine, proline, and hydroxyproline; little or no cystine)	Cartilage, connective tissue
(b) Keratins	Soluble in aqueous potassium bisulfate or thioglycolic acid solutions (contain cystine and are high in basic amino acids)	Hair, nails, skin

Simple Proteins Distinguished from the Solubility Group on the Basis of Amino Acid Content and Source

Protamines	Small size (MW $\simeq$ 5,000)	Sperm
Histones	Rich in basic amino acids	Cell nuclei

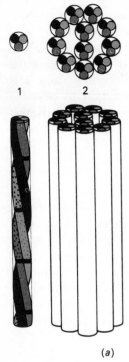

1   2

(a)

Amino acid	Whole wool[b]	α Keratose[c] 1
Cystine/2[i]	106	74.4
Aspartic acid	70.5	80.6
Threonine	61.7	48.0
Serine	112	90.3
Glutamic acid	109	131
Proline	85.2	42.7
Glycine	84.8	107
Alanine	61.4	56.3
Valine	39.0	58.0
Methionine	4.7	4.6
Isoleucine	24.6	31.5
Leucine	71.2	93.3
Tyrosine	38.6	43.3
Phenylalanine	26.7	31.5
Lysine	27.3	28.6
Histidine	7.1	6.1
Arginine	69.6	72.2
(NH_3)	—	(121)

[a] Residues per 1000 total residues.

0   50 Å

(b)

Figure 4.30   A scleroprotein, α keratin (α keratose). (a) A model of the molecular structure of α keratin. (1) Protofibril consisting of three α helices coiled into a rope. Within the axial repeat of 200 Å, three similar but not identical subunits are shown. Between each subunit there is a major interruption in electron density. (2) Microfibril containing 11 protofibrils, each consisting of a 3-strand rope. [R. D. B. Fraser, T. P. MacRae, and G. E. Rogers, *Nature*, **193**, 1,053 (1962).] (b) The amino acid composition of α keratose and whole wool. [S. Seifert and P. M. Gallop in *The Proteins* (H. Neurath, ed.), p. 317, Academic Press, New York, 1966].

Although proteins have a common basis in their amino acid composition, they are an extremely heterogeneous group of compounds. They differ in size, amino acid composition, non-amino acid components, solubility, and so forth. In attempts to facilitate communication concerning proteins, several classification schemes have been proposed. One of them, based on solubility, is shown in Table 4.12. The "average" protein of biochemical interest is usually an albumin or a globulin. The scleroproteins are highly specialized and contain a great deal of pleated sheet or helical structure. An example of a scleroprotein is given in Figure 4.30. The classification can be seen to be quite arbitrary. It is very difficult and indeed not worth the effort to attempt to classify newly isolated proteins using this scheme since it is likely that proteins exist which have solubilities intermediate between all the categories. The classification has been used, however, and the terms defined in it do occur in the literature.

TABLE 4.13   CONJUGATED PROTEINS CLASSIFIED BY NON-AMINO ACID COMPONENT

Protein	Prosthetic group	Linkage	Source or an example
Nucleoproteins (protamines or histone plus nucleic acid)	Nucleic acids	Ionic	Nucleus
Lipoproteins	Phospholipids Glycolipids Simple lipids Cholesterol	Hydrophobic	Most cells, blood, cell wall, membranes
Glycoproteins	Carbohydrate	Covalent	Most cells, cell walls
Mucoproteins	Hexosamine (Usually polymeric)		Sinovial fluid
Chromoproteins	A pigment—e.g., a heme	Covalent Ionic Hydrophobic	Hemoglobin, cytochromes
	Cobalamin		Diol dehydrase
	Nicotinamide Adenine dinucleotide		Many dehydrogenases
Metalloproteins	A metal—e.g., Zn^{2+}, Fe^{2+}	Coordinate-covalent	Alcohol dehydrogenase, Ferritin

A more useful classification is that based on the non-amino acid component of the protein shown in Table 4.13. There are two groups of proteins in this classification—the simple proteins, which contain only amino acids in peptide linkage, and conjugated proteins, which contain non-amino acid components. This classification is unsatisfactory because it is not possible to list all the known conjugated proteins under the headings listed. For example, acyl-carrier protein contains a covalently bound non-amino acid group which is neither a pigment nor a carbohydrate. However, the listing does convey some idea of the complexity of the group of compounds termed proteins.

The covalent linkages of the glycoproteins and mucoproteins present special problems in structure elucidation. There are two types of linkage known to occur between carbohydrate and a protein—an acetal linkage utilizing an alcohol residue on the protein and an *N*-glycosidic linkage involving a carbox-amide residue of a protein. Both are known to occur (Figure 4.31). These linkages can only be established by mild degradation of the protein with pro-teolytic enzymes followed by characterization of the carbohydrate structure and the linkage. Many glycoproteins are heterogeneous in the carbohydrate part but not in the polypeptide. There is no clear distinction made between glycoproteins and mucoproteins. The latter term is usually used for com-pounds having more carbohydrate than protein and may be applied to com-plexes between mucopolysaccharides (heteropolysaccharides containing amino sugars) and proteins.

The complexes between proteins and nucleic acids or lipids are not usually stabilized by covalent linkages. The conjugation between proteins and smaller

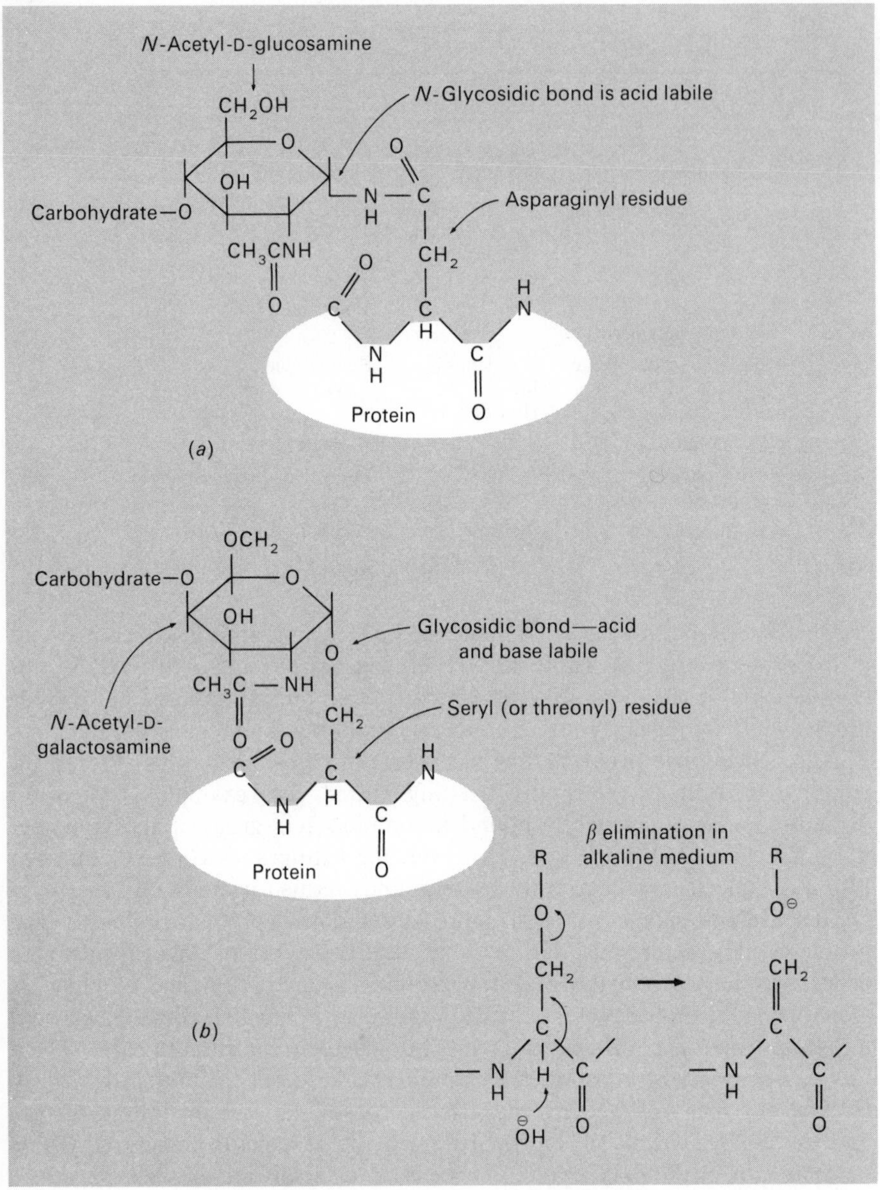

Figure 4.31   Covalent bonds in glycoproteins. (*a*) N-glycosidic bond. (*b*) O-glycosidic bond. Carbohydrate may contain glucosamine, galactosamine, galactose, mannose, fucose, sialic acid, and so forth. The linkage to protein may involve carbohydrates other than D-glucosamine and other carbohydrates may be linked to other positions in the molecule.

organic residues such as chromophoric groups varies with the particular conjugated protein; it may be covalent and therefore stable or noncovalent and allow the conjugated group to be removed under mild conditions. The catalytic proteins (enzymes) might be considered to represent a special case of the conjugated proteins. The enzymes form reversible conjugates with the materials upon which they act (substrates). There may be transient covalent bond formation between the enzyme and its substrate during this process or only noncovalent interactions may be involved. Among the most intriguing properties of the proteins are their enzymatic properties, and information regarding the composition catalytic site on the protein is frequently sought. An approach to obtaining such information can utilize the fact that when the enzyme is in the presence of its substrate groups, the reaction site may be masked by the substrate. If these groups are susceptible to substitution by chemical reagents or are observable spectroscopically, then studies in the presence and absence of substrate can aid in establishing the composition of the catalytic site (active site).

Many different metals are found in the metalloenzymes. In most, the location of the metal is not known, but it is reasonable to assume that it is held as a chelate since there are many ligands present in proteins, including the peptide bond. The metal atom provides a very powerful stabilizing force and many metalloproteins must be severely denatured before the metal can be removed.

For much of chemistry, the starting materials for synthesis or study are commercially available, and few texts devote space to discussion of sources of material. With few exceptions, pure proteins which are needed for studies on structure or function are obtained only on a do-it-yourself basis. Because of this, some discussion of the methods which can be applied to protein isolation and purification is in order.

Isolation and Characterization of Proteins

Except in rare circumstances in which a cell releases a protein into its environment, such as a bacterium releasing an enzyme to hydrolyze macromolecules, proteins are present inside cells. The average cell is large enough to contain some thousand different, protein molecules. Even those specialized cells which contain one protein in overwhelming proportion, such as the red blood cells which are 35 percent by weight hemoglobin (95 percent of the protein), contain many kinds of proteins in varying proportions. In the usual case, isolation requires the separation of molecules which differ only in detail in the various physical properties which are the bases for separation procedures. To get an idea of the kinds of proteins which may be encountered, look at Tables 4.12 and 4.13 which show classifications of proteins on the basis of content and solubility.

Molecules can be distinguished from one another on the basis of differences in size, solubility, electrical charge, density, vapor pressure, and adsorption (physical interactions). All of these except difference in vapor pressure are used to separate proteins. In addition, separation procedures can be applied analytically and serve as tests of purity; most are so used.

To isolate any substance, it is necessary to have some method of detecting the presence of that substance in a mixture. For simple compounds, tests for the presence of certain functional groups (aldehydes, olefins, and so forth) by chemical or physical means can be applied. With more complex materials such as proteins, all components of a mixture have almost identical properties. For example, almost all proteins contain all the amino acids and have properties which reflect this fact. They contain approximately 16 percent nitrogen, they have very similar spectral properties, and 1 mg/ml solution of most proteins will have an extinction coefficient at 280 nm of 1.0 (due to the presence of tryptophan and tyrosine). They contain histidine, arginine, cysteine, tryptophan, tyrosine, and phenylalanine, which can be tested for by simple chemical tests, and peptide bonds, which give the biuret test.

$$\begin{array}{c}\text{O=C}\diagup\text{NH} \\ \diagdown\text{HCR} \\ \text{HN}\diagup \\ \diagdown\text{C=O} \\ \text{RCH}\diagup\end{array} + Cu^{2+} \xrightarrow{\ OH^{\ominus}\ } \text{Biuret, intense blue chromophore}$$

These properties are useful since they allow the estimation of total protein in mixtures, but they cannot be used easily to distinguish between individual proteins. In many cases, however, proteins have a property which is unique, such as a structural (a colored cofactor) or a functional (enzymatic activity) feature. This feature can be used to follow the isolation of the protein, and isolation becomes a process of applying a treatment and determining whether it results in an enrichment of the desired protein.

Solubility. There is an enormous spread in the solubilities of proteins (Table 4.13). Note that the proteins of specialized tissues such as hair, feathers, horn, and hoof are produced in a fairly pure state and are so insoluble that their purification is relatively simple; however, studies of their molecular structure and reactivity are significantly complicated by this insolubility.

The globular proteins (albumins and globulins), on the other hand, are usually present as complex mixtures, and the separation of one from the group on the basis of solubility is more difficult although the approach most often used is simple. The mixture is placed in solution, and a second miscible solute is added which will cause the solubility of the proteins to be decreased. By experimentation, conditions are found such that the various components are deposited from solution by the addition of increased amounts of the

TABLE 4.14   ISOLATION OF RAT MUSCLE ALDOLASE

Reaction

$$
\begin{array}{c}
\text{O} \\
\parallel \\
H_2C-O-P-O_2^{2\ominus} \\
|\\
C=O \\
|\\
HO-CH \\
|\\
HC-OH \\
|\\
HC-OH \\
|\\
H_2C-O-P-O_2^{2\ominus} \\
\parallel \\
O
\end{array}
\quad \rightleftharpoons \quad
\begin{array}{c}
\text{O} \\
\parallel \\
H_2C-O-P-O_2^{2\ominus} \\
|\\
C=O \\
|\\
H_2C-O-H
\end{array}
\quad + \quad
\begin{array}{c}
HC=O \\
|\\
HC-OH \\
|\\
H_2C-O-P-O_2^{2\ominus} \\
\parallel \\
O
\end{array}
$$

D-Fructose 1,6-diphosphate   Dihydroxyacetone phosphate   D-Glyceraldehyde 3-phosphate
(FDP)

Fraction	Total protein (mg)	Specific activity μmole FDP cleaved/mg/min	Total activity
Crude extract	9.975	0.725	7,232
Precipitate, 40%	630	0.075	47
Supernate, 40%	9,450	0.833	7,872
Fraction, 40–52%	870	1.572	1,368
Supernate 52%	8,550	0.923	7,892
Fraction, 52–60%	840	8.957	7,524
Supernate, 60%	8,240	0.094	774

second solute. Among the more common second solutes is ammonium sulfate, which is very soluble in water. It is important that the second solute chosen does not denature the protein of interest, although if it denatures other proteins, it will probably alter their solubilities and may facilitate fractionation.

The isolation of proteins on the basis of differential solubilities is a strictly empirical procedure. Few, if any, rules can be given even though an enormous literature exists on the subject. A very simple example, the isolation of rat muscle aldolase by fractional precipitation, is given in Table 4.14.

Other agents which can cause changes in the solubility of certain proteins in a mixture can be used. For example, if some proteins are more sensitive to temperature, it may be possible to denature them while the desired protein is unaffected. Since denatured proteins usually possess quite different solubilities than native proteins, a fractionation may result.

It is worth noting then that when a protein is forced out of solution, it is often possible to obtain significant purification if it can be caused to crystallize. Although crystallinity per se cannot be used as a criterion of purity, it frequently facilitates isolation and makes purification a simpler task. Crystals of rat muscle aldolase obtained using the fractionation outlined in Table 4.14 are shown in Figure 4.32.

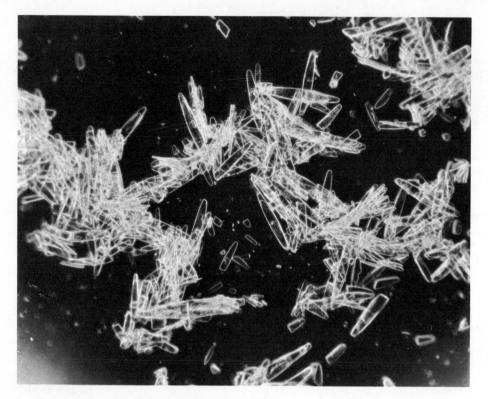

Figure 4.32   Crystals of rat muscle aldolase.

Size. Molecules can be sieved pretty much as rocks or potatoes are graded for size. The simplest devices resemble their macro counterparts in that they consist of membranes with pores of varying sizes through which solutes can diffuse or be forced if they are small enough. It is easy to see how very small molecules can be separated from very large ones by this process, which is termed *dialysis* (splitting with a membrane). The separation of molecules of fairly similar size depends very much on the homogeneity of the pore size in the membrane, however, and this technique of protein isolation is developing very rapidly at the time of this writing because of the increased availability of membranes having narrow ranges of pore size. The technique should also discriminate between molecules of the same size (molecular weight) but different shape; it is illustrated in Figure 4.33.

Separation on the basis of size has been extended by the development of granular materials which are porous (spongelike) and which are inert and insoluble in the usual aqueous systems. These materials can also be produced with pore sizes in a very narrow range so that molecules below a certain molecular weight can penetrate the pores and those having greater molecular weights are excluded. These materials can be used in columns to separate large (those

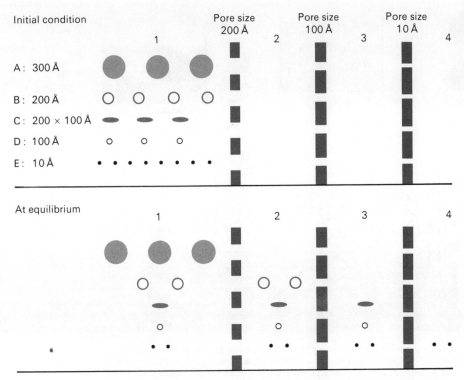

Figure 4.33 The separation of molecules on the basis of size using membranes. Initially all molecules are placed in compartment 1. At equilibrium they are distributed as shown in the lower diagram. Note that the rate at which equilibrium is achieved by various molecules will differ because of their size relative to pore sizes. In addition, the application of a driving force, such as an electric field, can increase the rate and alter the equilibrium distribution of molecules.

molecules which cannot enter the pores) from small (those molecules which can enter the pores). In addition, the smaller the molecules are which can penetrate the pores, the longer they are retained on the columns so that the technique is capable of finer fractionation (Figure 4.34). There are many types of porous materials (gels) available for exclusion chromatography, and the method is widely used in the purification of relatively small samples.

Electrical Charge. The polyelectrolyte character of proteins is the basis for several methods of fractionation which are widely used both analytically and preparatively.

There is a particular pH value at which a protein in a given system is electrically neutral (isoelectric point);† it may be associated with ions from the

---

† There is another condition in which a protein possesses electrical neutrality. If a pure protein is dissolved in pure water, only ions from the dissociation of the protein and the water will be present. The pH value under these circumstances is termed the *isoionic point.* Solutions of proteins at their isoionic points are prepared best by removing other ions by passage over mixed-bed ion-exchange resins—that is, a mixture of anion and cation exchange resins.

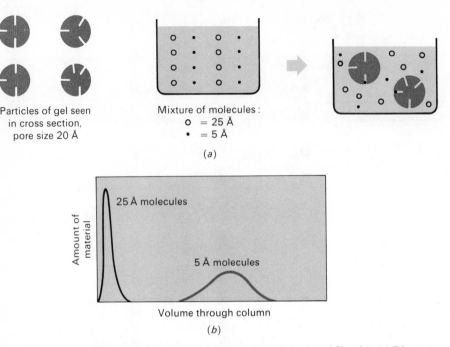

Particles of gel seen
in cross section,
pore size 20 Å

Mixture of molecules :
o = 25 Å
• = 5 Å

(a)

25 Å molecules

Amount of
material

5 Å molecules

Volume through column

(b)

Figure 4.34   The separation of molecules on the basis of size by gel filtration. (a) Diagrammatic representation of the basis for separation. When gel particles are added
to a solution containing molecules of different sizes, the 5Å molecules become
randomly distributed between the pores and the body of the liquid; the 25 Å
molecules are excluded from the pores. (b) Diagram of the separation that
could be achieved using a column of the gel. The small molecules are retarded
because they penetrate the gel and the large molecules come through with the
solvent front.

solvent, but it will have no tendency to migrate in a static electrical field. At
this pH value, the protein is also least soluble since the electrostatic repulsion
between molecules is zero.

At all other pH values, the protein will have a net charge and will tend to
migrate in an electric field (electrophoresis). Since different proteins have
different amino acid compositions, mixtures of proteins can frequently be
separated by electrophoresis in a buffer. The rate of migration depends on the
charge per unit of molecular weight. In some systems, electrophoretic mobility
and size can both be used to effect separation. Such a separation is shown in
Figure 4.35 in which the supporting medium, polyacrylamide gel, restricts the
motion of larger molecules. Electrophoresis can be applied to large-scale
isolation, but it is technically difficult and the method is used primarily as an
analytical tool. However, the ionic character of proteins can also be used to
effect separation on a larger scale using ion-exchange materials. These insoluble polystyrene or cellulose materials contain ionized, functional groups
such as $-SO_3^{\ominus}$, $-COO^{\ominus}$, $-NH_3^{\oplus}$, $-N^{\oplus}(R_3)$ which can exchange cations or
anions with those in the medium. The reaction is an equilibrium, and the

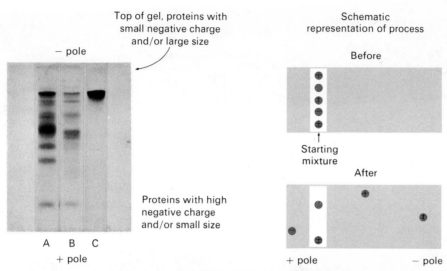

Figure 4.35 The separation of molecules on the basis of electrical charge and size by poly-acrylamide gel electrophoresis. Fractions from the purification of rat muscle aldolase are shown (see Table 4.14). A = crude extract. B = supernatant from crystallization step. C = crystals. Gel is formed in a tube by the polymerization of acrylamide. The gels are cleaned up by washing and electrophoresis before the addition of the protein. The protein is added in a narrow zone at the origin, and after separation by electrophoresis the gel is stained with a dye. The amount of protein in each band can be quantified spectrophotometrically.

position is decided by the relative affinities of the exchangeable ions for the functional group on the polymer. Using these compounds in columns allows even poorly exchanged ions to be retarded. Separations can be achieved on the basis of the varying affinities of proteins at a given pH value, and elution of proteins adsorbed to an exchange column can be achieved by altering the pH or increasing the ionic strength of the eluting solution (Figure 4.36). Both the size and electrical properties of proteins can serve as a basis for discrimination by using porous membranes with charges on the pores.

Density. The differences in density of proteins can be used to examine their purity and, on a relatively small scale, to separate them. Proteins can be caused to sediment in a high centrifugal field. The rate at which they do so and their position in the cell in which they are being sedimented depend on the strength of the applied field and the density of the supporting solvent. In Figure 4.37 is shown the distribution of a mixture of proteins in a medium through which they sediment.

The technique of ultracentrifugation has been widely applied to the determination of protein molecular weights and to the fractionation of cell organelles and proteins, especially the lipoproteins (complexes of lipid and protein). The lipoproteins are fractionated by flotation rather than sedimentation since they are less dense than the medium.

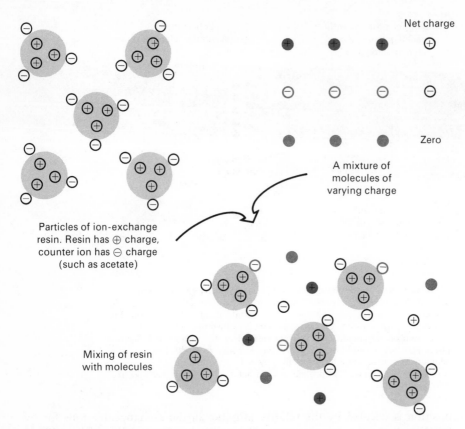

Figure 4.36   The separation of molecules on the basis of electric charge by ion-exchange chromatography. Molecules with a net charge are bound to the resin; others are not. In a column, the ⊖ molecules would be completely retained. They could be removed from the resin by increasing the concentration of anion in the medium, by adding a cation which is bound tightly to the protein, or by lowering the pH to decrease the ⊖ charge on the protein.

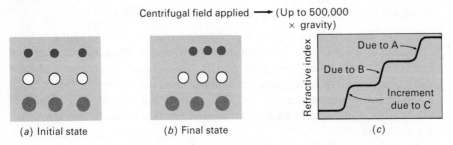

Figure 4.37   The separation of molecules on the basis of density by ultracentrifugation. (*a*) Molecules of varying size but the same molecular weight in a medium having the same density as the molecules of largest size. (*b*) The distribution achieved after centrifugation. (*c*) The separation can be observed by examining the refractive index through the cell. The refractive index is proportional to concentration.

Adsorption. Proteins adsorb to many surfaces and show differential properties in this regard. However, most of these tendencies are a nuisance rather than an assistance in purification. For example, the tendency of proteins to adsorb to glass, with concomitant denaturation, presents problems in cleaning glassware and in storing protein solutions. Reversible adsorption is possible and adsorption to calcium phosphate gels, for example, has been useful in purifying some proteins.

The design of adsorbants which possess functional groups for which a specific protein has affinity has received some attention. The preparation of an adsorbant having a substrate analog allows the adsorption of the enzyme. Elution with the substrate itself, with a competitive inhibitor or with solutions of low pH, caused the enzyme to be eluted. Many elegant examples of this technique have appeared recently, and it may become the most commonly used approach to the isolation of pure proteins.

Criteria of Purity. All the techniques described above can be used to evaluate the purity of a protein. It is almost impossible to establish purity beyond a doubt. The basic approaches fall into two categories. First, establish that the protein, when examined by physical processes, appears homogeneous. More than one technique should be used, and with each technique the conditions (pH, temperature, density, and so forth) should be varied. Second, repurify the protein by any of the techniques described, and determine its specific activity† before and after application of the technique. If the application of a number of purification processes does not alter the specific activity, the protein may be considered "pure."

REFERENCES

Anfinsen, C. B., Jr., M. L. Anson, J. T. Edsall, and F. M. Richards, ed., *Advances in Protein Chemistry*. This series appears annually and usually contains four to eight articles which review specific aspects of protein chemistry.

Hirs, C. H. W., ed., *Methods in Enzymology*, Vol. XI, Academic Press, New York, 1967. This volume deals with enzyme structure. It contains authoritative statements on many of the chemical procedures mentioned in this chapter. It is presently indispensable to workers in areas concerned with enzyme or protein structure.

Neurath, H., ed., *The Proteins*, 2nd ed., Academic Press, New York, 1963. This series of five volumes is intended to present complete discussions of the major topics of interest to protein chemists. For example, Volume I contains chapters on amino acid analysis and composition, the synthesis of peptides, intramolecular bonds, the determination of primary structure, and so forth. Other volumes contain articles on X-ray analysis, the size and shape of proteins in solution, and so forth. In general, the articles are written so that they are of use to a fairly wide audience.

Rich, A., and N. Davidson, eds., *Structural Chemistry and Molecular Biology*, W. H. Freeman and Co., San Francisco, 1968. This excellent volume was prepared in honor

---

† Specific activity is units of function per unit of mass. Units of function may be catalytic activity, color, radioactivity, physiological activity, and so forth.

of Linus Pauling whose pioneering studies on the peptide bond and protein structure form the basis for much of our present understanding. The book contains sections on the structure and chemistry of proteins, with contributions on X-ray diffraction, enzyme structure, thermodynamics of conformational transitions, exclusion diagrams, and spin labeling. The last technique utilizes free-radical-containing reagents as probes of protein structure. In addition, the volume contains articles on hydrogen bonding and water structure and authoritative discussions of many chemical principles.

FIVE | CARBOHYDRATES

Carbohydrates and their derivatives are found in almost all parts of the cell; they have both structural and functional roles. Polymeric carbohydrates are major components of cell walls in plants and microorganisms, and complex carbohydrate derivatives are found in membranes and in the nucleic acids which are responsible for the transfer of genetic information. The simpler carbohydrates are utilized for energy production in the cell, for the synthesis of noncarbohydrate constituents, and as components of many of the cofactors required in metabolic processes.

In this chapter an attempt is made to describe those aspects of carbohydrate chemistry which are important to the biochemical behavior of these compounds and which illustrate their chemical reactivities. The carbohydrates are unusual since in water many of them can exist in a variety of forms which have significantly different properties. Some of these forms are due to the establishment of equilibria between different compounds which are readily interconverted (tautomers); others are due to arrangements in space of the atoms of a molecule. These arrangements can be achieved by rotation about single bonds (conformers).

The first part of the chapter deals with the chemistry of simple model compounds and the complications which arise due to tautomerism. Some nomenclature is presented; without it the discussion becomes complex. The assignment of configuration and conformation to simple carbohydrates is then described, along with some of the approaches that have been made to these problems. A section on the kinds of reactions in which carbohydrates can participate is given that is intended to show the ways their structures can be manipulated, and in most cases an analogous biochemical reaction is listed. The treatment of polymeric carbohydrates is very brief, since to a large extent their chemistry is the sum of that of their components.

5.1 WHAT ARE CARBOHYDRATES?

The carbohydrates are a group of compounds which range in molecular weight from less than 100 to more than 1 million. Almost all of them contain groups which are derived from aldehyde or ketone functions. The smaller compounds, containing three to nine carbons and having at least one group derived from a carbonyl function, are called *monosaccharides*. The larger compounds are formed by condensation of the smaller ones through *acetal* linkages and range from dimers (*disaccharides*) to polymers containing many thousands of monosaccharide units (*polysaccharides*). Polymers are known which contain only one kind of monomer (*homopolysaccharides*) or contain several kinds of monomer (*heteropolysaccharides*). The simple carbohydrates contain carbon, hydrogen, and oxygen but many naturally occurring ones also contain phosphorus, sulfur, and (or) nitrogen. It is not possible to give a simple definition of what carbohydrates are, although this name itself represents the first attempt to define the composition of the members of this class as "hydrates of carbon." The definition of carbohydrates as polyhydroxy aldehydes or ketones or compounds derived from them is accurate enough for most purposes but emphasizes the carbonyl content of carbohydrates which is not readily apparent in most of the naturally occurring compounds. Most of the compounds which occur naturally are polyhydroxy *hemiacetals*, *hemiketals*, *acetals*, or *ketals* and a definition based on this fact would be more accurate.

5.2 HEMIACETALS, HEMIKETALS, ACETALS, AND KETALS

A hemiacetal is the product formed by the reaction of an aldehyde with an alcohol:

$$\begin{matrix} H \\ \diagdown \\ \diagup \\ R \end{matrix} C{=}O + HOR' \rightleftharpoons \begin{matrix} H \diagdown \diagup OH \\ C \\ \diagup \diagdown \\ R \quad OR' \end{matrix}$$

aldehyde   hemiacetal

A hemiketal is the analogous product formed by the reaction of a ketone with an alcohol. The reaction is catalyzed by H^+ and OH^+ and is readily reversible. Hemiacetals (or hemiketals) can undergo further acid-catalyzed reaction with alcohols to form acetals (or ketals) and water.

$$\underset{\substack{\text{hemiacetal}}}{\overset{\displaystyle \mathop{C}\limits^{\displaystyle H \quad OH}_{\displaystyle R \quad OR'}}{}} + \text{HOR}'' \overset{H^{\oplus}}{\rightleftharpoons} \underset{\substack{\text{acetal}}}{\overset{\displaystyle \mathop{C}\limits^{\displaystyle H \quad OR''}_{\displaystyle R \quad OR'}}{}} + H_2O$$

Note that in the *hemiacetal* the carbon derived from the carbonyl is asymmetric and, depending on the nature of R and R′, the hemiacetal will be a racemic mixture or an unequal mixture of both forms. The *acetal* may be symmetric or asymmetric depending on whether R′ and R″ are the same or different.

What will be the symmetry possibilities in the formation of hemiketals and ketals?

The reactions above show a carbonyl function in one molecule undergoing reaction with a hydroxyl in another, that is, an *inter*molecular reaction. *Intra*molecular reactions, where possible, occur more readily than intermolecular ones because the interacting groups are constrained to a smaller volume (the same molecule) and there is not such a large decrease in entropy as when two molecules react to form a single species. This generalization is very well illustrated by the carbohydrates where, except for the smallest members of the class, the compounds exist practically only in the form of internal hemiacetals. The carbohydrate D-ribose is represented below in all the forms in which it can exist. The representation used is that developed by Haworth. The plane of the ring is vertical to the plane of the paper with C-2 and C-3 nearest the observer. The substituents are then above or below this plane. If the crystalline form of D-ribose (**3**) is dissolved in a suitable solvent (H$_2$O), a very small proportion of

Haworth projections

the *aldehydo* form (**2**) is produced which by intramolecular hemiacetal formation gives rise to (**3**), (**4**), and (**5**).

At equilibrium the thermodynamically most favored form will predominate. Since the six-membered ring allows a greater freedom of motion of the component atoms (greater entropy) than the five-membered ring does and has fewer interactions between the substituents, the former is more abundant at equilibrium. The rate at which the five-membered ring forms from the straight chain is much greater than the rate at which the six-membered ring forms, however, since when the open chain is wiggling about in a random fashion the frequency with which the hydroxyl group on C-4 is situated so as to approach C-1 far exceeds the frequency with which the hydroxyl on C-5 occupies this position. The rates at which various ring sizes form when there are no bulky groups in the rest of the molecule to interfere are approximately

Ring size	5	6	3	7	4
Relative rate of formation	50,000	850	60	14	1

The data above is taken from a study involving the cyclization of ω-bromo-alkylamines but a similar variation has been shown with carbohydrates.

This effect, involving five- and six-membered rings, is seen when an acetal is formed. For example, when D-ribose is dissolved in methanol and acid is added, the hemiacetals rapidly disappear and an acetal (6) is formed; in time (7) and then (8) and (9) accumulate and at equilibrium about 75 percent of the product is present as (8) and (9).

The formation of acetals is reversible and, in the presence of a catalyst, the ratio of acetal to hemiacetal depends on the proportion of water in the medium.

5.3   NOMENCLATURE

Before discussing the behavior of carbohydrates in solution and the methods used to establish their structures, it seems advisable to outline the nomenclature of the monosaccharides since in doing so the complexity of the structural problem involved is also revealed. In addition, some familiarity with this nomenclature is essential to follow the discussion of structure and reactivity.

The following terms are useful in discussing the general aspects of carbohydrate chemistry.

(a) Carbohydrates are frequently referred to as *sugars*.

(b) Hemiacetals formed from *aldehydo* sugars are *aldoses* and hemiketals formed from *keto* sugars are *ketoses*.

(c) Aldoses containing a five-membered (tetrahydro*furan*) ring are *aldo-furanoses*; those containing a six-membered (tetrahydro*pyran*) ring are *aldopyranoses*.

(d) Acetals of sugars are *glycosides*; if they contain a five-membered ring, they are *furanosides*; if they contain a six-membered ring, they are *pyranosides*. The carbohydrate portion of a glycoside is the *glycone* and the alcohol portion is called the *aglycone*.

(e) The number of carbons in a sugar is indicated by a prefix; C_3 = triose, C_4 = tetrose, C_5 = pentose, C_6 = hexose, C_7 = heptose, C_8 = octose, and C_9 = nonose.

The following terms are used in discussing specific carbohydrates.

(a) Configurational prefixes. It is obvious that, to compare the arrangements of groups in different molecules properly, a formal agreement must exist with regard to the way in which the molecules are to be viewed. For carbohydrates, the convention developed by Emil Fischer is used; here the carbon chain is vertical with the lowest numbered carbon at the top. Numbering usually follows the convention that the most oxidized end of the molecule has the lowest number [see structures **(6)** and **(8)**].

When a compound has more than one asymmetric center, each center is viewed in turn, and a representation is constructed showing the relative positions of the substituents of the various asymmetric centers. These relative arrangements are referred to by trivial names as shown in Figure 5.1. It is important to note that the simplest aldose sugars having hydroxyl groups

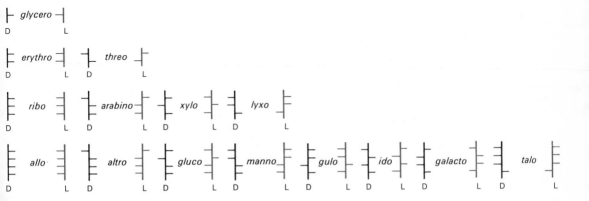

Figure 5.1   The relative arrangements of hydroxyl groups in the carbohydrates and the trivial names assigned to them. The set can be thought of as being built up one carbon at a time, each compound giving rise to two products as shown by the arrows. An enantiomeric set is derived from the L-*glycero* starting material.

on all but the first carbon and having one of these configurations are named by adding "se" to the configurational name.

The determination of configuration is rarely a great problem today, since all the configurations of the sugars have been established and authentic materials are available. However, the original establishment of the configuration of a sugar was a remarkable achievement. The method used is described in Section 5.6.

Configurational prefixes can be used even when the substituents are not hydroxyls and when they are not situated on adjacent carbons (see Section 5.4).

(b) Enantiomers (DL pairs) are compounds which are mirror images of each other; for example, the two forms of glyceraldehyde shown in Figure 5.2 are enantiomers (also see Chapter 2). They are designated D and L as shown on the basis of the location of the hydroxyl group relative to the carbon chain. All carbohydrates are considered to derive from one or the other of these compounds and belong to the D series or the L series depending on the configuration of the highest numbered asymmetric center (also see Figure 5.1 in which the enantiomeric pairs are shown). (Note: The designation D or L indicates nothing about the compound's optical activity.)

(c) Anomers: As with the simple acetals and hemiacetals, the glycosides [(6), (7), (8), and (9)] and aldoses [(1), (3), (4), and (5)] contain one more asymmetric center than the aldehyde (2) does. Carbon-1, the site of this asymmetry, is called the *anomeric* carbon and the hemiacetals and acetals which differ *only* in configuration at that carbon are called *anomers*; that is, (1) and (3), (4) and (5), (6) and (7), (8) and (9) are pairs of anomers. In order to designate a specific anomer, it is necessary to relate the configuration at the anomeric carbon to a reference configuration. Such a reference point is built into each

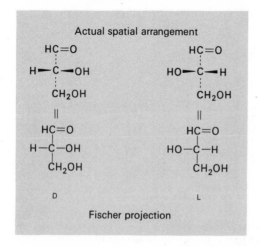

Figure 5.2

D- and L-Glyceraldehyde shown in stereoprojection and in the Fischer projection.

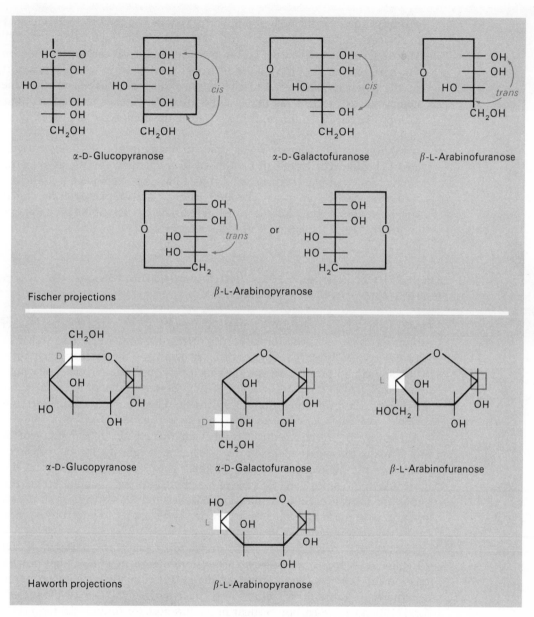

Figure 5.3   Fischer and Haworth projections of some common carbohydrates. In the Fischer projections, the *cis* and *trans* relationships between the anomeric hydroxyl and the highest numbered asymmetric carbon in the α and β forms, respectively, is shown. The same compounds are shown in Haworth projection; the colored square encloses the highest numbered asymmetric center and the open square, the anomeric carbon.

carbohydrate; it is the highest numbered asymmetric center which is related to the configuration of either D- or L-glyceraldehyde (Figures 5.1 and 5.2). The designation of the configuration at the anomeric carbon is made on the basis of the position of the hydroxyl (or aglycone which replaces it in the glycosides) relative to the group which establishes whether the compound is in the D or the L series. If in the Fischer projections these groups are *cis*, the compound is α; if *trans*, the compound is β (see Figure 5.3).

(d) *Epimers* are isomers which differ in configuration at a single asymmetric center. The term can be applied to a difference at any asymmetric center; for example, D-glucose and D-galactose are 4-epimers, D-arabinose and D-ribose are 2-epimers. The terms anomer and epimer describe similar differences in structure but the former is restricted to describing the anomeric center (C-1) and the latter to all other centers.

5.4   CONVENTIONS FOR REPRESENTING CARBOHYDRATES: FISCHER AND HAWORTH PROJECTIONS

Open-chain forms of carbohydrates are usually represented in the Fischer projection, whereas cyclic forms are represented using either the Fischer or Haworth conventions. To represent ring structures in the Fischer convention, the open-chain form is drawn and the ring form constructed from it. The position of the alcohol function involved in ring formation determines the side of the carbon chain on which the ring is drawn. The anomeric configuration is designated on the basis of (c) above, as shown in Figure 5.3.

The Haworth projection is an attempt at a more realistic representation of the ring forms of the sugars in which the placement of substituents can be shown with respect to the plane of the ring. When a Haworth projection is used, it is more difficult to state a simple rule for assigning anomeric configuration. It is important to realize, though, that both Fischer and Haworth projections are attempts to represent the same structure and some ability to interconvert these representations is desirable.

The two centers which are related by the term α or β are indicated in each structure in Figure 5.3; in Haworth projections there is no immediately obvious relationship between them. However, the most frequently encountered representations are those for a six-carbon D sugar in the pyranose (or pyranoside) form (10) and a five-carbon D sugar in the furanose (or furanoside) form (11).

α-D
10

β-D
11

In both cases the α group is below the plane and the β above the plane or the ring. The substituents which are to the right in the Fischer projection are down in the Haworth projection. These relationships are shown in Figure 5.3.

5.5 HIGHER CARBON SUGARS—MORE THAN FOUR ASYMMETRIC CENTERS

When more than four asymmetric centers are present, the carbohydrate is given two configurational prefixes; one for the four lowest numbered asymmetric centers and one for the rest of the molecule. The configuration of the highest numbered group is stated first, for example,

$$
\begin{array}{c}
\text{H—C=O} \\
\text{L-galacto} \left\{
\begin{array}{l}
\text{HO} \\
\text{OH} \\
\text{OH} \\
\text{HO}
\end{array}
\right. \\
\left.
\begin{array}{l}
\text{OH} \\
\text{OH}
\end{array}
\right\} \text{D-erythro} \\
\text{H}_2\text{COH}
\end{array}
$$

D-*erythro*-L-*galacto*octose

The configuration at the anomeric center in these "higher carbon" sugars is related to the enantiomeric relationship of the upper four asymmetric centers.

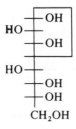

D-*erythro*-β-L-*galacto*furanooctose

5.6 CONFORMATIONAL ANALYSIS

In the discussion to this point Fischer and Haworth projections have been used. In arriving at an understanding of the chemistry of the carbohydrates, however, it is necessary to consider the molecules with bond distances, bond angles, space-filling characteristics, and so forth all as they really exist. Keeping all of these factors in mind at one time is very difficult and not always necessary, but failure to remind oneself that the usual conventions are merely conveniences and that reality is more complex can prove a costly oversight.

In this section the three-dimensional aspects of the carbohydrate ring systems are considered and the factors which contribute to their relative stabilities discussed. The most obvious contributions to the stability of the carbohydrate ring systems are made by the hydroxyl substituents and depend on interactions between them.

Analysis of the interaction between substituents requires consideration of the *conformations* available to the pyranose ring system. (A conformation is an arrangement in space of the atoms of a molecule which can be achieved by rotations about single bonds.) Since the valence-bond angle of oxygen is similar to that of tetrahedral carbon and the C—O bonds are only slightly shorter than C—C bonds, a pyranose ring is essentially equivalent to a cyclohexane ring in its conformational possibilities. As shown by Sachse, it is possible for the cyclohexane ring to exist in two different arrangements: the boat (**12**) and the chair (**13**) forms in which the valence-bond angles of the ring are entirely unstrained. The usual representation of a planar ring wrongly implies bond angles of 120 deg, which would result in a great deal of strain in six-membered rings.

The two strain-free forms (**12**) and (**13**) would be expected to have totally different properties, particularly with regard to interactions between various groups in the molecules. It is possible to evaluate these interactions by viewing parts of the molecule utilizing Newman projections. If an observer stands so that he is looking at carbon 1 of (**12**) in the direction of carbon 2, he will see (**14**).

He can repeat this process, moving around the molecule and sighting along each bond in turn, making six projections. Two will be like (**14**) and four will be like (**15**). If he carries out the same process on the chair form, all the projections will be similar to (**15**). Now in (**14**), an *eclipsed* arrangement, all the atoms are as close together as possible and interactions between them are maximal; whereas in (**15**), a *gauche* arrangement, the minimum interactions possible in a ring exist. On this basis alone we can conclude that a chair form will be more stable than a boat form. It is also important to note that the hydrogens on C-3 and C-6 in the boat form which are directed toward the center of the boat come closer together (1.8 Å) than the sum of their van der Waals radii (2.4 Å). The four hydrogens oriented parallel to the axis of the ring in the boat form (axial orientation) are also close enough together to produce an uncomfortable crowding. The difference in energy between boat and chair forms of cyclohexane is 5 to 6 kcal/mole, which means that at 25°C only 1 in 1,000 molecules exists in the boat form. Careful inspection of a model of cyclohexane will reveal that the boat form is quite mobile; conversion from one boat form to another is readily accomplished and there is an intermediate form (skew boat) in which the unfavorable interactions are somewhat relieved, but only to the extent of 1.6 kcal/mole.

In the pyranose ring system the presence of an oxygen (bond angle, 110 deg) presents no difficulty as far as strain is concerned; and since the oxygen atom bears no substituents, a little more room is available for those carried by carbon. The *p* electrons of oxygen are important in determining the position of groups at C-1, however, if those groups possess a strong dipole which can interact with those of the *p* electrons.

p orbitals of oxygen

If conformational projections are drawn for the D-glucopyranoses, it is found that a total of eight can be drawn both for the α and the β forms. Of the eight, two are chair forms and six are boat forms. On the basis of considerations of the conformers of cyclohexane, we can conclude that the boat forms will not be present to any great extent in a solution of glucose. This does not mean, however, that boat forms cannot exist nor that they are unimportant in reactions in which glucose participates. Of the two chair forms, one (**16**) has the bulky

β-D-glucopyranose

normal conformer (C1)

16

alternate conformer (1C)

17

groups at C-2, 3, 4, and 5 equatorially oriented and is termed the *normal* conformer (or C1 conformer) and the other (17) has them axially oriented and is termed the *alternate* conformer (or 1C conformer). Axially oriented bulky or polar groups on the same side of the ring constitute an arrangement which is highly unstable as compared to the equatorial one. An idea of the crowding which would exist in the chair form of β-D-glucose having axially oriented hydroxyl groups (**17**) is obtained by viewing the structure from above, outlining the atoms attached to the ring carbons which are above the ring with their single bond and van der Waals radii as shown in Figure 5.4. The interactions would appear even greater if the hydrogen atoms of the hydroxyl functions were included.

In most cases the most stable arrangement at the anomeric carbon is that one in which the dipoles of the ring oxygen and the anomeric substituent (—OR,

$$\underset{\text{O}}{\overset{\text{O}}{\parallel}}$$

O—CR, or halogen) are maximally separated—that is, the α anomer is favored in the normal conformer. This is termed the *anomeric effect.*

less stable   more stable

Solvation will influence the proportion of anomers; and neither the degree nor sites of solvation have been measured, so that the extent to which it affects the equilibrium between anomers cannot be determined. It is a fact, however, that in the case of α- and β-glucopyranose in water the ratio of anomers observed at 25°C reflects thermodynamic differences between solvated forms; solvation stabilizes the inherently less stable β form.

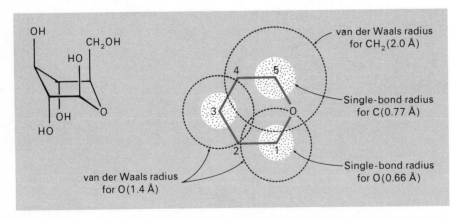

Figure 5.4   The alternate chair form of β-D-glucopyranose as viewed from above showing the van der Waals radii and single-bond radii for the groups above the ring.

5.7 ASSIGNMENT OF CONFIGURATION: STEREOISOMERS

The discussion above presumes that compounds of known structure are being dealt with, which is now usually the case. To establish the configuration of a carbohydrate is not always a simple matter. As an example, the configuration of D-glucose as established by Fischer is given here.

The following is not the precise proof used by Emil Fischer, but it serves to illustrate the type of reasoning he used. It is important to realize that the theory of LeBel and van't Hoff concerning the asymmetric carbon atom was not then (1891) as well accepted as it is today. Fischer's opening statement was "All previous observations in the sugar group are in such complete agreement with the theory of the asymmetric carbon atom that the use of this theory as a basis for the classification of these substances seems justifiable."

Fischer had two isomeric aldohexoses (glucose and mannose) and an aldopentose (arabinose). These were known to be linear polyhydroxy aldehydes, but their configurations were not known; at that time the names were entirely trivial.

It is possible to draw 16 six-carbon aldohexoses in two sets of enantiomers; only one set of enantiomers needs to be considered. Each set can be divided into groups of two which differ in configuration only at C-2.

$(\Delta = HC=O; \bullet = CH_2OH; \vdash = H-C-OH)$

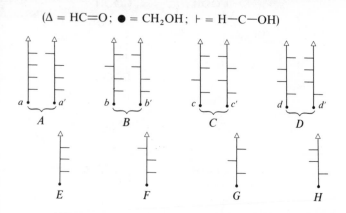

Fact. Glucose and mannose can be converted to the same osazone and osazone formation results in loss of asymmetry at C-2 only. Carbons 3 through 6 are not affected (see Section 5.15, V, E).

Conclusion. Glucose and mannose constitute a pair (*A, B, C,* or *D*).

Fact. Glucose and mannose can be oxidized by nitric acid to produce dibasic acids (CH_2OH and $HC=O$ converted to $COOH$) which are optically active.

Conclusion. Glucose and mannose are either pair *B* or pair *C* since one from each of the other pairs (*a* and *d*) would give an optically inactive dibasic acid.

Fact. Arabinose is oxidized to an optically active dibasic acid.

Conclusion. It is *F* or *H* or the enantiomer of *F* or *H* since *E* and *G* would give optically inactive dibasic acids.

Fact. Arabinose condenses with cyanide to give a pair of hexonic acids (HC=O of hexose oxidized to COOH, general name *aldonic* acid) which are epimeric at C-2 and which are enantiomers of the aldonic acids prepared from glucose and mannose.

Conclusion. Arabinose is the enantiomer of *F*. It cannot be the enantiomer of *H* since it would give rise to the pair *D* which was eliminated earlier. This also eliminated pair *C* which would arise from *G*. Glucose and mannose are pair *B*. If the pentose is an L sugar, the hexoses are D. This was assumed to be the case.

Fact. The dibasic acid from glucose can form a lactone (an internal ester) in which only one of the COOH groups has reacted, as shown below. The lactone can be reduced to a CH$_2$OH group to give an aldonic acid which is different from the aldonic acid produced from glucose itself by mild oxidation.

Conclusion. Only *b* could give rise to a different aldonic acid since *b'* gives rise to a dibasic acid with end-to-end symmetry. Therefore *b* represents glucose.

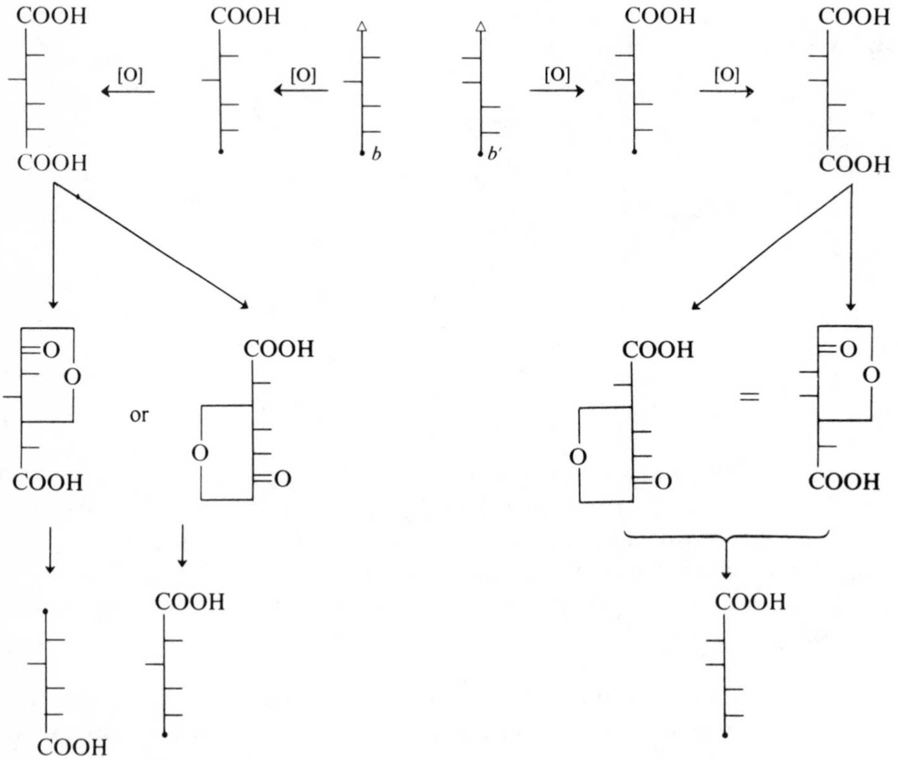

5.8 ASSIGNMENT OF CONFIGURATION: ANOMERS

The determination of anomeric configuration poses a difficult problem and usually requires that both α and β forms be prepared. When both are available, the anomeric assignment can usually be made on the basis of an empirical rule developed by C. S. Hudson which states that the anomer having the more positive specific rotation in the D series is the α anomer. Conversely, the anomer with the more negative specific rotation in the L series is the α anomer.†

The only exceptions so far noted to this rule are certain unsaturated glycopyranosyl esters and a 2,4-dinitrophenylamino derivative of glucosamine.

Hudson also proposed that the molecular rotation (specific rotation × molecular weight) of a carbohydrate is the sum of the rotations due to its component asymmetric centers. He demonstrated that this proportion held very well for the contribution of the anomeric carbon in a large number of glycosides. If the rotation of glycosides is considered to be due to the sum of the contribution of the anomeric carbon (A) and the rest of the molecule (B), then the rotation of one anomer will be $+A$, $+B$ and of the other $-A$, $+B$ (Figure 5.5). The sum of the rotations of the two anomers will then be $2B(+A + B - A + B)$ and the difference will be $2A$. Hudson evaluated the data from a large

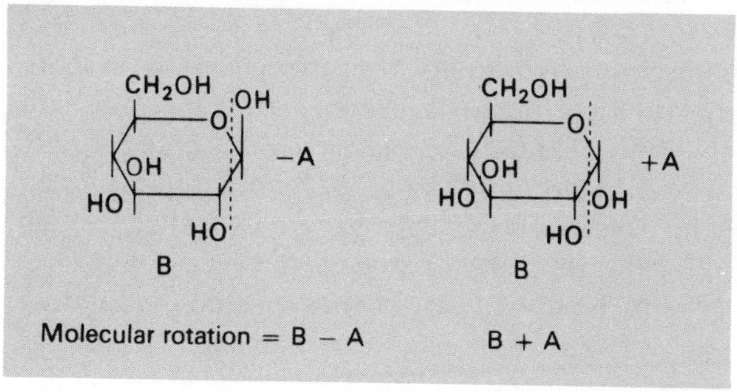

Molecular rotation = B − A   B + A

Figure 5.5

The division of a pyranose sugar into A and B parts for application of the Hudson isorotation rules.

number of compounds and showed that to a resonable approximation A and B are independent of each other. It is possible, therefore, to assign a tentative anomeric configuration to a glycoside when the carbohydrate component is known. The contribution of the B portion of the molecule is subtracted from the molecular rotation of the glycoside and the value of A so obtained is compared with those reported for other anomeric carbons bearing similar aglycones. For example, the value of A for methyl glycopyranosides is approximately 18,500 deg (Table 5.1). A methyl pyranoside of xylose which has $M_D = +25,240$ must be the α form since the B portion of xylopyranosides contributes $+7,250$. The difference $(M_D - B)$ is $+17,990$ and since the α form is that with the higher

† The α-D-compound is the mirror image of the α-L compound.

TABLE 5.1   2A VALUES FOR METHYL GLYCOSIDES[a]

	M_D	M_D	Sum (2B) $\alpha + \beta$	Difference (2A) $\alpha - \beta$
D-Glucoside	+30,630	−6,300	+24,330	+36,930
D-Galactoside	+37,380	−80	+37,300	+37,300
D-Xyloside	+25,240	−10,740	+14,500	+35,980
L-Arabinoside	+2,840	+40,260	+43,100	−37,260
			Average for 2A	36,950

[a] "Correlations of Optical Rotary Power and Structure in the Sugar Group," C. S. Hudson, *Rapports sur les Hydrats de Carbone (Gluides), Union Internationale de Chimie, Paris*, pp. 59–78 (1931).

rotation, this must be the α form. This is a trivial example, but it illustrates the approach.

As one might expect, this simple rule is not perfect and there is some effect of the structure of the carbohydrate (B portion) on the rotation of the anomeric carbon (A portion). In particular, changes in configuration at carbon-2 affect the value of A and the value for methyl mannosides is 14,450 deg. Similarly the value of B is somewhat affected by the nature of the group at the anomeric carbon. For glucosides the average value is about 12,000 deg with a range from 9,850 to 16,800 deg.

The determination of anomeric configuration by nmr spectroscopy is more direct and depends on two facts. First, the hydrogen on the anomeric carbon gives a signal at significantly lower field (approximately $\tau = 4.2$ ppm for fully acetylated pyranose sugars in chloroform solution). Secondly, the projected angle between the anomeric hydrogen and the hydrogen on C-2 influences the shape of the signal observed for the anomeric hydrogen. In Figure 5.6 projections are shown along the 2C—1C bond of a pair of anomers and the splitting of the signal for the anomeric hydrogen atom in cycles per second. In this case the ring of the sugar is represented as it exists with all angles equal to 109 deg (see Section 5.10).

When the hydrogens at C-2 and C-1 are *trans* as in Figure 5.6(a), they interact strongly and the coupling constant $(J_{1,2})$ is large, approximately 8 Hz. When the angle between the hydrogens at C-2 and C-1 is small however, as in Figure 5.6(b), $J_{1,2}$ is smaller, approximately 3 Hz. In those cases where the hydrogen at C-2 is disposed as in Figure 5.7(c) and (d), the angle between it and the anomeric hydrogen is always approximately 60 deg and the coupling constant cannot be used to evaluate the anomeric form. In this case the fact that equatorial hydrogens usually absorb at lower field (smaller values of τ) can be used to assign configuration. Some care is necessary, however, since other groups in the molecule can alter chemical shift values (τ values) significantly. Neither can the value for the coupling constant $(J_{1,2})$ in all cases be equated with the observed splitting of the signal for the anomeric proton. In some cases, "virtual long-range

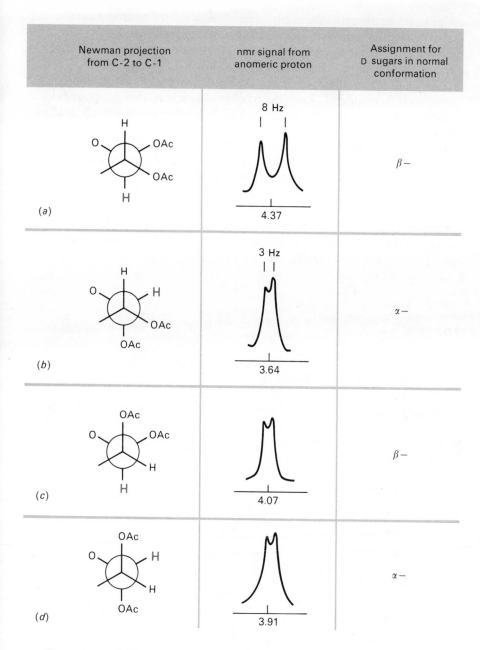

Figure 5.6   The nmr signals of the anomeric protons of pyranoses showing the dependence of splitting on the configuration at carbon-2. (a) and (b) 2-Hydroxyl equatorial. (c) and (d) 2-Hydroxyl axial.

coupling"† can alter the effective interaction between the hydrogens at C-1 and C-2 so that the signal from the former cannot be interpreted in a simple way. The presence of virtual long-range effects usually broadens the peaks and in some circumstances their presence can be allowed for.

X-Ray diffraction studies on crystals can establish the relative positions of all the atoms in a compound. Enough carbohydrates and related compounds have been studied to verify that the configurations inferred from less direct procedures are correct.

The use of enzymes which hydrolyze glycoside bonds as reagents to establish the anomeric configuration of glycosides is particularly useful with polysaccharides but has also been applied to simple glycosides. The technique depends on the demonstration that an enzyme is specific for a particular type of bond (α or β) prior to using it to infer the presence of that bond. Most of the glycoside-cleaving enzymes which have been isolated are also quite specific for the glycone portion of the glycoside. For example, maltase cleaves α-glycosidic bonds of D-glucopyranosides most readily and the α-glycosidic bonds of other D-pyranosides much more slowly.

5.9   KETOSES

A great deal of the foregoing discussion of aldoses applies equally well to ketoses, but one or two problems arise which need clarification. The two most common ketoses are fructose (**18**) and ribulose (**19**). Obviously fructose is a trivial name, and ribulose is an incorrect name. Fructose contains three asymmetric carbons and has the *arabino* configuration. Ribulose has the *erythro* configuration. The correct names of the two compounds are D-*arabino*-hexulose and D-*erythro*-pentulose; since the keto group occupies the 2 position,

$$
\begin{array}{ccc}
\text{H}_2\text{COH} & \text{H}_2\text{COH} & \text{H}_2\text{COH} \\
| & | & | \\
\text{C}=\text{O} & \text{C}=\text{O} & \text{HOCH} \\
| & | & | \\
\text{HOCH} & \text{HCOH} & \text{C}=\text{O} \\
| & | & | \\
\text{HCOH} & \text{HCOH} & \text{HCOH} \\
| & | & | \\
\text{HCOH} & \text{H}_2\text{COH} & \text{HCOH} \\
| & & | \\
\text{H}_2\text{COH} & & \text{H}_2\text{COH} \\
\end{array}
$$

	19	
18	D-ribulose	**20**
D-fructose	D-*erythro*-pentulose	D-*arabino*-3-hexulose

it is not necessary to locate it by number. For ketoses with the carbonyl group

---

† Virtual long-range coupling occurs when the proton at C-2 is strongly coupled to the proton at C-3. This effect is transmitted to the proton at C-1 and the signal from that proton is perturbed.

elsewhere, however, the name must specify the position, e.g., D-*arabino*-3-hexulose (**20**).

The completely trivial name D-fructose is firmly established and will ordinarily be used, but names such as D-ribulose and L-xylulose are totally incorrect because they imply a configurational relationship which does not exist. They are, however, in wide use by biochemists.

A difficulty frequently arises in representing ketoses by use of Haworth and Fischer projections. For example, the correct representations of β-D-fructo-furanose are (**21**) and (**22**). A number of texts show the Fischer projection (**23**)

| **21** | **22** | **23** |
| β-D-fructofuranose | β-D-fructofuranose | α-D-fructofuranose |

for β-D-fructofuranose, which when one tries to make a model, turns out to be the α anomer. The difficulty arises because the α and β forms of the ketoses are derived from the corresponding aldoses having one less carbon by replacing the anomeric H with CH_2OH. The incorrect representations arise because the ring projection is created from the open-chain projection without consideration of this fact.

5.10 ANOMERS: CHEMICAL AND PHYSICAL PROPERTIES

It is important to realize that anomers, although derivable from the same aldehyde sugar and frequently readily interconvertible, are chemically and physically separate and distinct compounds. This point is well illustrated by α-

TABLE 5.2 PROPERTIES OF α- and β-D-GLUCOPYRANOSE

| Property | D-*Glucopyranose* | |
	α	β
Specific rotation, $[\alpha]_D^{25}$	+112 deg	+18.7 deg
Melting point	146°C	150°C
Solubility in H_2O (g/100 ml)	82.5 at 25°C	154 at 15°C
Rate of oxidation by bromine	1	250
Rate of oxidation by glucose oxidase	>150	1

and β-D-glucopyranose. Both of these compounds are crystalline and a super-saturated aqueous solution of D-glucose will deposit either form if innoculated with seed crystals. As seen from Table 5.2, the two compounds have distinctly different properties.

Optical Rotation

Few aldoses have been isolated in more than one crystalline anomeric form. On dissolution of one of the crystalline hemiacetals in a suitable solvent, a change in the optical rotation of the solution with time can usually be observed. This change is termed *mutarotation* and indicates that isomers, other than the one dissolved, are being formed. In the case of D-glucose this change appears to reflect a simple equilibration between the α- and β-pyranose forms [Figure 5.7(a)]. Other sugars show a much more complicated mutarotation, however, indicating that forms of different ring size, or aldehydo forms, may be occurring

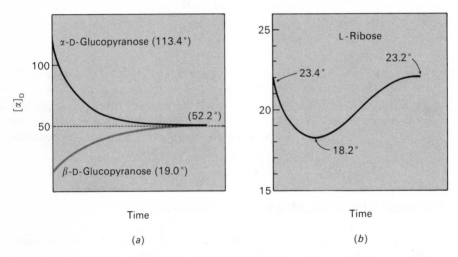

Figure 5.7   The mutarotation of some representative monosaccharides. (*a*) D-Gluco-pyranoses. (*b*) L-Ribose.

in the process [Figure 5.7(b)]. The activation energy for mutarotation is between 13 and 17 kcal/mole. For interconversion involving the formation of furanose forms, the reactions are more rapid (energies of activation, approximately 13 kcal/mole), whereas interconversions of pyranose forms are slower (energies or activation, approximately 17 kcal/mole). At equilibrium the composition of the mixture is determined by the relative thermodynamic stabilities of the various forms in the solvent chosen.

In the case of a simple equilibration of α- and β-pyranose forms, the process appears to be first-order and the integrated rate equation has the form

$$k_1 + k_2 = \frac{1}{t} \ln \frac{\alpha_0 - \alpha_\infty}{a_t - \alpha_\infty}$$

where k_1 is the rate constant for the reaction $\alpha \rightarrow \beta$ and k_2 is the rate constant for the reaction $\beta \rightarrow \alpha$. The terms α_0, α_t, and α_∞ are the rotation at times $0, t$, and ∞. When $\log (\alpha_0 - \alpha_\infty/\alpha_t - \alpha_\infty)$ is plotted versus time, a straight line is obtained with slope equal to $(k_1 + k_2)/2.303$ (Figure 5.8). The same slope is obtained whether the data for the mutarotation of the α or of the β form are used if the reaction is a simple equilibration. If other forms are present, the slope is not constant.

The mutarotation of β-L-ribopyranose [Figure 5.7(b)] probably reflects the rapid formation of furanose forms superimposed on the slower formation of more stable pyranose forms. The equilibrium value of the specific rotation is similar to the initial value; but whereas the latter reflects the presence of a single component, the former is due to the presence of a mixture. The proportion of each ring form and anomer present at equilibrium is determined by its relative stability, which is determined by the number and kind of steric and polar interactions present in the solvated molecule.

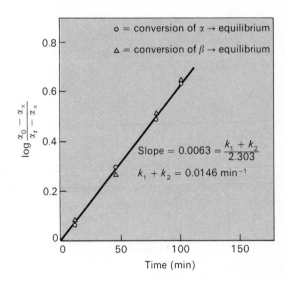

Figure 5.8
The determination of the sum of the rate constants for the mutarotation of D-glucopyranoses.

Melting Point

The small difference in melting point between α- and β-D-*gluco*-pyranose cannot be considered significant. It is generally true, however, that for carbohydrates the forms having higher melting points have more intermolecular hydrogen bonding. This is especially noticeable with the alditols (acyclic compounds obtained from the aldoses by reduction of the aldehyde group).

Solubility

Presumably β-D-glucopyranose fits into water structure (that is, solvated by water) more readily than the α form. The conformations of the anomers are described in Section 5.6 and it can be seen that the β form could fit very well into the ice-like structure of water shown in Figure 2.4.

Bromine Oxidation

It has been demonstrated clearly that the very slow rate of oxidation of α-D-glucopyranose reflects its rate of conversion to the β form followed by oxidation under the conditions of the reaction rather than an actual oxidation of the α anomer. The reaction has been proposed to require attack by $Br^{\oplus}$ at the anomeric oxygen and the development of the steric situation shown in Figure 5.9,

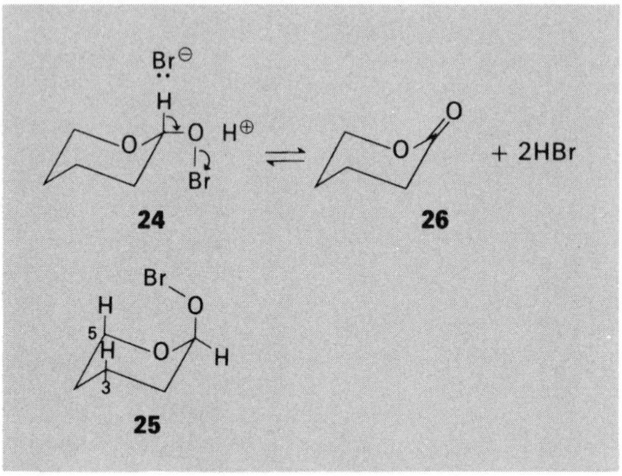

Figure 5.9

The mechanism of oxidation of β-D-glucopyranose by bromine.

(**24**) (H and Br must be *antiparallel*). With the α anomer (**25**) this arrangement is hindered by the hydrogens attached to carbons 3 and 5 of the ring. The fact that bromine oxidation results in the formation of δ-lactones such as (**26**) rather than acyclic acids indicates that the aqueous solution of glucose does indeed contain predominately pyranose forms. The difference in rate of oxidation between the tautomers having equatorial anomeric hydroxyl groups and those having axial anomeric hydroxyl groups has been used to establish the anomeric composition of mixtures of sugars.

Glucose oxidase

This is an enzyme which converts β-D-glucopyranose to the corresponding lactone. It is highly specific and is used as a very definitive test for the presence of D-glucose. The small rate observed for the α anomer probably reflects the rate of mutarotation rather than a slow oxidation.

5.11   ASSIGNMENT OF CONFORMATION: CUPRAMMONIUM COMPLEXES

Since conformational isomers are interconvertible by rotation about single bonds, the energy barriers between conformers are usually small, at least when compared to the energy barriers between other types of isomers. To determine the conformational preferences of a compound requires that it be examined by a technique which does not alter the equilibrium between conformers. This requirement would seem to rule out the use of most chemical processes as tools for conformational analysis since in all cases a complex is formed which might easily alter the normal distribution of conformations.

TABLE 5.3   CUPRAMMONIUM COMPLEX FORMATION BETWEEN VICINAL HYDROXYL GROUPS[a]

Conformation	Resistance Change	Molecular Rotation Change
	50–70	0
I II	50–70	I + 2,000 II − 2,000
	0	0
	0	0

[a] R. E. Reeves, "Cuprammonium-Glycoside Complexes," *Advances in Carbohydrate Chemistry*, **6**, 107 (1951).

In the crystalline state the conformation can be determined unequivocally by X-ray crystallography. Even though X-radiation is sufficiently powerful to break chemical bonds, the technique can be used because the crystal structure holds each molecule relatively rigidly and the radiation is largely diffracted, not absorbed.

In solution the problem is more difficult. The molecules are not oriented; with carbohydrates, tautomers (α- and β-furanose and pyranose forms) as well as conformational isomers can be present. Despite the drawbacks inherent in studies involving complex formation (chemical reaction), such methods have been important in developing present ideas and methods and warrant some discussion.

Some early studies utilized the fact that properly situated pairs of hydroxyl groups separated by less than 3.5 Å form a complex with cuprammonium ions. When adjacent hydroxyl groups are eclipsed, a very stable complex is formed; at 60 deg a complex is present, but at any projected angle above this no complex is observed (Table 5.3). Complexing results in an increase in the resistance of the cuprammonium solution. The cyclic complex formed by hydroxyls at 60 deg to each other is usually asymmetric and has a large specific rotation. Observation

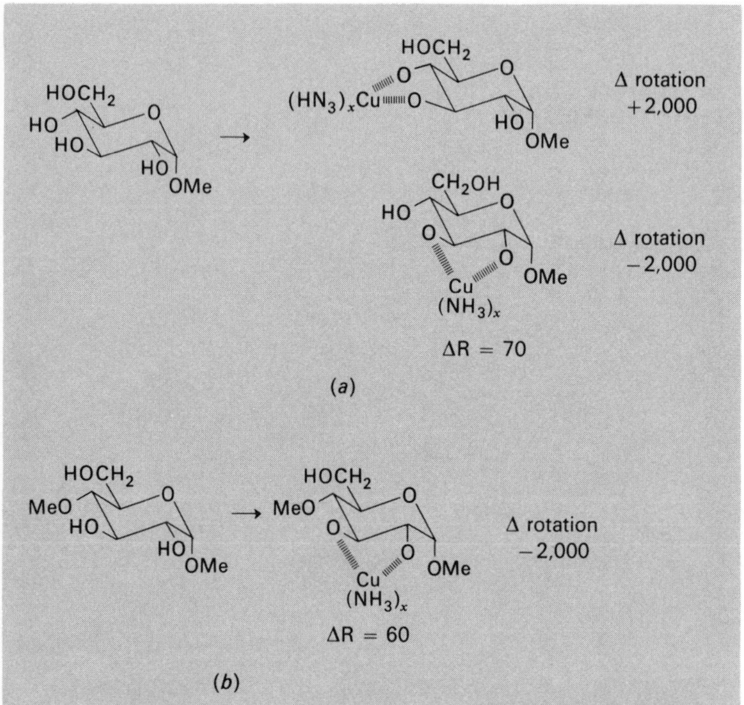

(a)

(b)

Figure 5.10   Cuprammonium complex formation with methyl-D-glucopy-ranoside and methyl 4-O-methyl-D-glycopyranoside.

of the rotational changes and conductivity changes allowed both the extent and site of complex formation in the molecule to be evaluated. For example, methyl α-D-glucopyranose forms a complex which, by conductivity measurements, appears to involve one pair of hydroxyls. However, the rotational change is quite insignificant. It appears that the solution contains an equimolar mixture of two complexes, one involving the 2- and 3-hydroxyl groups and having a large positive rotation and the other involving the 3- and 4-hydroxyls and having a large negative rotation as shown in Figure 5.10(a). This deduction was confirmed by studies on model compounds in which one of the OH groups was blocked, removing its ability to complex; for example, methyl 4-O-methylglucose cannot form a complex involving the 4-OH group and gives a large positive rotation change in cuprammonium solutions but the same change in resistance as methyl α-D-glucopyranose [Figure 5.10(b)].

On the basis of these studies and the assumption that the pyranose ring exists as one of the strain-free chair or boat forms, it was possible to assign arbitrary units to the factors which decrease the stability of the pyranose ring system (Table 5.4). By summing the factors for a given conformation, its stability

TABLE 5.4   INSTABILITY FACTORS[a]

	Arbitrary units
Axial OH	1
Axial CH$_2$OH	2
Axial CH$_2$OH + axial OH at C-1 or C-3	2.5
Δ2 Effect (see Figure 5.11)	2.5

[a] R. B. Kelly, "Modification of Reeve's Instability Factors," *Canadian Journal of Chemistry*, **35**, 149 (1957).

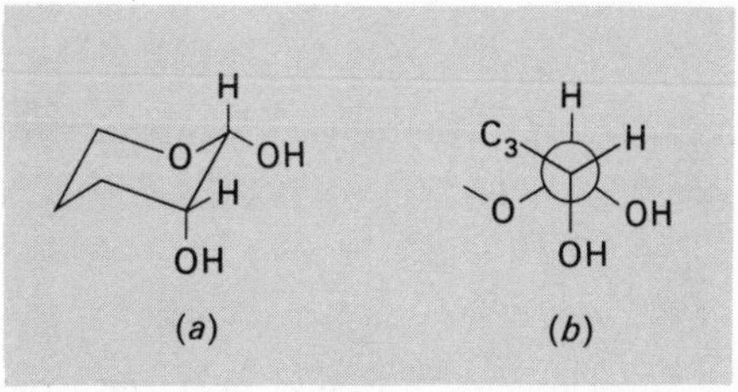

Figure 5.11

The Δ-2 effect. A highly unstable arrangement if hydroxyls at positions 1 and 2 relative to the ring oxygen due to dipole-dipole interactions. (a) Shown in stereoprojection. (b) Shown in Newman projection.

relative to another conformation can be estimated. When differences of one unit or less are obtained, a completely reliable prediction of conformational preference cannot be made. Examples of the use of these factors are given in Figure 5.12.

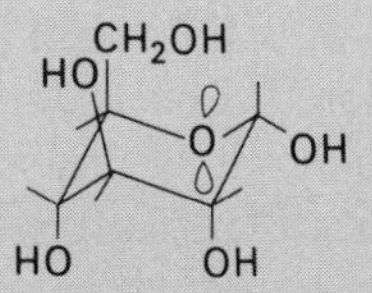

Alternate form

I Instability units

3-Axial OH	3
Axial CH_2OH	2
Axial CH_2OH + axial 3 OH	2.5
Δ-2 effect	2.5
Total	10

II Interaction energies

		kcal/mole
(a) Due to axial substituents above the ring	H versus OH	0.45
	H versus CH_2OH	0.90
	OH versus CH_2OH	1.60
(b) Due to axial substituents below the ring	HO versus OH	1.90
	2x OH versus O	0.70
(c) Due to equatorial groups	None	
	Total	5.55

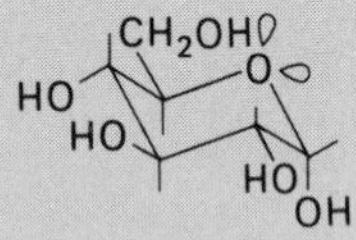

Normal form

I Instability units

1-Axial OH	1
Total	1

II Interaction energies

		kcal/mole
(a) Due to axial substituents above the ring	None	
(b) Due to axial substituents below the ring	2 H versus OH	0.90
(c) Due to equatorial groups (starting at 1)		
	ax OH versus eq OH	0.35
	eq OH versus eq OH	0.35
	eq OH versus eq OH	0.35
	eq OH versus eq CH_2OH	0.35
	Total	2.30

Conclusions: I—Normal form much more stable.   II—Normal form much more stable. $\Delta G° = -2.25$ kcal/mole. At 25°C K_{eq} = [normal]/[alternate] $\cong$ 40, and 97% of molecules are in the normal conformation.

Figure 5.12   The prediction of the conformational preference of α-D-glucopyranose using instability factors and interaction energies.

5.12   BORATE COMPLEXES OF THE CYCLITOLS

A similar approach to the conformational analysis of the sugars involves evaluation of the effects of interactions of nonbonded groups on the stability of tridentate borate complexes (Figure 5.13) of the cyclitols.† From the equilibrium constants for complex formation and analysis (using molecular models) of the interactions between hydroxyl groups in the free cyclitol and in the complexes, it is possible to calculate the interaction energies listed in Table 5.5 for carbohydrates in water. These are generally applicable to carbohydrates and can be used to calculate differences between conformational and configurational isomers as shown in Figure 5.13.

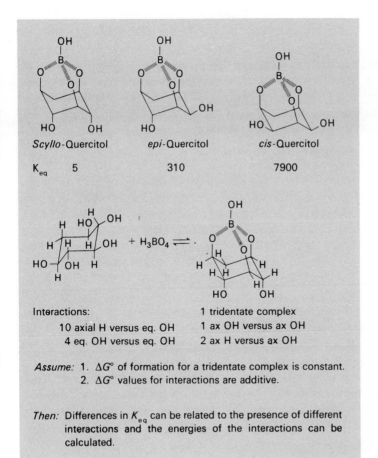

Assume: 1. $\Delta G°$ of formation for a tridentate complex is constant.
2. $\Delta G°$ values for interactions are additive.

Figure 5.13

Then: Differences in K_{eq} can be related to the presence of different interactions and the energies of the interactions can be calculated.

Tridentate complex formation between borate and some cyclitols and the calculation of interaction energies.

† The cyclitols are hydroxylated cyclohexane derivatives which have stable ring systems and which have been used as models for carbohydrates.

TABLE 5.5   INTERACTION ENERGIES (kcal/mole) BETWEEN SUBSTITUENTS ON CYCLOHEXANE OR PYRANOSE RINGS IN WATER[a]

	$axOH$	$axCH_3 (or\ CH_2OH)$	$eq.\ OH$
axH	0.45	0.9	0.0
axOH	1.90	1.6	0.35
eq OH	0.35	0.35	0.35
eq CH_3 (or CH_2OH)	0.35	—	0.35
Ring oxygen p electrons	0.35		
Anomeric effect			
(interaction between the ring oxygen and an			
equatorial OH or OR)	0.55		
Δ-2 effect (see Figure 5.11)	0.45		

[a] E. L. Eliel, N. L. Allinger, S. J. Angyal, and G. A. Morrison, *Conformational Analysis*, John Wiley & Sons, Inc., New York, 1966, p. 355.

Using the equation $\Delta G^0 = -RT \ln K$, the proportion of α-pyranose form in the equilibrated aqueous solutions of a number of carbohydrates has been calculated and compared to the proportion found by rotation and by oxidation with bromine (Table 5.6). In many cases the agreement is excellent between

TABLE 5.6   PROPORTION OF α PYRANOSE (%) IN AQUEOUS SOLUTION

Carbohydrate	Calculated from Values in Table 4.4	Found by Rotation	Found by Bromine Oxidation	Found by nmr
Glucose	36	32	37.4	36
Mannose	68	68.8	68.9	67
Galactose	36	29.6	31.4	27
Lyxose	73	76	79.7	71
Xylose	36	34.8	32.1	33

[a] S. J. Angyal, "The Composition and Conformation of Sugars in Solution," *Angewandte Chemie, International Edition*, **8**, 157 (1969), with permission of Verlag Chemie, GmbH.

calculated and experimental results. In other cases (for example, galactose) there is only an approximate fit. It is probable, in the case of galactose and others like it, that furanose forms are present as an appreciable proportion of the equilibrated solution.

5.13   GLYCOSIDE HYDROLYSIS

The bulk of carbohydrate in nature is polysaccharide in which monosaccharides are joined by glycosidic bonds. Structural studies as well as studies of the biochemical behavior of the sugars inevitably require consideration of the cleavage of the glycosidic bond.

Two plausible mechanisms for glycoside hydrolysis have been proposed; both are shown in Figure 5.14. Mechanism *a* involves protonation of the oxygen of

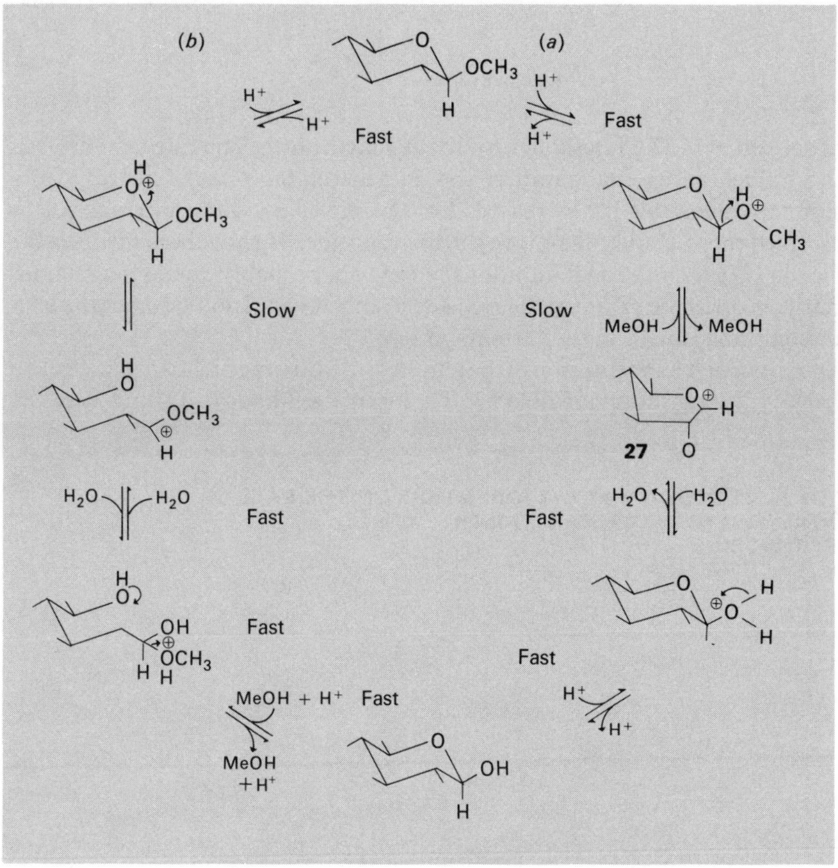

Figure 5.14   Possible mechanisms of glycoside hydrolysis. (*a*) involves a cyclic carbonium ion; (*b*) involves a ring-opened carbonium ion.

the aglycone and the formation of a cyclic carbonium-oxonium ion (**27**) which has carbon atoms 1, 2, and 5 and the ring oxygen in a plane. Mechanism *b* is initiated by protonation of the ring oxygen and involves an acyclic carbonium-ion intermediate. Both routes are unimolecular in the rate-determining step. At present the bulk of the experimental evidence favors Mechanism *a*, although there has been no definitive proof offered. Methyl glycopyranosides in which the more stable anomer (in terms of nonbonded interactions) has the aglycone in an equatorial orientation are hydrolyzed more readily than those with an

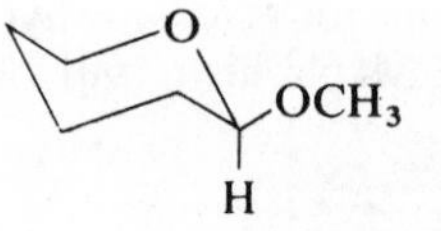

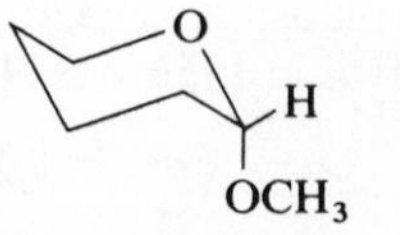

2   1

Rate of hydrolysis

axial substituent. The rates differ by about a factor of 2. The rate of hydrolysis
is also influenced by configuration and, in general, the rates of hydrolysis in-
conformational strain is increased. For example, α-D-galactopyranoside is
hydrolyzed more readily than α-D-glucopyranoside. If hydrolysis involves the
cyclic ion (27), then the configurational effects are probably due to interactions
which develop as the protonated glycoside loses methanol and assumes the half-
chair conformer found in the carbonium ion.

Alteration of the substitution of the tetrahedropyran ring has a drastic effect on
hydrolysis rates, as shown in Table 5.7. The loss of the 2-hydroxyl group increases

TABLE 5.7   EFFECT OF SUBSTITUTION AT C-2 AND AT C-5 ON THE RATE OF
ACID-CATALYZED HYDROLYSIS OF METHYL α-D-
GLUCOPYRANOSIDE[a]

Compound		Relative Rate
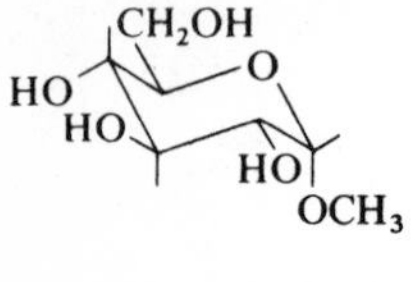		1
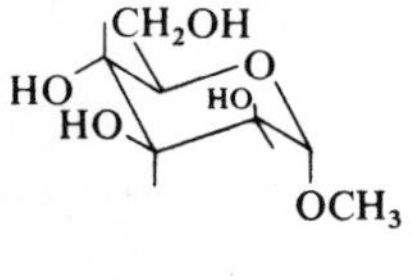	2-Epimer	2.4
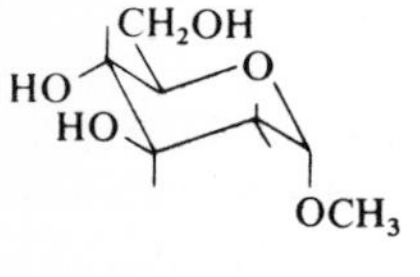	2-Deoxy	2,090
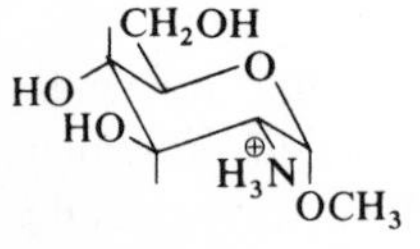	2-Amino-2-deoxy	0.1

TABLE 5.7 EFFECT OF SUBSTITUTION AT C-2 AND AT C-5 ON THE RATE OF ACID-CATALYZED HYDROLYSIS OF METHYL α-D-GLUCOPYRANOSIDE *(cont.)*

Compound		Relative Rate
	2-Acetamido-2-deoxy	3.0
	5-CH$_2$OH replaced by H (methyl α-D-xylopyranoside)	4.5
	6-Deoxy	5.0
	6-*O*-Methyl	0·7
	6 Position oxidized (methyl α-D-glucopyranosiduronic acid)	0.9
	6-Amino-6-deoxy	0.1

[a] J. N. BeMiller, Acid-Catalyzed Hydrolysis of Glycosides, *Advances in Carbohydrate Chemistry,* **22**, 25 (1967), with permission of Academic Press, Inc.

the rate enormously; this is partly due to relief of steric interactions in achieving the transition state but more so to loss of the electron-withdrawing action of the hydroxyl group. Electron-withdrawing groups in the 2 position decrease the basicity of the oxygen of the aglycone (and the ring oxygen) and therefore decrease the concentration of the conjugate acid. In addition, and perhaps more important, they make it more difficult to develop the carbonium ion at carbon 1. The increased effectiveness of an ammonium group at C-2 (as compared to the hydroxyl group) attests to the importance of the inductive effect. The conversion of the 2-amino group to an acetamido group decreases its electron-withdrawing power and increases the rate of hydrolysis considerably.

Alteration of the substituents at carbon 5 produces much smaller changes in hydrolysis rate, and it is difficult to know whether the observed effects are due to alterations in the steric or inductive properties of the groups. Glycosides of uronic acids are generally somewhat more stable than those of the parent sugar but the effect is not large. The greater separation of carbon 6 from the site of protonation (at the aglycone oxygen) probably insulates the latter from inductive effects. The five fold increase in rate of the 6-deoxy derivative is surprisingly large, although the methyl group should be electron donating and the compound is best compared to the methyl α-D-xylopyranoside.

5.14   FURANOSES AND FURANOSIDES

The nitrogen furanosides of D-ribose and 2-deoxy-D-ribose (properly 2-deoxy-D-*erythro*pentose) are found in all living systems in the nucleic acids. There are few other abundant naturally occurring furanose forms. Much less is known of the furanose forms of sugars since in those cases where pyranose rings can form they are more stable and predominate. In the case of the four-carbon sugars, the furanose rings are the only ones that can exist but these sugars have not been extensively studied.

The furanosides are less stable than the corresponding pyranosides but where the potential for the formation of both exists, the furanosides are formed faster. If the conditions of the reaction permit the establishment of equilibrium, then the pyranosides will predominate. The rates of formation of methyl pentofuranosides are given in Table 5.8 along with the proportions of the various products at equilibrium. It is clear that the rate of glycoside formation does not relate to the composition of the equilibrium mixture. The rates reflect only the differences between the standard free energies of the reactants and the transition states and are due in part to nonbonded interactions in the latter.

Relatively little attention has been given to the conformational analysis of furanoses (-osides). The ring has been shown not to be planar and can exist in an *envelope* or *twist* conformation which are deisgnated E and T, respectively. In the E conformer one atom is out of the plane of the ring, and it is designated by a superscript or a subscript, depending on whether it is above or below the plane. Structure (**28**) shows a E^3 conformer; structure (**29**) shows a E_3 conformer. A similar convention is used for the T forms and a T_2^3 conformer

TABLE 5.8 THE RELATIVE RATES OF FORMATION OF METHYL PENTOFURANOSIDES AND THE EQUILIBRIUM COMPOSITION OF THE GLYCOSIDATION REACTION[a]

Compound	First Product	Relative Rate of Glycosidation	Furanosides α	Furanosides β	Pyranosides α	Pyranosides β
Lyxose (HC=O, HO, HO, —OH, CH₂OH)	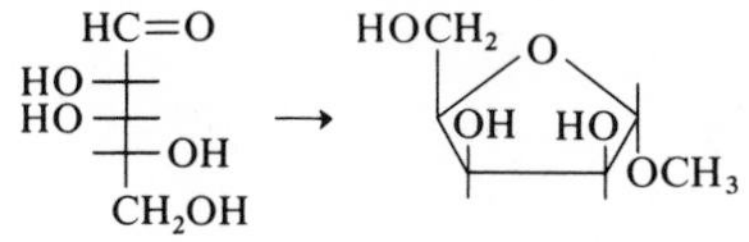 HOCH₂ ring, OH HO, OCH₃	1	1.4 T_2^3 or E^3	0 T_2^3	88.3	10.3
Arabinose (HC=O, HO, —OH, —OH, CH₂OH)	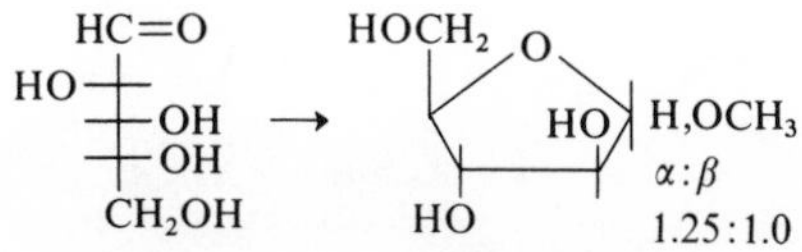 HOCH₂ ring, HO, H,OCH₃, α:β 1.25:1.0	1.9	21.5 T_2^3 ot T_3^2	6.8 E_2	24.5	47.2
Xylose (HC=O, —OH, HO, —OH, CH₂OH)	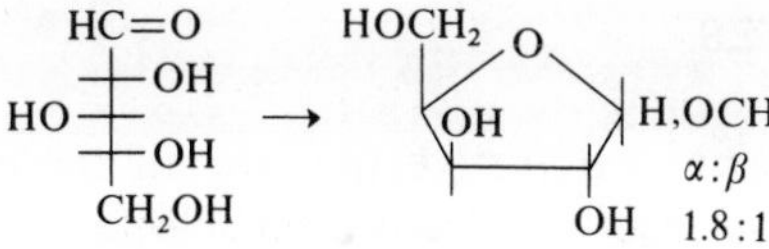 HOCH₂ ring, OH, H,OCH₃, α:β 1.8:1.0	2.5	1.9 T_3^2	3.2 T_3^2 or E_3	65.1	29.8
Ribose (HC=O, —OH, —OH, —OH, CH₂OH)	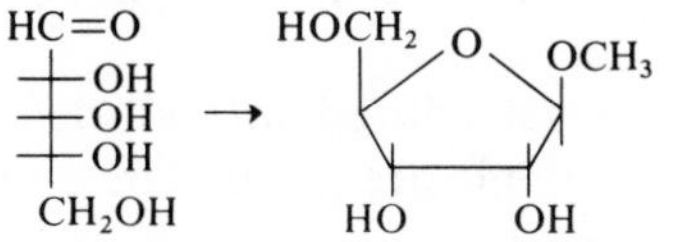 HOCH₂ ring, OCH₃, HO, OH	12.4	5.2 E^3	17.4 E^3	11.6	65.8

[a] C. T. Bishop and F. P. Cooper, *Canadian Journal of Chemistry*, **41**, 2743 (1963).

is shown in (**30**). The interactions between hydrogens, hydroxyl groups,

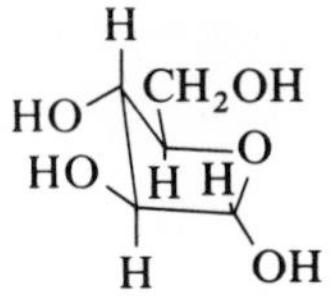

28 E^3

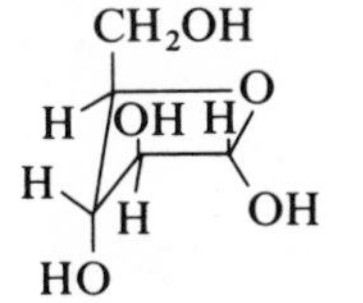

29 E_3
α-D-arabinofuranose

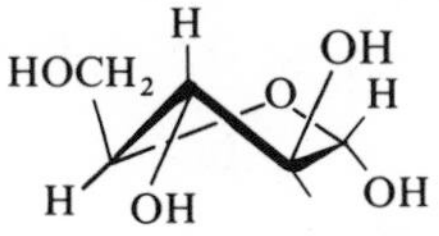

30 T_2^3

hydroxymethyl groups, and aglycone groups can be analyzed much as those in the pyranose ring. This analysis has been performed based on the ratio of the methyl pentofuranosides at equilibrium. The conformations assigned on this basis are given in Table 5.8; they should be considered tentative assignments.

The furanose ring itself is less stable than the pyranose ring but the difference is difficult to quantitate. For tetrahydrofuran-2-ol (**32**) and tetrahydropyran-2-ol (**31**), the difference in stability (relative to the corresponding ω-hydroxy-aldehydes) has been evaluated at only 0.4 kcal/mole. The difference in the highly substituted carbohydrates is probably larger (see Table 5.8 and compare the proportions of furanosides to pyranosides at equilibrium).

$$HC=O \quad \rightleftharpoons \quad (ring) \quad\quad HC=O \quad \rightleftharpoons \quad (ring)$$

CH₂OH   **31**   CH₂OH   **32**

11.4%   88.6%   6.1%   93.9%

5-hydroxypentanal   4-hydroxybutanal

5.15 REACTIONS OF CARBOHYDRATES

The following list of reactions is not exhaustive nor are all the conditions given for carrying them out. These reactions have been selected to illustrate the methods frequently used to characterize carbohydrates or to synthesize those which are not otherwise available.

In many cases new nomenclature is introduced without definition. In all cases a general name and a specific example are given. The correct usage of nomenclature can be implied from them.

To help orient the reader to the relevance of the examples chosen, one or more of the ways in which each reaction is used are listed. For most reactions an analogous biochemical process is described. Although the analogies are by no means perfect, they illustrate the fact that most of the problems faced by the chemist are faced by some biological species and that biological catalysts utilize processes adapted to their chemical task in much the same way as the chemist does.

I. Reduction

General reaction:

Aldose or ketose ⟶ alditol

CH₂OH
OH
HO
OH
OH
CH₂OH

Specific examples:

$$\text{D-Glucose} \longrightarrow \text{D-glucitol}$$
$$\text{D-Fructose} \longrightarrow \text{D-glucitol} + \text{D-mannitol}$$

Reagents: Sodium borohydride, hydrogen + catalyst (Pt, Raney Ni).
Use: To correlate the configuration of carbohydrates.
Biochemical process:

D-*Erythro*pentulose 5-phosphate   L-ribitol 1-phosphate
(ribulose 5-phosphate)   (D-ribitol 5-phosphate)

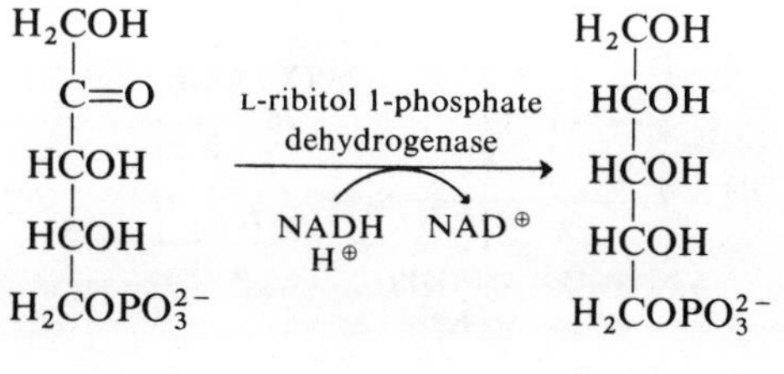

L-ribitol 1-phosphate
dehydrogenase

NADH   NAD$^\oplus$
H$^\oplus$

Note that the enzyme produces only *one* of the possible products.

II. Oxidation

A. Aldehyde to acid

General reaction:

$$\text{Aldose} \longrightarrow \delta\text{-aldonolactone} \longrightarrow \text{aldonic acid} \longrightarrow \gamma\text{-aldonolactone}$$

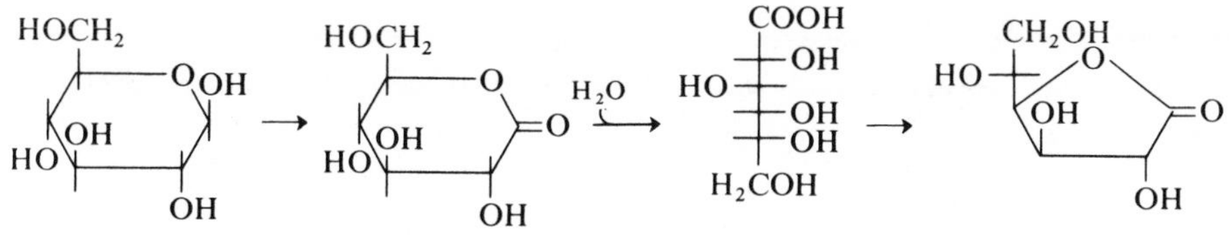

Specific examples:

β-D-Glucopyranose $\longrightarrow$ D-glucono-δ-lactone $\longrightarrow$ D-gluconic acid $\longrightarrow$ D-glucono-γ-lactone

Reagents: Bromine at pH 6.0 specifically oxidizes aldoses with equatorial anomeric hydroxyl groups; the other anomer must rearrange to the form which can be oxidized. The δ-aldonolactone is hydrolyzed to the aldonic acid. Nitric acid is less specific; it is not known whether a lactone is an intermediate and the reaction can continue to give the dibasic saccharic acid (both carbon-1 and carbon-6 are oxidized to acids). Hypoiodite in alkaline solution oxidizes aldoses to aldonic acids quantitatively.

Use: Correlation of configuration and the preparation of synthetic intermediates.

Analogous biochemical process:

Glucose 6-phosphate ⟶ 6-phosphoglucono-δ-lactone ⟶ gluconic acid 6-phosphate

Two enzymes are required, the first carries out a reaction similar to that carried out by bromine and the second opens the lactone ring so that the next enzyme in the biochemical sequence does not have to depend on the uncatalyzed hydrolysis for its substrate. Note that the biochemical oxidation is essentially the reverse of the biochemical reduction—both involve nicotinamides as coenzymes.

B. Involving the carbonyl group with cleavage of the carbon chain

General reaction:

Aldoses or ketoses ⟶ enediol ⟶ aldonic acids, saccharinic acids, and other organic acids

D-Glucose $\xrightarrow{\text{with oxygen}}$ formic acid + D-arabinonic acid

Specific examples:

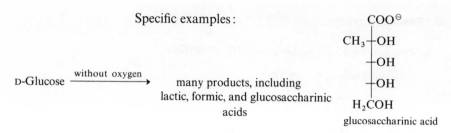

$$\begin{array}{c} COO^{\ominus} \\ CH_3 - OH \\ -OH \\ -OH \\ H_2COH \end{array}$$

D-Glucose $\xrightarrow{\text{without oxygen}}$ many products, including lactic, formic, and glucosaccharinic acids

glucosaccharinic acid

Reagents: In the case shown in which the aldonic acid is formed, the reagents are dilute alkali and molecular oxygen and the products shown are formed in good yield. In the absence of oxygen, internal oxidation-reduction occurs which produces an as yet incompletely analyzed mixture of products.

Use: These reactions have been studied to elucidate the behavior of carbohydrates in alkali and for the synthesis of some interesting derivatives.

Analogous biochemical process:

Catechol $\longrightarrow$ cis,cis-muconic acid

$$\text{[structure: catechol]} \quad + \quad \begin{array}{c} O^* \\ \| \\ O^* \end{array} \xrightarrow{\text{catechol-1,2-oxygenase}} \text{[structure with } COO^{\ominus}, COO^{\ominus}]$$

There are a number of enzymes in soil bacteria which function to oxygen-ate aromatic compounds so as to initiate their breakdown for the provision of energy. The reaction is analogous to a periodate oxidation, discussed in Part IX. In nature, however, carbohydrate chains are altered in length by enzymes which catalyze aldol-type reactions (Part VII) or by oxidative decarboxylation.

C. Reducing sugar tests

General reaction:

Aldose or ketose or compounds which can be converted to them by heating in alkaline medium $\longrightarrow$ mixed oxidation products and a color change in the oxidizing agent

Reagents: Alkali and an oxidizing agent such as $Cu^{2\oplus}$ (used in the Benedict's and Fehling's tests), $Ag^{\oplus}$ (Tollens test), dinitrosalicylate, hypoiodite (NaOI, gives a quantitative yield of the corresponding aldonic acid), and $Fe(CN_6)^{3\ominus}$ (Hagedorn–Jensen method). Other tests do not involve the reducing properties of the sugars.

III. Rearrangements

A. Induced by alkali under anaerobic conditions

General reaction:

Aldose or ketose $\longrightarrow$ enediol $\longrightarrow$ epimeric aldoses and ketoses

Specific example:

$$
\text{D-Glucose} \longrightarrow
\begin{array}{l}
\text{D-mannose } 2.5\% \\
\text{D-glucose } 63.5\% \\
\text{D-fructose } 31.0\% \\
+ \text{ degradation products}
\end{array}
$$

Reagents: Dilute alkali. The reaction is called the Lobry de Bruyn-Alberda van Ekenstein transformation. It is improbable that an equilibrium between the 2-epimers, glucose, mannose, and fructose has ever been achieved since the enolization of fructose can lead to a 3,4-enediol which can lead to a 3-keto compound. The reaction can continue down the chain. Additional rearrangements are catalyzed by more concentrated alkali. Pyridine has been used to give a "controlled" rearrangement of the pentoses.

Use: To establish 2-epimeric relationships of some sugars.

Analogous biochemical process:

D-Glyceraldehyde 3-phosphate $\longrightarrow$ dihydroxyacetone phosphate

The reaction of interest is catalyzed by an isomerase. The sequence shown has not been proved but constitutes a logical interpretation of the observed events which include exchange of a proton with the solvent.

B. Induced by acid and alkali—Mutarotation

General reaction:

Aldose or ketose ⇌ anomeric aldose or ketose

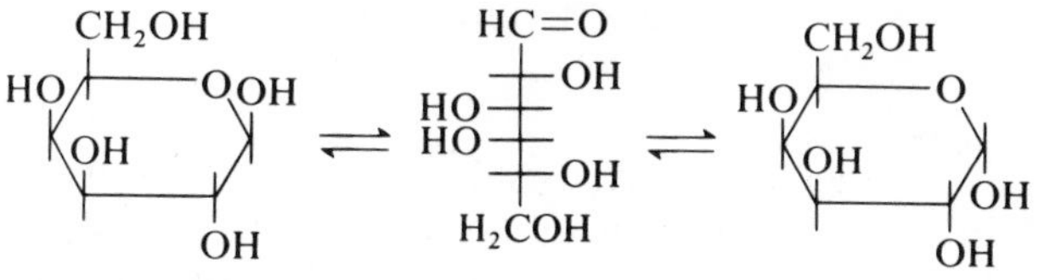

Specific example:

β-D-Galactopyranose   ⇌   α-D-galactopyranose

Reagents: Dilute acid or alkali. Catalysis is approximately 4,000 times more efficient by OH^- than by H^+ and is even more efficient by the difunctional catalyst 2-hydroxypyridine which acts by a push-pull mechanism as an acid and a base.

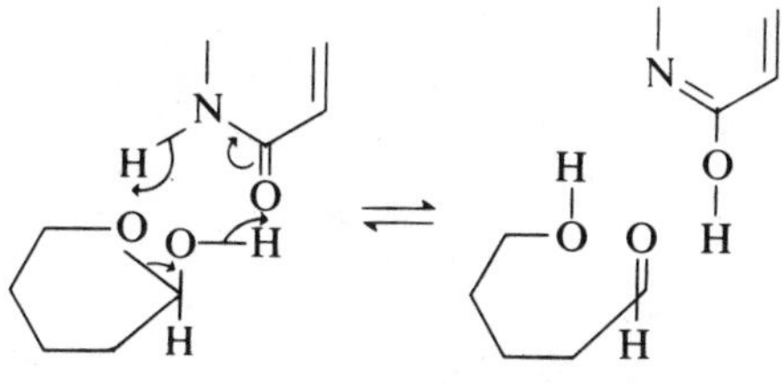

Use: The direction of mutarotation can be used to determine anomeric configurations (see Figure 5.7).

Analogous biochemical process: The conversion shown above is catalyzed by a fungal enzyme which is an extremely effective catalyst. (It also catalyzes the mutarotation of glucose, lactose, and maltose.)

IV. Dehydration

General Reaction:

Aldohexose or ketohexose $\xrightarrow{H^+}$ 5-hydroxymethyl furfural

Aldopentose or ketopentose $\xrightarrow{H^+}$ furfural

Specific example:

α-D-Fructopyranose ⟶ 5-hydroxymethyl furfural

Reagents: Concentrated strong acids such as sulfuric acid. The products condense with phenols to give colored substances; the reaction serves as the basis for several qualitative and quantitative tests; for example, Molisch test: α-Naphthol-purple color, a general test for pentoses and ketoses.

Seliwannoff test: Oricinol-red color, a test for ketoses.

Analogous biochemical process:

$$\text{D-Glyceric acid 2-phosphate} \rightleftharpoons \text{phosphoenolpyruvate}$$

Enolase is highly specific and the reaction is readily reversible—unlike the chemical process described.

V. Addition reactions of the carbonyl group

A. Cyanide (Kiliani synthesis)

General reaction:

Aldose or ketose ⟶ 2-epimeric nitriles ⟶ 2-epimeric aldonic acids

Specific example:

$$\text{D-Arabinose} \longrightarrow \begin{array}{l} \text{D-gluconic acid} \\ \text{D-mannonic acid} \end{array}$$

Reagents: Aqueous sodium cyanide. The hydrolysis of the nitriles occurs very readily and in some cases no specific treatment is required. The proportions of the 2-epimers in the product depends on the configuration of the starting material, but the epimer having the hydroxyl group at C-2 *trans* to that at C-4 usually predominates.

Uses: Establishing configurational relationships between sugars and extending the carbon chain.

B. Nitromethane (Sowden-Fischer synthesis)

General reaction:

Aldose or ketose $\longrightarrow$ 2-epimeric *aci*-nitroalcohols $\longrightarrow$ 2-epimeric aldoses

$$\text{HC}{=}\text{O} + \text{CH}_2\text{NO}_2^{\ominus} \longrightarrow \begin{array}{cc} \text{HCNO}_2 & \text{HCNO}_2 \\ | & | \\ \text{HCOH} & \text{HOCH} \\ | & | \end{array} \longrightarrow \begin{array}{cc} \text{HC}{=}\text{O} & \text{HC}{=}\text{O} \\ | & | \\ \text{HCOH} & \text{HOCH} \\ | & | \end{array}$$

Specific example:

$$\text{L-Arabinose} \longrightarrow \underset{80\%}{\text{L-glucose}} + \underset{20\%}{\text{L-mannose}}$$

Reagents: Nitromethane and alkali in methanol or water. The *aci*-nitro-alcohol salt is converted to the aldehyde by adding a solution of the compound in 2 N sodium hydroxide slowly to 16 N sulfuric acid. The proportion of epimers is quite variable; in the case cited, L-arabinose gives 80 percent or more of the L-gluco compound.

Use: Extension of the sugar chain. The synthesis of rare sugars such as L-glucose and L-mannose. The nitroalcohols can also be dehydrated to give the 1,2-olefins, which are susceptible to attack at C-2 by nucleophiles such as methoxide anion. This reaction has been used to synthesize 2-O-methyl-D-ribose—a naturally occurring methylated sugar.

Analogous biochemical process: Biochemical systems have many mechanisms available for the addition of one-carbon units to acceptor molecules. None of them is strictly analogous to the cyanohydrin or nitromethane synthesis. For example, glycine is converted to serine by the transfer of the methylene group from 5,10-methylene-5,6,7,8-tetrahydrofolic acid.

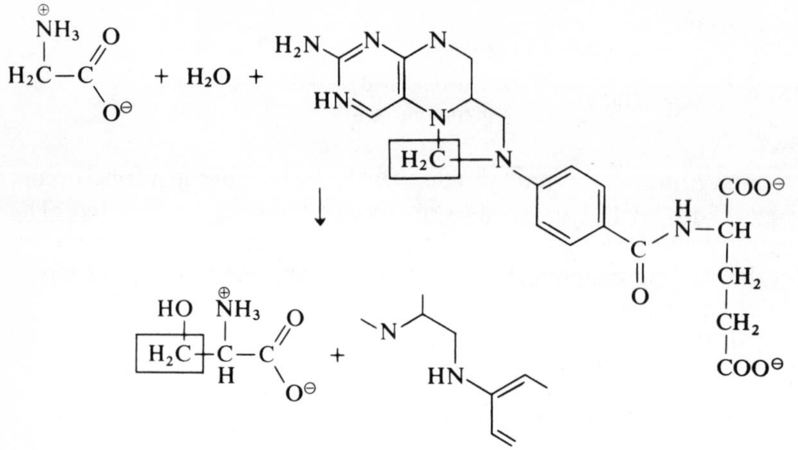

C. Amines

General reaction:

Aldose or ketose + amine $\rightleftharpoons$ carbinolamine $\rightleftharpoons$ Schiff base $\rightleftharpoons$ glycosyl amine

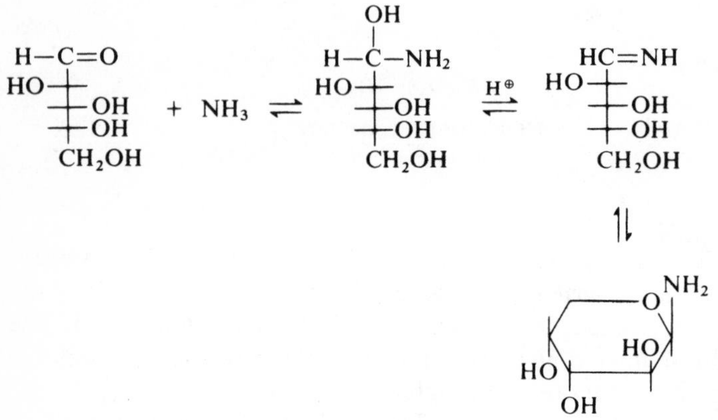

Specific example:

D-Arabinose + liquid NH_3 $\longrightarrow$ D-arabinosylamine
D-Arabinose + NH_2OH $\longrightarrow$ D-arabinose oxime

Reagents: Treatment with amines or hydroxylamine in a suitable solvent (alcohols, water, and so forth). Glycosylamines can rearrange via Schiff bases and 1,2-enediols to the corresponding *keto* form.

This is the Amadori rearrangement.
Use: The synthesis of biologically interesting compounds.
Analogous biochemical process:

1. Dihydroxyacetone phosphate + NH_2-aldolase $\rightleftharpoons$ Schiff base

This represents the first step in the synthesis of fructose 1,6-diphosphate from dihydroxyacetone phosphate and D-glyceraldehyde 3-phosphate by the enzyme aldolase.

2. D-Fructose 6-phosphate + L-glutamine $\longrightarrow$ D-Glucosamine 6-phosphate + L-glutamate

D. Thiols

General reaction:

Aldose or ketose + thiol $\longrightarrow$ thio-hemiacetal $\longrightarrow$ dithioacetal

Specific example:

D-Galactose + ethanethiol $\longrightarrow$ D-galactose diethyl dithioacetal

Reagents: The reaction is acid-catalyzed, usually concentrated hydrochloric acid is used but zinc chloride has been employed.

Use: The dithioacetal can be oxidized with peracids to the corresponding disulfone which is very susceptible to alkaline degradation, yielding an aldose with one fewer carbon atoms and a *bis* (alkylsulfonylmethane).

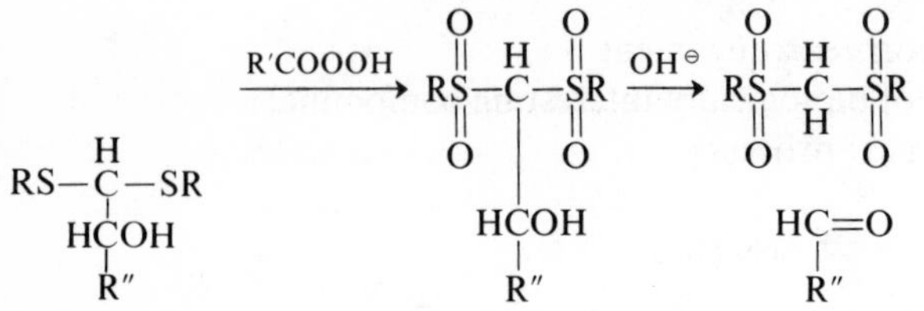

$$
\begin{array}{c}
\text{H} \\
\text{RS}-\overset{|}{\underset{|}{\text{C}}}-\text{SR} \\
\text{HCOH} \\
\text{R}''
\end{array}
\xrightarrow{\text{R}'\text{COOOH}}
\begin{array}{c}
\text{O} \quad\quad \text{O} \\
\|\quad \text{H} \quad\| \\
\text{RS}-\overset{\|}{\underset{\|}{\text{C}}}-\text{SR} \\
\text{O} \quad | \quad \text{O} \\
\text{HCOH} \\
\text{R}''
\end{array}
\xrightarrow{\text{OH}^{\ominus}}
\begin{array}{c}
\text{O} \quad\quad \text{O} \\
\|\quad \text{H} \quad\| \\
\text{RS}-\overset{\|}{\underset{\|}{\text{C}}}-\text{SR} \\
\text{O} \quad \text{H} \quad \text{O} \\
\text{HC}=\text{O} \\
\text{R}''
\end{array}
$$

Analogous biochemical processes:

D-Glyceraldehyde 3-phosphate $\xrightarrow[\text{dehydrogenase}]{\substack{\text{glyceraldehyde} \\ \text{phosphate}}}$ enzyme-bound thiohemiacetal

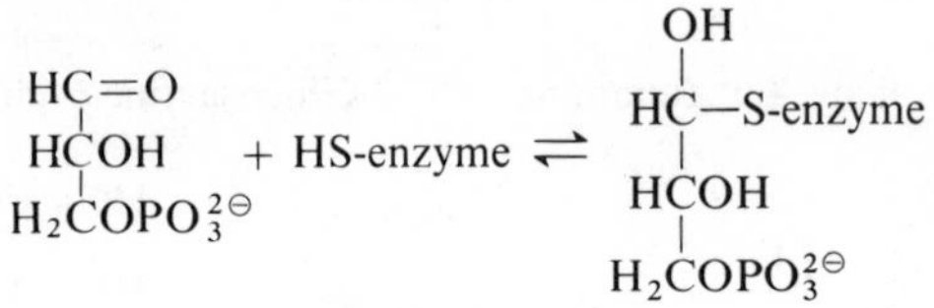

$$
\begin{array}{c}
\text{HC}=\text{O} \\
\text{HCOH} \\
\text{H}_2\text{COPO}_3^{2\ominus}
\end{array}
+ \text{HS-enzyme} \rightleftharpoons
\begin{array}{c}
\text{OH} \\
| \\
\text{HC}-\text{S-enzyme} \\
\text{HCOH} \\
\text{H}_2\text{COPO}_3^{2\ominus}
\end{array}
$$

The thiohemiacetal is proposed as the first step in the conversion of D-glyceraldehyde 3-phosphate to 1,3-diphosphoglyceric acid. In the next step the thiohemiacetal is oxidized to the corresponding thioester. Shortening of the carbon chain of sugars is accomplished by oxidative decarboxylation.

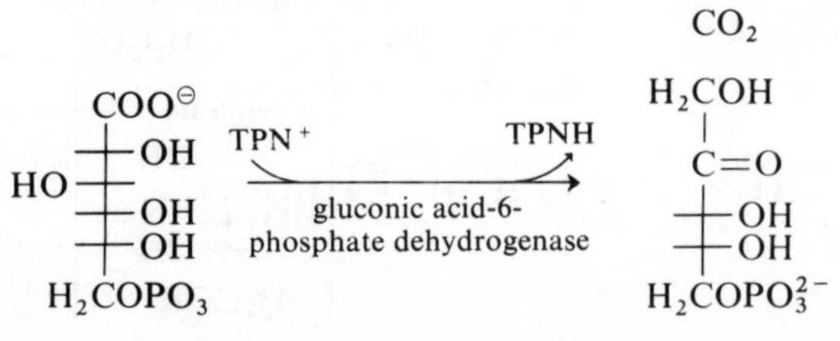

$$
\begin{array}{c}
\text{COO}^{\ominus} \\
\text{HO}-\text{+}-\text{OH} \\
-\text{OH} \\
-\text{OH} \\
\text{H}_2\text{COPO}_3
\end{array}
\xrightarrow[\substack{\text{gluconic acid-6-}\\\text{phosphate dehydrogenase}}]{\text{TPN}^+ \quad\curvearrowright\quad \text{TPNH}}
\begin{array}{c}
\text{CO}_2 \\
\text{H}_2\text{COH} \\
\text{C}=\text{O} \\
-\text{OH} \\
-\text{OH} \\
\text{H}_2\text{COPO}_3^{2-}
\end{array}
$$

E. Hydrazines

General reaction:

$$\text{Aldose or ketose} \longrightarrow \text{hydrazone} \longrightarrow \text{osazone}$$

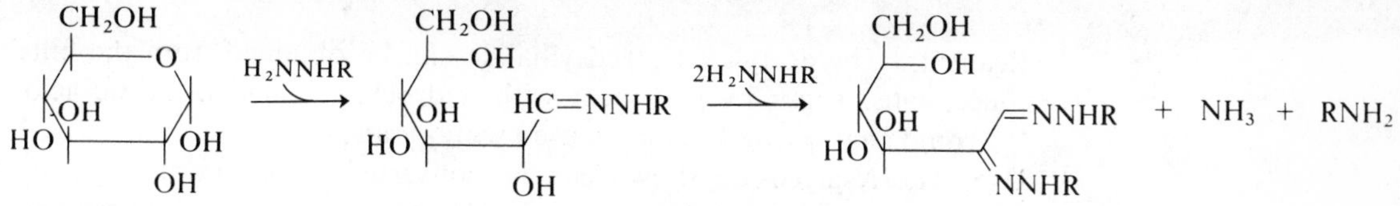

Specific example:

D-Glucose
or
D-mannose + ϕNHNH$_2$ $\longrightarrow$ D-glucose phenylosazone
or
D-fructose

Reagents: Usually phenylhydrazine.

Use: The phenylosazones have characteristic crystal structures and the different sugars react at very different rates. Since the asymmetry of the carbons adjacent to the carbonyl group is destroyed, sugars which are epimeric at C-2 give the same osazone although they may do so at different rates.

VI. *Glycoside formation*

General reaction:

Acetylated glycosylhalide $\longrightarrow$ acetylated glycoside

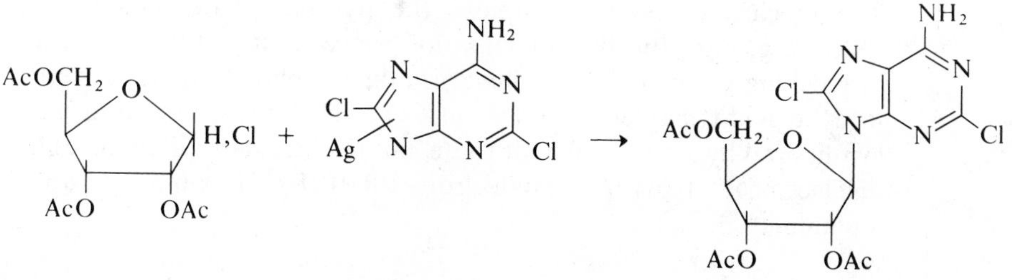

Specific example:

Tri-*O*-acetyl-D-ribofuranosylchloride + 2,8-dichloroadenine silversalt ⟶

9-(tri-*O*-acetyl-β-D-ribofuranosyl)-2,8-dichloroadenine

Reagents: The acetylated glycosylhalide can be obtained from the fully acetylated sugar by treatment with hydrochloric acid in acetic acid. Bromo analogs are frequently used with pyranose sugars.

Use: The *N*-glycoside shown can be converted to naturally occurring adenosine and this type of synthesis is used to prepare nucleosides and their analogs. The acetylglycosylhalides will react with any suitable nucleophile to produce a glycoside (ROH, RSH, and so forth).

Analogous biochemical processes:

1.

5-Phosphoribosyl pyrophosphate (PRPP)   orotate   orotidine synthetase

orotidine 5-phosphoate   pyrophosphate

The biochemical system employs the pyrophosphate moiety as a "leaving group" and the condensation is accomplished in the absence of blocking groups. The product, orotidine 5'-phosphate, is converted uridylic acid (Chapter 7) by loss of CO_2. The purine nucleosides are synthesized by a quite different route, but the first step still involves the displacement of pyrophosphate from PRPP by the amido group of glutamine.

2. Glycosides are synthesized by transfer of a glycosyl residue from donor to an acceptor. Many types of acceptors are found but among the most common are the hydroxyl groups of other carbohydrates. For example, glycogen synthetase catalyzes the following:

Uridine diphosphate glucose + glycogen $\longrightarrow$ glucose-glycogen

UDPG

VII. Aldol condensation

General reaction:

Aldehyde or ketone $\longrightarrow$ ketose

Specific example:

D-Glyceraldehyde
or $\longrightarrow$ D-fructose + D-sorbose + DL-dendroketose
dihydroxyacetone

$$
\begin{array}{l}
\text{HC=O} \\
\text{HCOH} + B \\
\text{H}_2\text{COH}
\end{array}
\begin{array}{c}
\text{BH} \\
\rightleftharpoons \\
\text{BH}
\end{array}
\begin{array}{l}
\text{HC=O} \\
{}^{\ominus}\text{COH} \\
\text{H}_2\text{COH}
\end{array}
\rightleftharpoons
\begin{array}{l}
\text{H}\overset{\ominus}{\text{C}}\text{OH} \\
\text{C=O} \\
\text{H}_2\text{COH}
\end{array}
\begin{array}{c}
\text{BH} \\
\rightleftharpoons \\
\text{BH}
\end{array}
B +
\begin{array}{l}
\text{H}_2\text{COH} \\
\text{C=O} \\
\text{H}_2\text{COH}
\end{array}
$$

D-glyceraldehyde   dihydroxyacetone

D-glyceraldehyde   dihydroxyacetone

$$
\begin{array}{l}
\text{H}_2\text{COH} \\
\text{C=O} \\
\text{HCOH} \\
\underset{\ominus}{} \\
\text{H—C}\overset{\curvearrowright}{\underset{}{=}}\text{O} \\
\text{HCOH} \\
\text{CH}_2\text{OH}
\end{array}
\qquad
\begin{array}{l}
\text{H}_2\text{COH} \\
\text{C=O} \\
\text{HCOH} \\
\underset{\ominus}{} \\
\text{HOCH}_2\text{—C—CH}_2\text{OH} \\
\overset{}{\underset{\text{O}}{\|}}
\end{array}
$$

$H^{\oplus}$ $\downarrow H^{\oplus}$

$$
\begin{array}{l}
\text{H}_2\text{COH} \\
\text{C=O} \\
\text{HOCH} \\
\text{HCOH} \\
\text{HCOH} \\
\text{CH}_2\text{OH}
\end{array}
\;+\;
\begin{array}{l}
\text{H}_2\text{COH} \\
\text{C=O} \\
\text{HCOH} \\
\text{HOCH} \\
\text{HCOH} \\
\text{CH}_2\text{OH}
\end{array}
\qquad
\begin{array}{l}
\text{H}_2\text{COH} \\
\text{C=O} \\
\text{HCOH} \\
\text{HOCH}_2\text{—COH} \\
\text{CH}_2\text{OH}
\end{array}
$$

D-fructose   D-sorbose   DL-dendroketose

Use: The reaction has been studied mainly to ascertain its mechanism. Analogous biochemical process:

Dihydroxyacetone phosphate + D-glyceraldehyde 3-phosphate $\rightleftharpoons$ D-fructose 1,6-diphosphate

$$
\begin{array}{l}
\text{H}_2\text{COPO}_3^{2\ominus} \\
\text{C=O} \\
\text{H}_2\text{COH}
\end{array}
\;+\;
\begin{array}{l}
\text{HC=O} \\
\text{HCOH} \\
\text{H}_2\text{COPO}_3^{2\ominus}
\end{array}
\;\overset{\text{aldolase}}{\rightleftharpoons}\;
\begin{array}{l}
\text{H}_2\text{COPO}_3^{2\ominus} \\
\text{C=O} \\
\text{HOCH} \\
\text{HCOH} \\
\text{HCOH} \\
\text{H}_2\text{COPO}_3^{2\ominus}
\end{array}
$$

The reaction produces only fructose 1,6-diphosphate and appears to involve the intermediate formation of a Schiff base between the enzyme and dihydroxy-acetone phosphate (see Part V,C above). Schiff base formation promotes aldol condensations in model systems.

VIII. Reactions of the alcohol groups

A. Ester formation

General reaction:

Alcohol $\longrightarrow$ ester

Specific example:

β-D-Glucopyranose $\longrightarrow$ penta-O-acetyl-β-D-glucopyranose

Reagents: Acetic anhydride in pyridine at 0°C for the reaction shown. At higher temperatures a mixture of tautomeric pentaacetates is produced. Esters are also formed using acetylchloride, benzoylchloride, or other acyl halides. Most of the bulky esterifying agents react most readily with primary hydroxyl groups, for example, benzoyl chloride, diphenylphosphorochloridate (see Section 7.4), and p-toluenesulfonyl chloride ($CH_3C_6H_4SO_2Cl$). Amino groups are also acylated under similar conditions.

Uses: Intermediates in the synthesis of glycosides and carbohydrate derivatives.

Biochemical example:

Acetyl coenzyme A + choline $\longrightarrow$ acetylcholine

$$CH_3C\!\!\overset{O}{\underset{SCoA}{\big<}} + HOCH_2CH_2\overset{\oplus}{N}(CH_3)_3 \longrightarrow CH_3\overset{O}{\overset{\|}{C}}{-}OCH_2CH_2\overset{\oplus}{N}(CH_3)_3$$

The carbonyl carbon of the acyl group of acetyl coenzyme A is susceptible to nucleophilic attack by $HO{-}R$, $H_2N{-}R$, $HS{-}R$, $O_3PO{-}R$, and so forth. Acetyl CoA serves in most biological acetylation reactions.

B. Ether formation

General reaction:

Alcohol $\longrightarrow$ ether

$$+ \quad CH_3I \quad \longrightarrow$$

Specific example:

Methyl-β-D-ribofuranoside $\longrightarrow$ methyl 2,3,5-tri-O-methyl-β-D-ribofuranoside

Reagents: Methyl iodide or dimethylsulfate and an agent to promote ionization of the alcoholic group and to accept the acid released. Methylations occur smoothly with methyl iodide in dimethylsulfoxide using dimethylsulfoxide anion as an acid acceptor. The methyl ethers are stable under acidic and alkaline conditions. Other alkyl halides can be used and benzyl chloride with powdered potassium hydroxide in an inert solvent is used to make benzyl ethers which are stable to acid and alkali but which can be removed by hydrogenolysis on palladium catalysts.

Triphenylmethyl ethers (trityl) are formed using triphenyl chloromethane in pyridine. The reaction is quite specific for primary hydroxyl groups and the trityl group is easily removed with dilute acid or by hydrogenolysis.

triphenylchloro   'trityl ether'   triphenyl
methane   carbinol

Uses: Methyl and other stable ethers are used in proving structures of glycosides. Benzyl and trityl ethers are used as blocking groups in synthetic studies.

Analogous biochemical process:

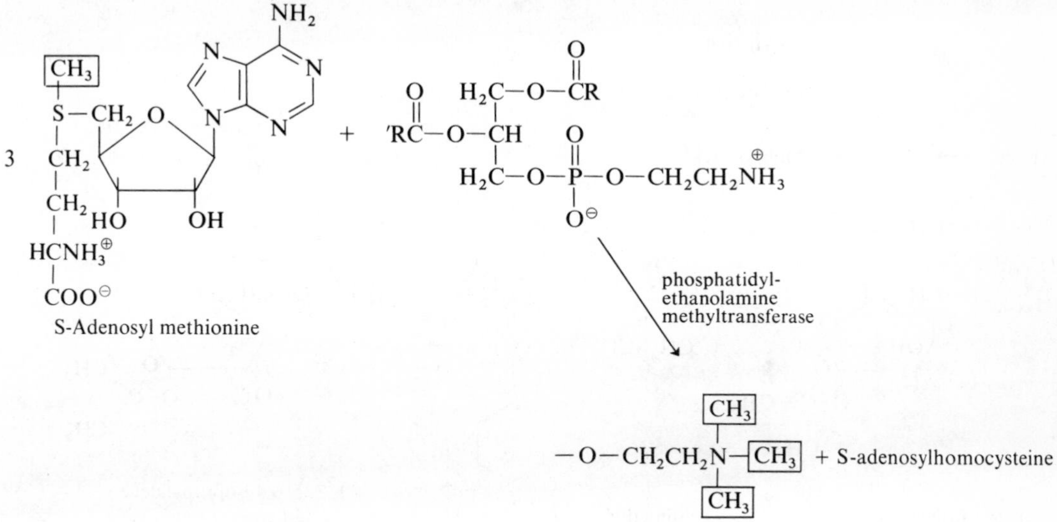

S-Adenosyl methionine

phosphatidyl-
ethanolamine
methyltransferase

$$-O-CH_2CH_2N-\boxed{CH_3} + \text{S-adenosylhomocysteine}$$

S-Adenosyl methionine is utilized for the methylation of many biologically important compounds such as tRNA in which some of the bases and some of the riboses are methylated.

C. Acetals and ketals

General reaction:

Pair of hydroxyl groups + aldehyde or ketone $\longrightarrow$ cyclic acetal or ketal

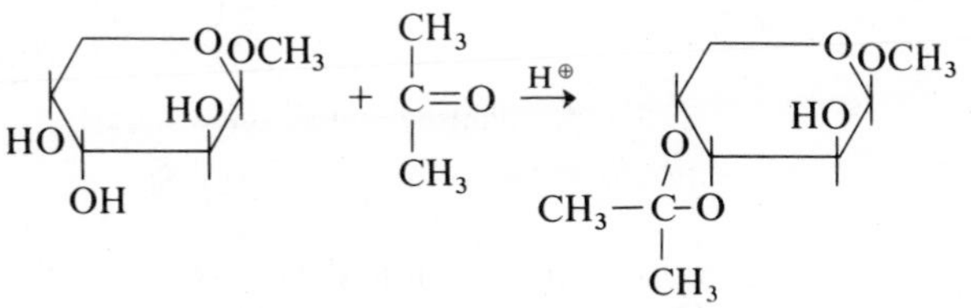

Specific example:

Methyl β-D-arabinopyranoside + acetone $\longrightarrow$ methyl 3,4-O-isopropylidene-β-D-arabinopyranoside

Reagents: Acetone and other ketones plus sulfuric acid, or Lewis acids such as copper sulfate, and zinc chloride give cyclic ketals usually with five-membered dioxolane ring systems. Aldehydes (acetaldehyde, formaldehyde, benzaldehyde) give cyclic acetals; six-membered rings are common. The reaction is reversible and the groups can be removed by acid hydrolysis.

Uses: In synthetic sequences.

D. Oxidation

General reaction:

Alcohol $\longrightarrow$ aldehyde, ketone, or acid

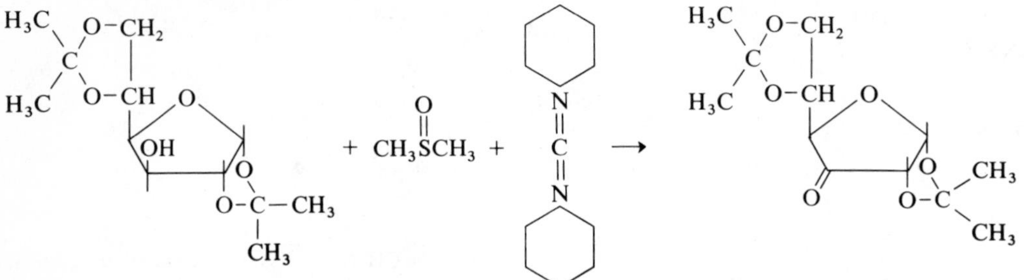

| 1, 2; 5, 6-Diisopropylidene α-D-glucofuranose | dimethyl sulfoxide | dicyclohexyl carbodiimide | 1, 2; 5, 6 diisopropylidene -3-keto-α-D-*ribo*- hexofuranose |

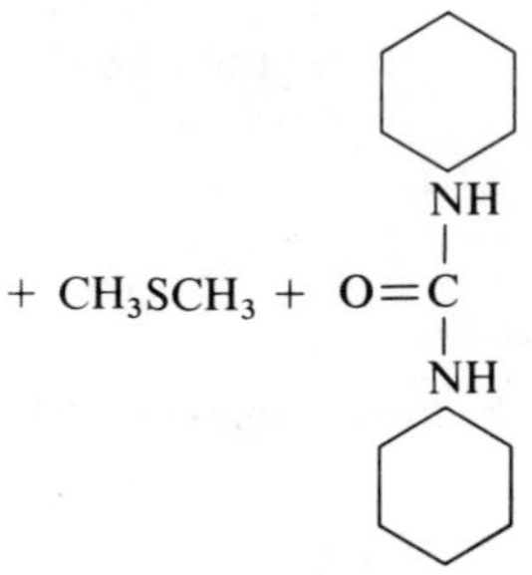

dimethyl sulfide   dicyclohexyl urea

Reagents: Those shown can oxidize isolated secondary hydroxyl groups to ketones and isolated primary hydroxyl groups to aldehydes. Dimethyl sulfoxide plus acetic anhydride or phosphorous oxychloride are capable of similar oxidations.

Uses: In the synthesis of rare sugars. For example, the reduction of the keto compound shown above with sodium borohydride gives an allose derivative almost quantitatively. The attack by hydride ion takes place from the least hindered side of the molecule.

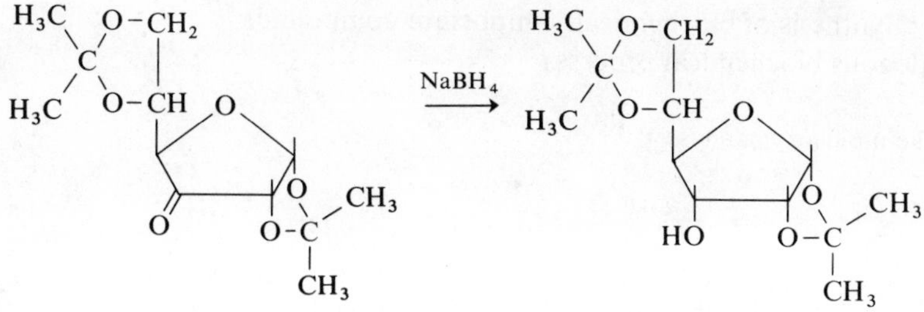

D-allose derivative

Analogous biochemical process (see Part I: Reduction): Many sugar alcohols and cyclitols are oxidized by enzymes utilizing the nicotinamide dinucleotides DPN$^+$ and TPN$^+$. They are usually fairly specific and do not oxidize all alditols.

E. Reduction

General reaction:

Primary alcohol $\longrightarrow$ methyl group

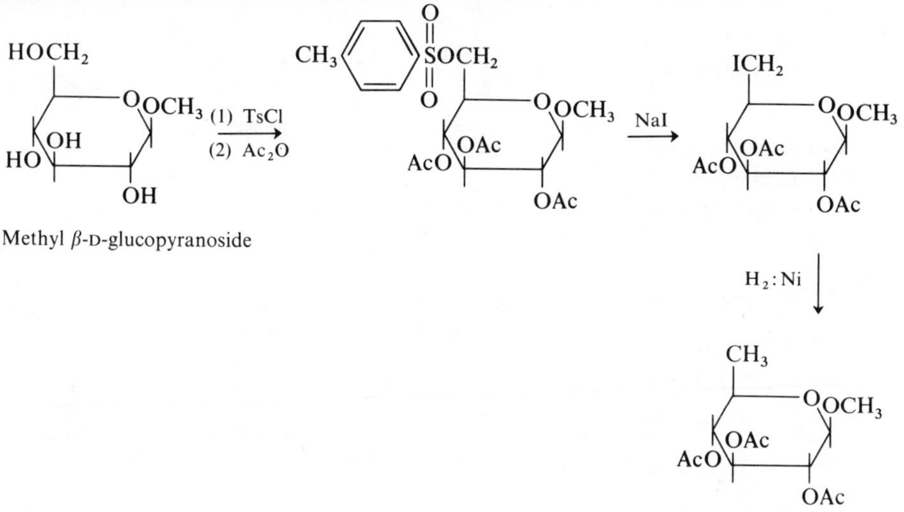

Methyl β-D-glucopyranoside

(1) TsCl
(2) Ac$_2$O

NaI

H$_2$:Ni

Methyl 2,3,4-tri-O-acetyl-6-deoxy-β-D-glucopyranoside

Reagents: The starting tosyl-acetyl glycoside is prepared from the glycoside by treatment with 1 molar equivalent of p-toluenesulfonyl chloride in pyridine in the cold followed by acetylation. The tosyl function is readily replaced by treatment with sodium iodide in hot acetone and the iodo compound is reduced catalytically to the 6-deoxy derivative.

Use: Synthesis of biochemically important compounds.

Analogous biochemical process:

GDP mannose $\longrightarrow$ GDP-4-keto-6-deoxymannose

The enzyme action is highly specific for the substrate shown and the reaction appears to be an intramolecular oxidation reduction.

IX. Periodate Oxidation

General reaction:

Adjacent oxidizable functions:

Alcohol or amine $\longrightarrow$ aldehyde or ketone

Aldehyde or ketone $\longrightarrow$ carboxylic acid

Carboxylic acid $\longrightarrow$ carbon dioxide

Specific example:

$$H_2C=O$$
$$+$$
$$5HCOOH$$
$$+ 5IO_3^{\ominus}$$

with H_2COH, HO, OH, OH, H,OH $+ 5IO_4^{\ominus} \longrightarrow$

D-Glucose—Note formate esters are formed during the oxidation and are slowly hydrolyzed to yield the acid and alcohol.

$$C_6H_5CH \quad O-CH_2 \quad \quad H,OH + 2IO_4^{\ominus} \longrightarrow$$

$$HC=O$$
$$\quad OH \quad HC\ C_6H_5 +$$
$$H_2C \quad O$$

$$2HCOOH$$
$$+$$
$$2IO_3^{\ominus}$$

4,6-*O*-benzylidene-D-glucopyranose   2,4-*O*-benzylidene-D-erythrose

This reaction gives D-erythrose, which although naturally occurring is rare.

Reagents: Sodium metaperiodate ($NaIO_4$) in aqueous solution is most frequently used but periodic acid (H_5IO_6) has found some applications.

Uses: The elucidation of carbohydrate structures (see Section 5.17). A complicating factor in the use of periodate is "overoxidation" in which groups which are between strong electron-withdrawing functions can be hydroxylated so as to be oxidized subsequently in a normal fashion.

$$HC=O \quad HC=O \quad \xrightarrow{IO_4^-} \quad HC=O \quad \xrightarrow{\underset{H_2O}{H_2O}} \quad HC=O$$
$$H-C-R \rightleftharpoons C-R \quad HO-C-R \quad HO-C-R$$
$$HC=O \quad HCOH \quad HO-C-OH \quad HC=O$$
$$\quad\quad\quad\quad\quad\quad\quad\quad\quad\quad H$$

Type of structure   Subject to
subject to "overoxidation"   normal oxidation

Analogous biochemical reaction: See Part IIB: Oxidation. The reaction shown there is similar to that occurring in periodate over-oxidation.

5.16 AMINO SUGARS

Amino sugars are widely distributed in nature and among the most common are D-glucosamine (2-amino-2-deoxy-D-glucose), D-galactosamine, and the sialic acids (**33**). The last are derivatives of 2-amino-2-deoxy-D-mannose (mannosamines). The compounds are frequently *N*-acetylated.

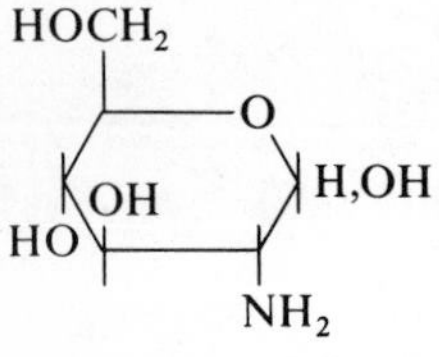

D-glucosamine 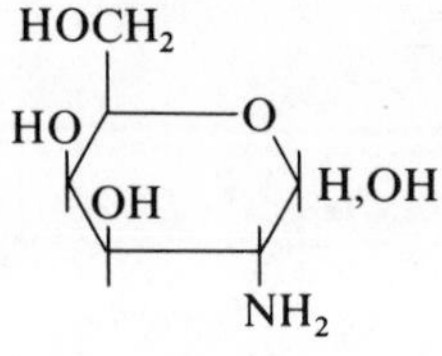   D-galactosamine

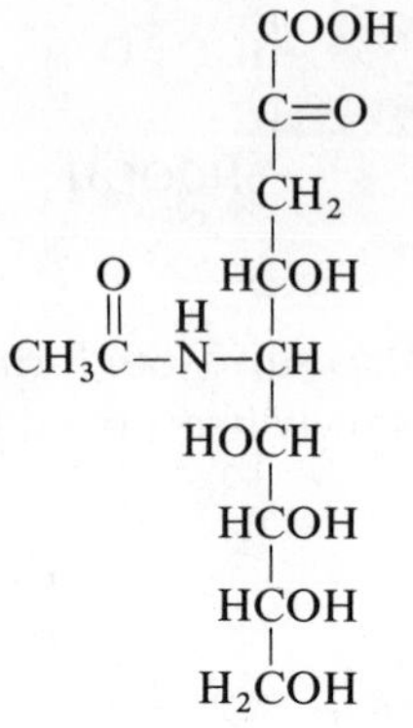

N-acetylneuraminic acid
(a sialic acid)
33

Many antibiotics have been found to contain sugars in which various hydroxyls are replaced by amino or methylated amino functions. Bacterial cell walls are characterized by their content of muramic acid (**34**) in their mucoproteins.

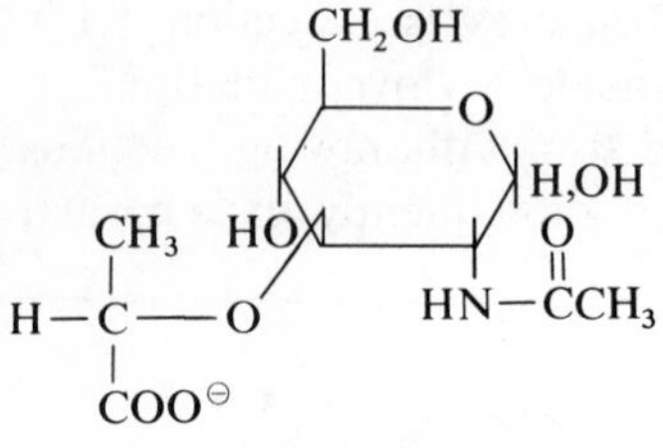

muramic acid
34

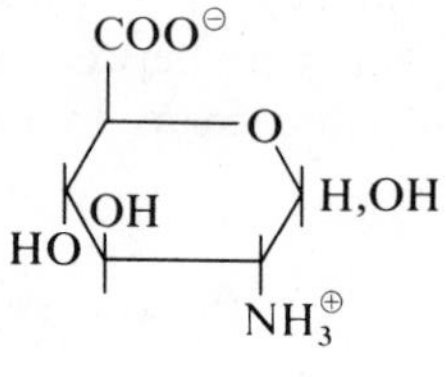

35

The occurrence of 2-amino-2-deoxy-D-glucuronic acid (**35**) and the *galacto* and *manno* isomers in bacterial polysaccharides has been demonstrated. The list of amino sugars in natural material is growing rapidly and their importance in structural and functional polymers is becoming increasingly apparent (Table 5.9).

5.17   DEOXY SUGARS

The most widely distributed deoxy sugar is 2-deoxy-D-ribose (**36**) which, if correctly named, would be known as 2-deoxy-D-*erythro*-pentose. It is found

TABLE 5.9   SOME BIPOLYMERS CONTAINING AMINO SUGARS

Polymer	Monomeric Units	Linkage	Source
Chitin	N-acetyl-D-glucosamine	β-(1–4)	Invertebrate exoskeleton
Hyaluronic acid	D-glucuronic acid and N-acetyl-D-glucosamine	β-(1–3) β-(1–4)	Synovial fluid substance
Chondroitin sulfate A or Chondroitin 4-sulfate	D-glucuronic acid and N-acetyl-D-glacatasamine-4-sulfate	β-(1–3) β-(1–3)	Cartilage
Chondroitin sulfate B or Dermatansulfate	L-iduronic acid and N-acetyl-D-galactosamine-4-sulfate	β-(1–3) β-(1–4)	Cartilage
Heparin	D-glucuronic acid, 2-O-sulfate, and N-acetyl-D-glucosamine N, 6-O-sulfate	α-(1–4)	Many tissues

36

in all forms of life as a constituent of DNA (deoxyribonucleic acid). It is present as N-glycosides, which are considerably more labile to hydrolysis than those of the fully hydroxylated forms (see Chapter 7). Other naturally occurring deoxy sugars seem to be of much more limited distribution and occur as components of polysaccharides, glycoproteins, or glycosides. They are sometimes found in nature methylated, for example, 6-deoxy-3-O-methyl-D-galactose (D-digitalose).

TABLE 5.10   A NUMBER OF NATURALLY OCCURRING DEOXY SUGARS

Deoxy Sugar Trivial Name	Systematic Name	Representative Source
L-Rhamnose	6-Deoxy-L-mannose	Bacterial cell wall, vegetable mucilages
L-Fucose	6-Deoxy-L-galactose	Human milk, blood group substances, seaweed
D-Fucose	6-Deoxy-D-galactose	Plant glycosides
D-Quinovose	6-Deoxy-D-glucose	Plant glycosides
Digitoxose	2,6-Dideoxy-D-*ribo*-hexose	Plant glycosides
Paratose	3,6-Dideoxy-D-*ribo*-hexose	Bacterial antigen
Tyvelose	3,6-Dideoxy-D-*arabino*-hexose	Bacterial antigen
Colitose	3,6-Dideoxy-L-*xylo*-hexose	Bacterial antigen
Abequose	3,6-Dideoxy-D-*xylo*-hexose	Bacterial antigen
Chalcose	3-O-Methyl-4,6-dideoxy-D-glucose	Chalcomycin, an antibiotic

In many cases they are responsible for the antigenic specificity of the polymers which contain them. A few of the naturally occurring deoxy sugars are listed in Table 5.10.

5.18 OLIGOSACCHARIDES AND POLYSACCHARIDES

There is no clear line of demarcation between oligo- and polysaccharides ("oligos," a few; "polys," many). They are condensation products of monosaccharides, linked by glycosidic bonds. The two groups of compounds can be arranged in homologous series, whose lowest members are disaccharides (that is, dimers of monosaccharides).

Although most oligo- and polysaccharides have a terminal monomer present as a "reducing sugar," the contribution of this portion to the properties of the molecule decreases as the size of the polymer (degree of polymerization) increases so that polysaccharides do not behave as reducing sugars although most di-, tri-, and small oligosaccharides do. A notable exception is sucrose, a dimer of fructose and glucose with both anomeric hydroxyls involved in the glycosidic linkage and which is, therefore nonreducing. This finding with respect to the gradual diminuation of reducing sugar with increasing polymerization can be generalized to the extent that in a homologous series the properties of the polymers approach those of the highly polymerized materials as chain length

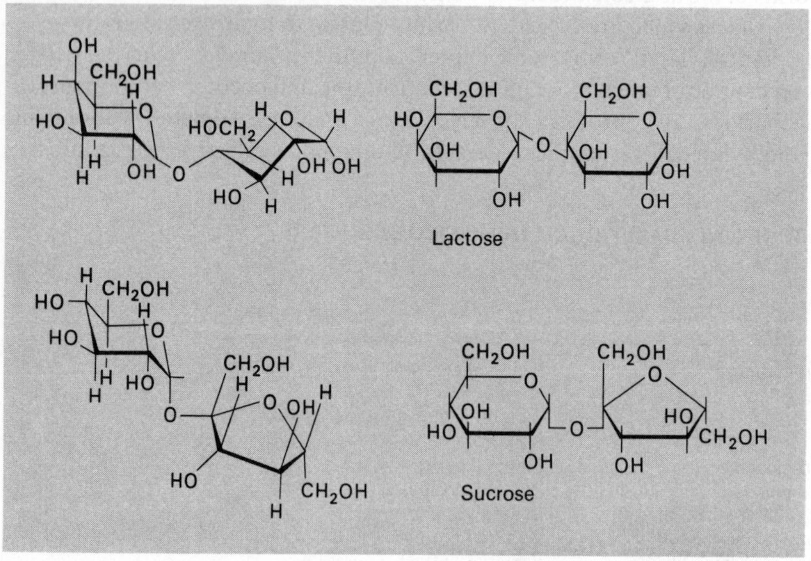

Figure 5.15 The structure of lactose and sucrose in stereorepresentation and in Haworth projection.

increases. This relationship holds for molecular rotation, melting point, and chromatographic behavior.

The structures of two naturally occurring disaccharides are given in Figure 5.15 and the proof of structure of maltose is given below to illustrate the methods which have been applied to this problem. The structure is proved if the findings listed in steps 1 through 4 are evaluated. However, the additional findings (5 and 6) confirm the proof and serve as examples of approaches which can be used on more complex polymers.

Proof of structure of maltose:

1. Maltose is a reducing sugar.

2. Hydrolysis gives only D-glucose.

3. Reduction of the saccharide to a substituted alditol, followed by complete methylation and hydrolysis, yields equimolar proportions of 1,2,3,5,6-penta-*O*-methyl-D-glucitol   and   2,3,4,6-tetra-*O*-methyl-D-glucopyranose which can be characterized by comparison with authentic materials. This clearly demonstrates that maltose is a disaccharide and that the glucopyranosyl moiety is attached to the 4-hydroxyl of the reducing end. It also shows that the reducing end has a pyran ring.

H_2COCH_3
|
$HCOCH_3$
|
CH_3OCH
|
$HCOH$
|
$HCOCH_3$
|
H_2COCH_3

1,2,3,5,6-Penta-*O*-methyl-D-glucitol (from the reducing end)

CH_2OCH_3

CH_3O OCH_3 H,OH

OCH_3

2,3,4,6-Tetra-*O*-methyl-D-glucopyranose (from the nonreducing end)

4. The only aspect of the structure to be evaluated is the anomeric configuration of the glucopyranosyl moiety. The fact that the rotation of maltose decreases during hydrolysis has been used as an indication of the α configuration of the glucosidic bond. The enzyme maltase which cleaves maltose very rapidly has been shown to cleave only α-glucosides and this constitutes the best evidence that maltose has an α-glycosidic linkage. Nmr spectroscopy can be used in certain cases to establish anomeric configuration. In general axial hydrogens on the pyranose ring absorb at higher magnetic field than equatorial hydrogens do. For example in cellobiose [4-(βD-glucopyranosyl)-D-glucopyranose] the hydrogen atom attached to the anomeric carbon of the glucopyranosyl residue absorbs at $\tau = 5.38$ ppm and is axial. The anomeric proton of the glucopyranosyl residue of maltose absorbs at $\tau = 4.58$ ppm and therefore is equatorial. The coupling (J) between the anomeric hydrogen atom and that at carbon-2 can also give

information since if they are *trans* antiparallel, the value of J is large (~ 8 Hz); if they have a *gauche* arrangement, J is small (<4 Hz) (Figure 5.6). Thus, if a glucosyl residue is in the normal chair form, the β configuration (which has H-1 and H-2 in the *trans* antiparallel arrangement) has a large splitting of the signal from the anomeric proton and the α configuration shows a small splitting. A splitting of less than 4 Hz is observed.

Maltose = 4-*O*-(α-D-glucopyranosyl)-β-D-glucopyranose

5. If the methyl glycoside of maltose is completely methylated and then hydrolyzed, two products are obtained in equimolar amounts. They are 2,3,4,6-tetra-*O*-methyl-D-glucose (37) and 2,3,6-tri-*O*-methyl-D-glucose (38)

(by comparison with knowns). This establishes the size of the ring in the glucopyranosyl portion and limits the possibilities for the ring size and intermonomer linkage in the reducing moiety. The linkage involves either the 4 or 5 position and the ring is either furan or pyran.

6. Periodate oxidation proceeds smoothly with overoxidation occurring. The final consumption of periodate is 11.0 molar equivalents with the formation of 9.0 formic acids, 2.0 formaldehydes, and 1.0 carbon dioxide. Overoxidation is consistent only with a pyran ring in the *reducing* moiety, giving the aldehyde (39) as an intermediate. Initial oxidation of a furanose substituted at the 5 position would lead to the aldehyde (40) which is not susceptible to further oxidation.

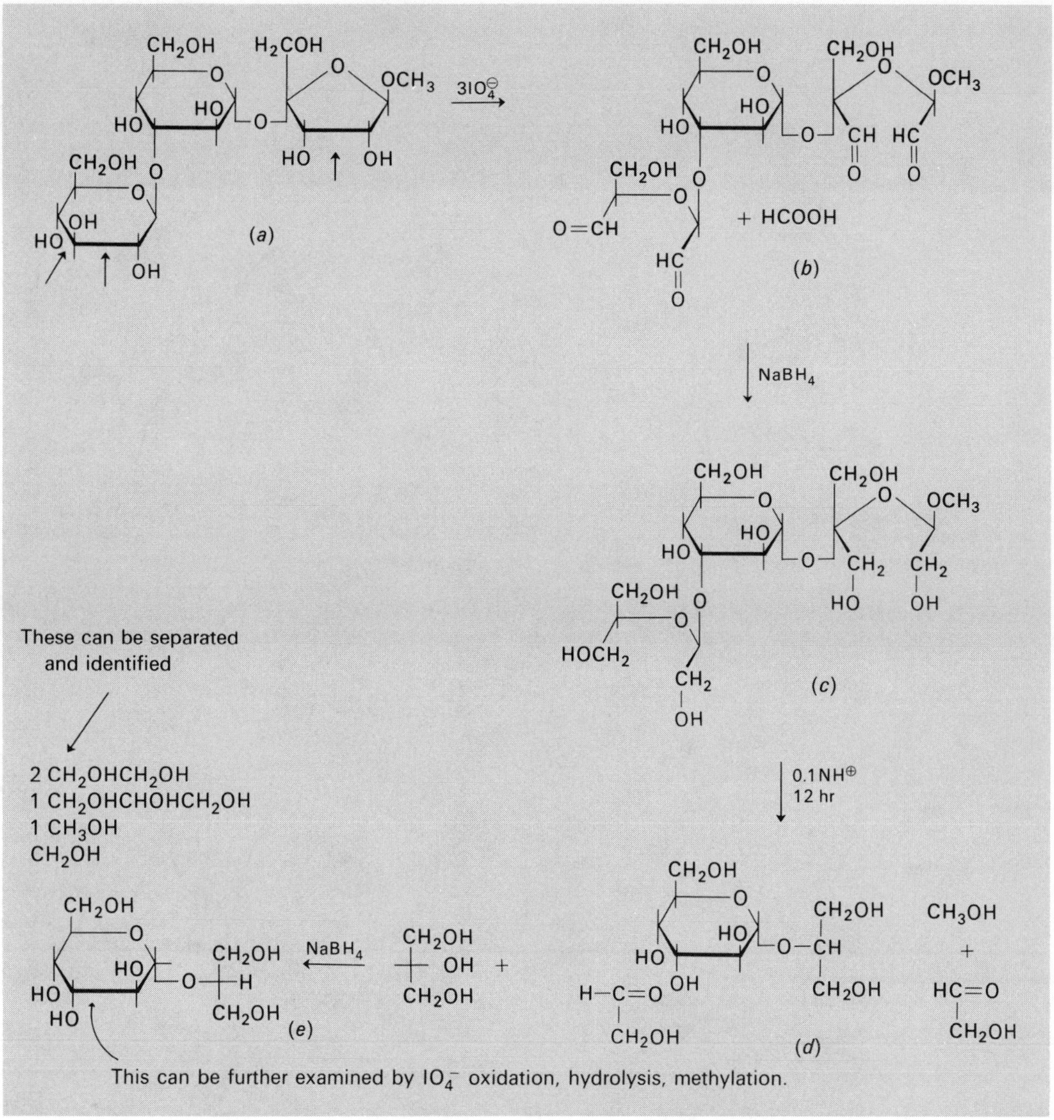

Figure 5.16   The degradation of an oligosaccharide by periodate oxidation—sodium borohydride reduction—hydrolysis—sodium borohydride reduction. (*a*) The nonreducing starting material showing position of periodate oxidation. The number of moles of periodate consumed can be determined. (*b*) The oxidized product. Formic acid production is measured and indicates the presence of hydroxyls on three consecutive carbons. (*c*) The product of borohydride reduction in which the acetal bonds in the acyclic portions are much more acid-labile than the normal glycosidic bond. (*d*) The product of acid hydrolysis. (*e*) The second reduction produces 2 ethylene glycol molecules, 1 glycerol and 1 methanol which can be easily identified and quantified by chromatographic procedures. Much of the parent structure can be deduced from the amount of periodate consumed and the final products.

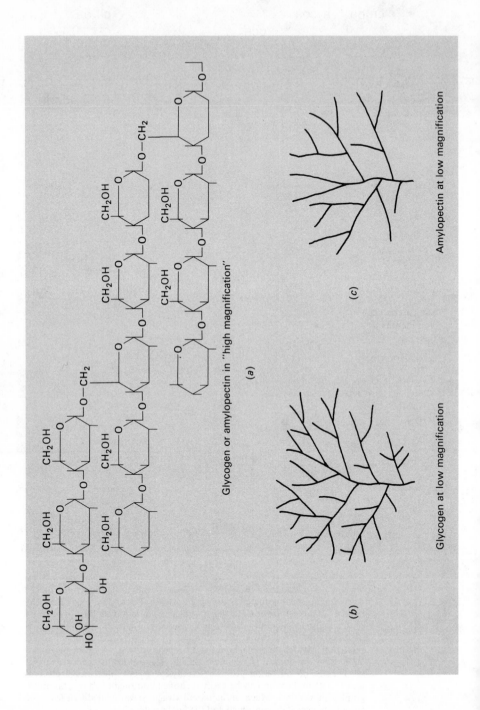

Glycogen or amylopectin in "high magnification"

(a)

Glycogen at low magnification

(b)

Amylopectin at low magnification

(c)

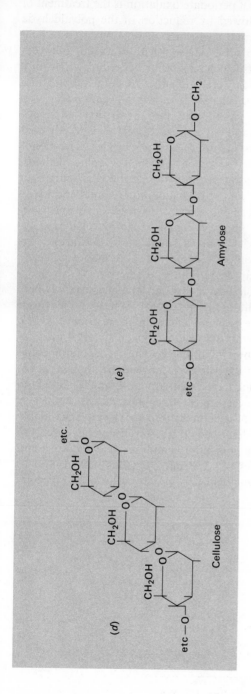

Figure 5.17 Amylopectin, amylose, glycogen, and cellulose. (*a*) A portion of a glycogen or amylopectin showing glycosidic bonds involved (α 1,4 and α 1,6). (*b*) Highly branched structure of glycogen. (*c*) Less highly branched structure of amylopectin. (*d*) A portion of cellulose structure showing β 1,4 linkages. (*e*) The α 1,4 linkages of amylose.

An important extension of the use of periodate oxidation is the treatment of a polysaccharide with periodate followed by reduction of the polyaldehyde with sodium borohydride. The polymer so formed is very sensitive to acid hydrolysis in those portions where an oxidation had occurred and acid hydrolysis yields a mixture of products which may be further reduced to readily identifiable alditols and residues which were originally resistant to oxidation. The degradation of a hypothetical trisaccharide is shown in Figure 5.16.

Permethylation followed by hydrolysis to yield partially methylated mono-saccharides, as described for maltose and in Section 5.15, VII, B, is widely used in establishing the composition and structure of polysaccharides. The mono-saccharides can be separated and quantitated by chromatographic methods. The successful application of this method requires the synthesis of samples of authentic materials for comparison with the isolated compounds and many partially methylated monosaccharides have been synthesized for this purpose.

Polysaccharides can be classified under two main headings; "homopoly-saccharides," which in hydrolysis yield only a single monosaccharide, and "heteropolysaccharides," which are composed of more than one monosaccharide species. The generic name *glycan* is applied to these polymers.

Homopolysaccharides of great biological interest are the glucans (polymers of glucose), starch, glycogen, and cellulose (Figure 5.17). Starch and glycogen are the main chemical energy reserve of plants and animals, respectively. The main structural features are glycopyranosyl residues linked by α-1,4 linkages (as in maltose) and branches formed by α-1,6 linkages. The degree of branching in glycogen (a branch point every 8 to 10 residues) is much higher than in starch, the latter being known in forms which have essentially no branches (amylose) and in forms which have a branch every 20 to 30 residues (amylopectins). A large number of variations of starch structures from amylose to the highly branched amylopectins have been described. Both starch and glycogen are very large, with molecular weights ranging into the millions. Neither are found in nature as single molecular species but are heterogeneous with respect to molecular weight and are treated as single substances only because the molecules

TABLE 5.11   SOME HOMOPOLYSACCHARIDES[a]

Name	Source	Monomer	Linkage	Form
Laminaran	Seaweeds	D-Glucose	α-1,3	Linear
Mannan	Ivory nut	D-Mannose	1,4	Linear
Inulin	Artichoke	D-Fructose	β-1,2	Linear
Xylan	Most plants (corncobs)	D-Xylose	β-1,4	Linear
Floridean starch	Red algae	D-Glucose	1,4 and 1,3	Branched
Yeast mannan	Yeast	D-Mannose	1,2; 1,3; and 1,6	Branched
Galactan	Beef lung	D-Galactose	?	Branched

[a] R. L. Whistler and C. L. Smart, *Polysaccharide Chemistry*, Academic Press, Inc., New York, 1953.

differ from one another only in degree of polymerization, which has little effect on most of the chemical and physical characteristics of the compound.

Cellulose, which is the major constituent of the supporting structures of plants, is a linear polymer of glucopyranosyl residues linked β-1,4 (Figure 5.17). The relatively small difference in structure between amylose and cellulose (α-1,4 linkage to β-1,4 linkage) produces a marked difference in physical and chemical properties but an even more striking difference in biological properties. Whereas starch is almost universally acceptable as a foodstuff, cellulose is digested only by some microorganisms.

Many other homopolysaccharides are known. A few are listed in Table 5.11.

Heteropolysaccharides have been isolated from many sources and differ widely in constitution and function. A number of polymers containing amino sugars are listed in Table 5.9 and further examples of heteropolymers are given in Table 5.12.

TABLE 5.12   SOME HETEROPOLYSACCHARIDES[a,b]

Name	Composition	Linkage	Form
Neutral gum tragacanth	L-Arabinose	?	Branched
	D-Galactose	1,4	
Pneumococcus Type III polysaccharide	D-Glucose	1,3'	Linear
	D-Glucuronic acid	1,4'	
Blood group substances	D-Galactose	?	Branched
	D-Glucosamine		
	L-Fucose		
Polysaccharide from *M. tuberculosis*	D-Arabinose	?	Branched
	D-Mannose		
	L-Rhamnose		
	D-Glucosamine		

[a] See various articles in *Advances in Carbohydrate Chemistry*, Reference 1.
[b] M. Stacey and S. A. Barker, *Polysaccharides of Micro-organisms*, Oxford University Press, 1960.

REFERENCES

Stanek, J., M. Cerny, J. Kocourek, and J. Pacak, *The Monosaccharides*, Academic Press, Inc., New York, 1963. An extensive treatment of the chemistry of the monosaccharides with many references. A useful coverage of the literature up to 1962.

Methods in Carbohydrate Chemistry (R. L. Whistler and M. L. Wolfrom, eds.), Academic Press, Inc., 1962. There are five volumes: Vol. I. "The Analysis and Preparation of Sugars", Vol. II. "Reactions of Carbohydrates," Vol. III. "Cellulose," Vol. IV. "Starch," and Vol. V. "General Polysaccharides." The series is invaluable to the person interested in synthetic carbohydrate chemistry or structure elucidation. Each reaction has been dealt with by an expert in the area and the careful editing and referencing make this a prime source.

Advances in Carbohydrate Chemistry (M. L. Wolfrom, ed.), Academic Press, Inc., New York. This annual publication first appeared in 1945. It usually contains 8 to 10 reviews of areas of interest to carbohydrate chemists, with topics ranging from commercially important aspects of starch to discussions of conformational analysis and the synthesis of biologically important compounds such as mononucleotides. This series is the first place to look for a good review.

PHOSPHATE
SIX | ESTERS

6.1 INTRODUCTION

Phosphorus is present in all living tissues. In soft tissues it accounts for from 0.1 to 0.26 percent of the net weight and in bone and teeth, for 5.0 and 13.0 percent, respectively. In all cases it is present in the form of phosphates ($H_2PO_4^{\ominus}$ and $HPO_4^{2\ominus}$), mono- or dialkyl phosphates or pyrophosphates, or their esters. In these forms the phosphates account for from 3.0 to 6.0 percent of the organic materials in most tissues.

More important than the fact that phosphate esters are indigenous is the fact that they are involved in almost every aspect of cellular function. They are responsible for energy transfer from nutrients to cellular synthetic processes; they are utilized in information storage and the replication and translation of stored information, to differentiate between storage and metabolically active forms of cellular components, and as sources of energy for nerve transmission and muscle contraction.

The phosphate esters are found as derivatives of carbohydrates, lipids, proteins, and nucleosides, and they are discussed in the appropriate chapters in this book. It is intended in this chapter to describe some of the general properties of phosphates and their esters.

6.2   ORTHOPHOSPHATES AND PYROPHOSPHATES

Orthophosphoric acid is an acid of intermediate strength. In aqueous solution it ionizes fairly completely. At pH values above 4.0 the proportion of H_3PO_4

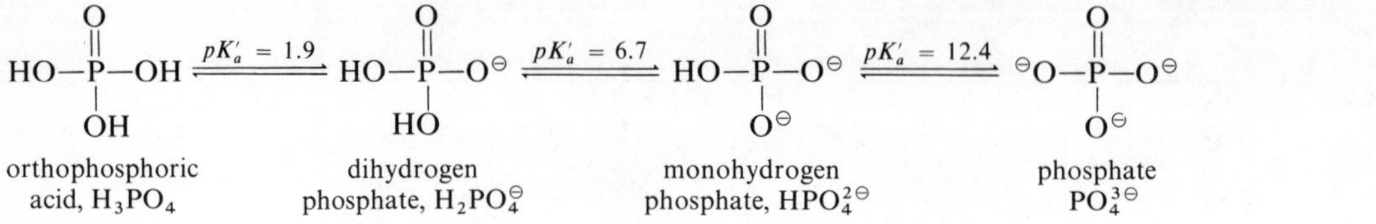

orthophosphoric acid, H_3PO_4 $\xrightarrow{pK'_a = 1.9}$ dihydrogen phosphate, $H_2PO_4^{\ominus}$ $\xrightarrow{pK'_a = 6.7}$ monohydrogen phosphate, $HPO_4^{2\ominus}$ $\xrightarrow{pK'_a = 12.4}$ phosphate $PO_4^{3\ominus}$

present in solution is negligible.†

The second and third ionization constants of orthophosphoric acid are 2×10^{-7} and 3.6×10^{-13}, respectively (corresponding to pK'_a values of 6.7 and 12.4). The pH of most tissues is in the range of 7.0 to 7.4; thus the ratio of $HPO_4^{2\ominus}$ to $H_2PO_4^{\ominus}$ varies between 2 to 1 and 5 to 1 and the proportion of $PO_4^{3\ominus}$ is negligible. The abbreviation P_i is frequently used to designate the mixture of orthophosphate ions present in aqueous solution.

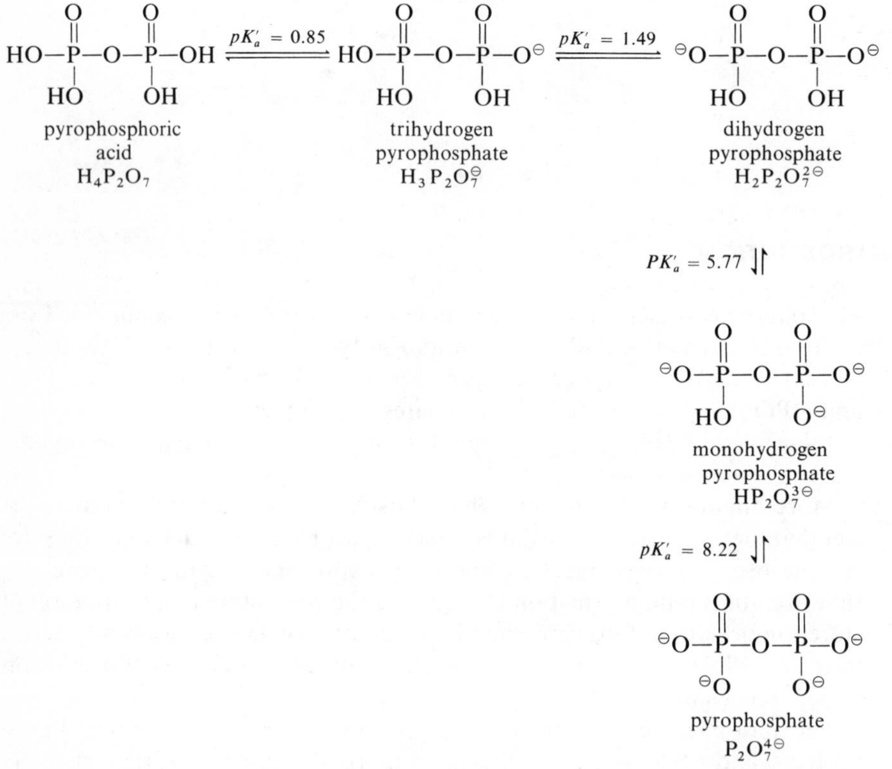

pyrophosphoric acid $H_4P_2O_7$ $\xrightarrow{pK'_a = 0.85}$ trihydrogen pyrophosphate $H_3P_2O_7^{\ominus}$ $\xrightarrow{pK'_a = 1.49}$ dihydrogen pyrophosphate $H_2P_2O_7^{2\ominus}$

$PK'_a = 5.77$ ↕

monohydrogen pyrophosphate $HP_2O_7^{3\ominus}$

$pK'_a = 8.22$ ↕

pyrophosphate $P_2O_7^{4\ominus}$

† The proportion of HA present when pH = pKa' + 2 is 1 percent and it can usually (but not always) be neglected. See Section 4.1A.

The structure and pK'_a values of pyrophosphoric acid (measured in solutions 0.1 N or less) are shown above. The first two protons are readily dissociated, while the second pair is approximately equivalent to the second proton of orthophosphoric acid. There are several important points to be considered. The loss of a proton produces a species which is capable of rehybridization of bonds and which is negatively charged. Both factors make loss of the second proton more difficult since it is more difficult to separate $H^\oplus$ from the negative environment and there are an increasing number of sites to which the proton can return. Pyrophosphate esters retain the characteristics of a strong acid. This is also found with esters of orthophosphate, where mono- and dialkyl esters behave as strong acids, as shown below.

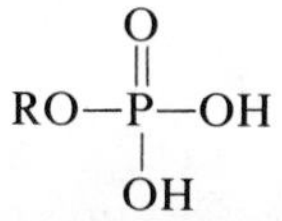

monoester	diester	symmetrical diester	unsymmetrical diester
pK'_a 1.5 and 6.7	1.5	1.0 and 2.0	1.0 and 6.0

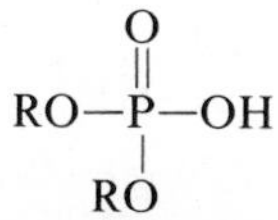

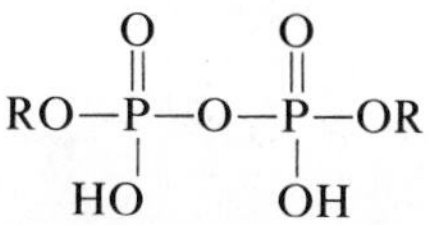

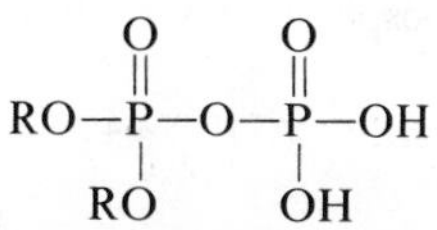

6.3   PHOSPHATE TRANSFER POTENTIALS

Pyrophosphate possesses an important chemical reactivity; it is an acid anhydride, and in common with other anhydrides it has a tendency to undergo solvolysis. More generally it can be thought of as having a potential to transfer an orthophosphate group to a suitable acceptor. Phosphate transfer reactions

TABLE 6.1   PHOSPHATE TRANSFER POTENTIALS[a]

Donor	Potential $\Delta G^{\circ\prime}$ kcal/mole
Enolpyruvate phosphate	−12.8
Creatine phosphate	−10.5
Acetyl phosphate	−10.1
ATP	−6.9
Pyrophosphate	−6.6
ADP	−6.4
Glucose-1-phosphate	−5.0
3-Phosphoglycerate	−3.1
Fructose-1-phosphate	−3.1
Glycerol-1-phosphate	−2.3

[a] F. H. Carpenter, *J. Am. Chem. Soc.*, **82**, 1,111 (1960). T. C. Bruice and S. J. Benkovic, *Bioorganic Mechanisms*, Vol. II, W. A. Benjamin, Inc., New York, 1966, pp. 176–180.

are important biochemically and it is useful to define a scale of phosphate transfer potentials in which all phosphate donors are compared in their ability to donate to water. The values being compared are standard free energies of hydrolysis at pH 7.0 and 25°C in the presence of 0.01 M $Mg^{2\oplus}$ ($\Delta G^{\circ\prime}$). A larger negative value indicates increased ability to transfer a phosphate group to water.

Theoretically a compound at the top of Table 6.1 can donate its phosphate to the acceptor portion of any compound below it in the table and generate the corresponding phosphate (Figure 6.1). The reaction can be thought of as the sum of two other reactions:

Enolpyruvate phosphate + H_2O $\longrightarrow$ "pyruvate" + phosphate + 12.8 kcal/mole
ADP + phosphate $\longrightarrow$ ATP + H_2O − 6.9 kcal/mole

Enolpyruvate phosphate + ADP $\longrightarrow$ "pyruvate" + ATP + 5.9 kcal/mole

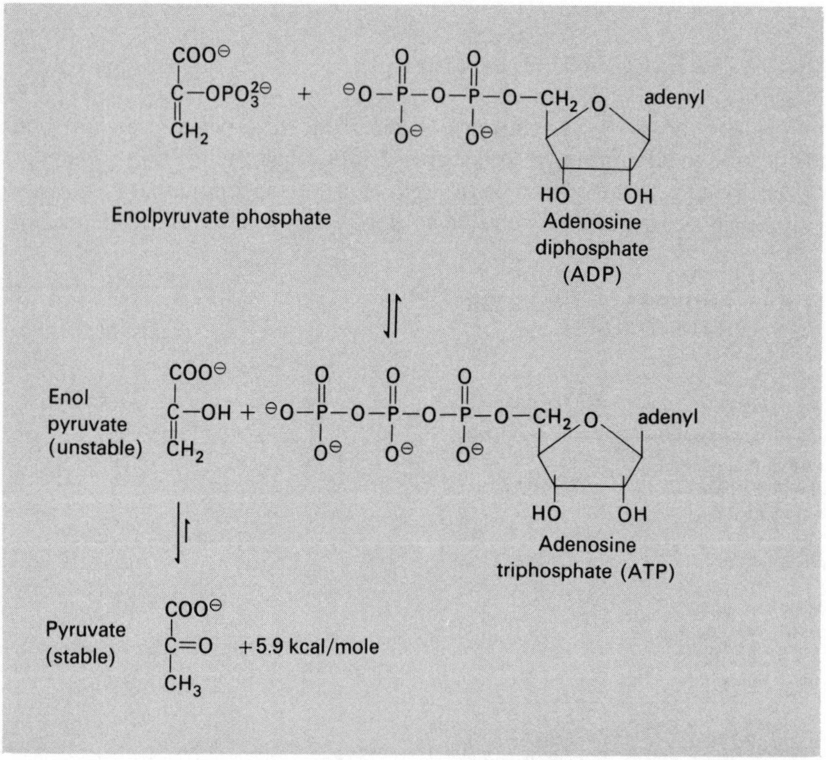

Figure 6.1   The reaction of enolpyruvate phosphate with adenosine diphosphate.

In this case a major proportion of the energy released is actually due to the ketolization of "enolpyruvate" to form pyruvate which has $\Delta G^{\circ\prime} = -8$ kcal/mole.

Biochemically, phosphate transfer reactions involve ATP (or some other "high-energy phosphate"†) as the donor and are enzyme-catalyzed. The reaction inevitably involves exchange of protons with the medium. A net uptake or release can occur depending on the pK_a' values of the product phosphate and the pH of the medium.

$$\text{ROH} + \text{ATP} \rightleftharpoons \text{ROP} + \text{ADP} \pm \phi\text{H}^{\oplus}$$

The dependence of ΔG° observed for the reaction

$$\text{ATP} + \text{H}_2\text{O} \longrightarrow \text{ADP} + P_i + \phi\text{H}^{\oplus}$$

where ϕ can be positive or negative is given by

$$\Delta G_{\text{obs}} = \Delta G^{\circ\prime} + \phi RT \ln[\text{H}^{\oplus}]$$

and the effect of pH on ΔG° is shown in Figure 6.2.

Clearly the $[\text{H}^{\oplus}]$ term is very important in this transfer reaction and is frequently important in the transfer of phosphate groups to other acceptors.

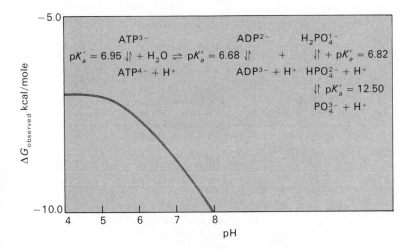

Figure 6.2   The effect of pH on ΔG_{obs} for the hydrolysis of ATP. All the ionizations shown in the figure must be considered and all the equilibria must be individually satisfied [F. H. Carpenter, *J. Am. Chem. Soc.*, **82**, 1,111 (1960). Copyright © 1960 by the American Chemical Society. Reprinted by permission of the copyright owner.]

† "High-energy phosphates" can be defined as those with $\Delta G^{\circ\prime}$ of hydrolysis > 6.0 kcal/mole. The term *does not* refer to the energy required for bond dissociation but to the net difference in energy between reactants and products.

6.4 SYNTHESIS:

A. Phosphate Esters

The organic synthesis of phosphate esters is frequently accomplished by utilizing phophoryl group donors in nonaqueous systems. The general reaction can be written

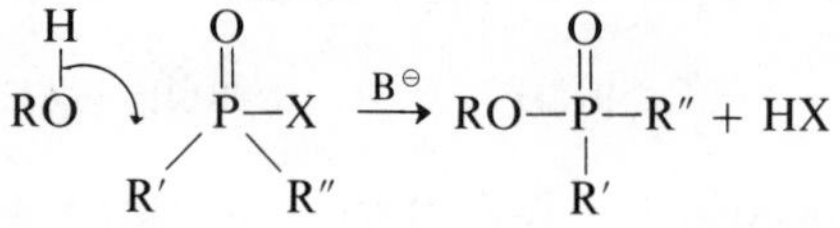

$$
\underset{R'}{\overset{H \quad O}{RO \searrow \underset{\diagdown}{\overset{\parallel}{P}}-X}} \xrightarrow{B^{\ominus}} RO-\overset{O}{\underset{R'}{\overset{\parallel}{P}}}-R'' + HX
$$

The leaving group X is usually chloride and R' and R'' are varied to accomplish specific synthetic goals (see Table 6.2). The reaction is subject to base catalysis

$$\overset{O}{\overset{\parallel}{}}$$
TABLE 6.2 PHOSPHORYLATING AGENTS $XPR'R''$

R'	R''	X	Name and comment
Cl—	Cl—	—Cl	Phosphorous oxychloride, Mild conditions; $-20°C$, multiple reaction possible[a]
C_6H_5O—	C_6H_5O—	—Cl	Diphenyl phosphoro-chloridate. Groups can be removed with H_2:Pt[b]
p-NO$_2$C$_6$H$_4$O—	Cl—	—Cl	p-Nitrophenyl phosphoro dichloridate. Group can be removed with OH⁻ and can be used to synthesize diesters[c]
p-NO$_2$C$_6$H$_4$O—	p-NO$_2$—C$_6$H$_4$O—	—OP(OC$_6$H$_4$NO$_2$—p)$_2$	Tetra-p-nitrophenyl pyrophosphate a powerful reagent. Groups removable by alkali[d]
N=CH$_2$CH$_2$O—	HO—	—OH	β-Cyanoethylphosphate in the presence of dicyclohexyl carbodii-mide C$_6$H$_{11}$N=C=NC$_6$H$_{11}$. Cyanoethyl group can be removed by dilute OH[e]

[a] E. Fischer, Ber., **47**, 3,193 (1914).
[b] P. Brigl and H. Muller, Ber., **72**, 2,121 (1939).
[c] A. F. Turner and H. G. Khorana, J. Am. Chem. Soc., **81**, 4,651 (1959).
[d] J. G. Moffat and H. G. Khorana, J. Am. Chem. Soc., **79**, 3,741 (1957).
[e] P. T. Gilham and G. M. Tener, Chem. and Ind. (London, 542 (1959); also G. M. Tener, J. Am. Chem. Soc., **83**, 159 (1961).

and is frequently carried out in the presence of an organic amine, such as pyridine, as a catalyst and acid acceptor.

Two other methods of synthesis should be mentioned. The first involves a reversal of roles between the phosphate and the alkyl residue in which a phosphate anion attacks the alkyl residue in a typical nucleophylic displacement (S_N1 or S_N2). The second is specific for the synthesis of sugar 1-phosphates and

$$RX + {}^{\ominus}O-\underset{\underset{R'}{|}}{\overset{\overset{O}{\|}}{P}}-R'' \longrightarrow RO\underset{\underset{R'}{|}}{\overset{\overset{O}{\|}}{P}}-R'' + X^{\ominus}$$

involves fusion of the corresponding acetate with pure orthophosphoric acid (mp 42°C).

α-D-glucopyranose
pentaacetate

β-D-glucopyranose
-1-phosphate

The synthesis of phosphate esters of biological interest is frequently difficult since the acceptor group is part of a multifunctional molecule and other potentially reactive groups must be protected. This blocking must be done so that the groups introduced for protection can be removed under mild conditions which do not disrupt the desired structure.

A single example of the synthesis of a phosphate ester is given in Figure 6.3. The synthesis involves many steps to produce the compound to be phosphorylated and a number of steps to remove the blocking groups from the carbohydrate and the phosphate groups.

Many more complex phosphate esters have been synthesized. The nucleotides and polynucleotides synthesized by Khorana and co-workers beautifully illustrate the complexity and utility of such work (Section 8.3A). The latter because the compounds synthesized were used to demonstrate the validity of the genetic code.

B. Pyrophosphate Esters

Pyrophosphate esters can be synthesized in some cases by obvious extension of the techniques used for the preparation of monophosphates. For example, the treatment of an alkyl halide with triethylammonium tribenzyl pyrophosphate

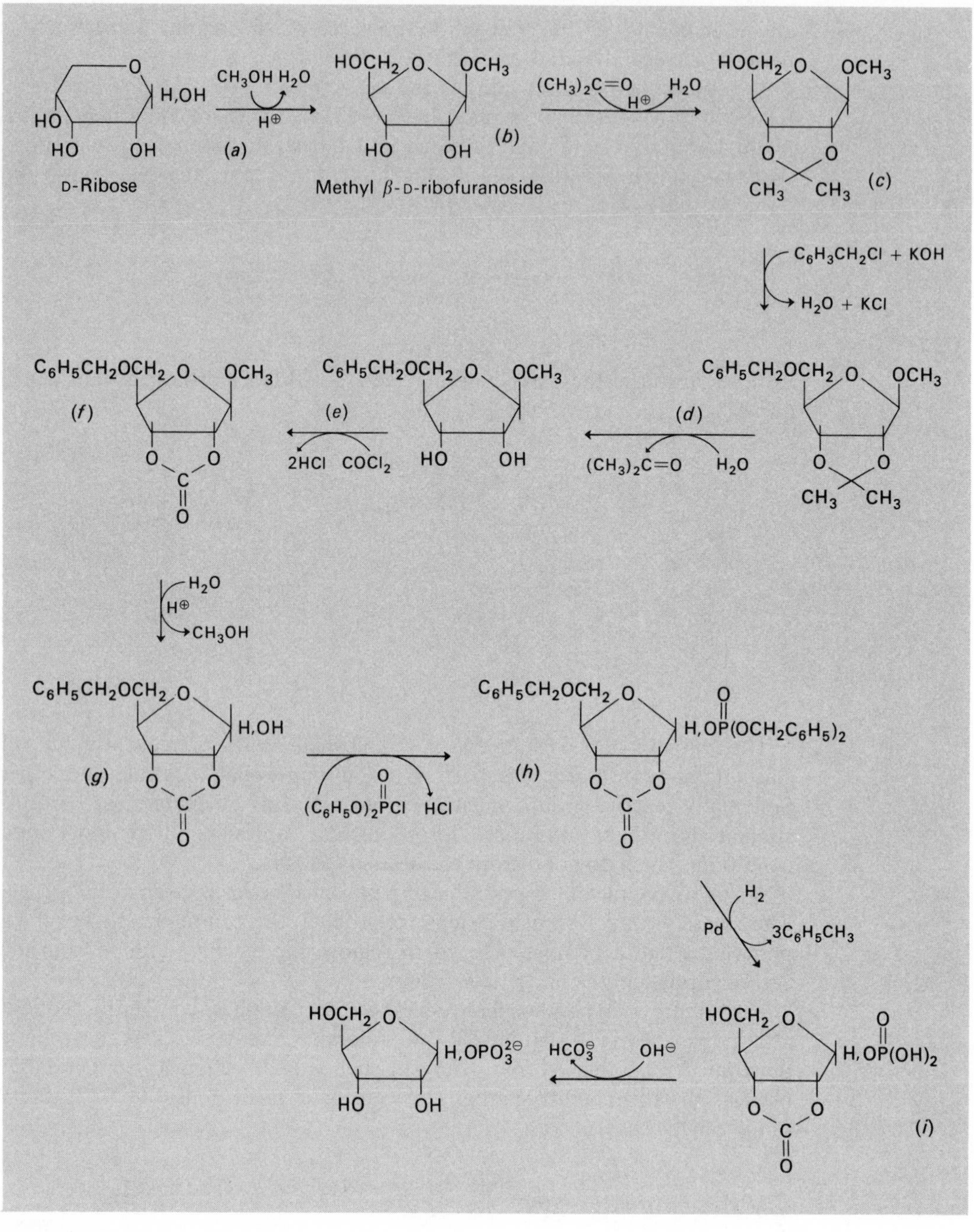

D-Ribose   (a)

Methyl β-D-ribofuranoside   (b)

(c)

$C_6H_3CH_2Cl$ + KOH

H_2O + KCl

(f)   (e)   (d)

2HCl COCl$_2$ (CH$_3$)$_2$C=O   H$_2$O

H_2O
$H^\oplus$
CH_3OH

(g)   (h)

$(C_6H_5O)_2PCl$ HCl

H_2
Pd $3C_6H_5CH_3$

$HCO_3^\ominus$ $OH^\ominus$

(i)

Figure 6.3

The synthesis of ribofuranosyl phosphate. The phosphate group is acid-labile, base-stable. The problem is to make a biologically important compound containing a furanosyl ring from a starting material, ribose, which exists in the pyranose form. The first steps in the sequence must be designed to produce a furanose form with a blocking group at position 1 which can be removed to allow phosphorylation at that position. If the hydroxyl group at position 5 is blocked, the possibility of pyranosyl ring formation will be eliminated. (a) Ribose is primarily in the pyranose form in solution. It reacts rapidly with methyl alcohol to give methyl furanosides which are acid-labile and base-stable. (b) Adjacent *cis* hydroxyl groups react readily with acetone to give a cyclic ketal which is very acid-labile (more so than the methyl glycoside) and base-stable. (c) The 5-hydroxyl can be converted to a benzyl ether which is acid- and base-stable. This blocking group at OH-5 eliminates the possibility of the compound reverting to the pyranose form. The benzyl group can be removed by hydrogenolysis (H_2 + Pd) and this can be done near the end of the synthesis. (d) The isopropylidene group can now be removed so that it can be replaced with an acid-stable, base-labile group (a cyclic carbonate is chosen) so that the group covering the 2- and 3-hydroxyls can be removed after the phosphate is in position and so that the 1-hydroxyl can be uncovered in the next step. Since the desired phosphate is acid-labile, the group at the 2,3 position must be removable with alkali. Mild acid hydrolysis is used so as not to unblock the 1-hydroxyl at this step. (e) The cyclic carbonate is formed using phosgene. (f) The cyclic carbonate is acid-stable and the methyl group can be removed to allow phosphorylation. (g) The furanose form is a mixture of anomers. It can be phosphorylated with a number of different reagents but one must be chosen which is compatible with the present structure. Dibenzyl phosphorochloridate is used so that the blocking groups on the phosphate and the 5-OH can be removed simultaneously. (h) The phosphate is a mixture of anomers. The catalytic removal of benzyl groups is accomplished in methanolic solution with Pd as a catalyst. (i) The carbonyl group is removed with alkali. If the 2,3-isopropylidene group used in the first steps had been retained to this point, its removal with acid would have resulted in loss of the phosphate group.

can yield the corresponding pyrophosphoryl derivative after removal of the benzyl groups. Much more elegant synthetic methods were required for the synthesis of nucleotide coenzymes such as coenzyme A. This synthesis was accomplished using adenosine 2′,3′-phosphate 5′-phosphoromorpholidate in which the 5′-phosphate group is susceptible to attack by a nucleophylic species and D-pantetheine 4-phosphate was used as that species (Figure 6.4).

Coenzyme A has also been synthesized utilizing anionic displacement of diphenyl phosphate from a diesterified pyrophosphate (Figure 6.5). This type of reaction appears to be general and has been used to synthesize uridine diphosphate glucose and other unsymmetrical pyrophosphate diesters.

C. Cyclic Phosphates

In the syntheses described above, cyclic phosphate diesters were used as blocking groups. They also occur naturally and in the course of degradative procedures. Synthesis can be accomplished by the several routes shown in Figure 6.6. Route A utilizes POCl$_3$ and a suitable diol, followed by treatment with water; B uses phenyl phosphorodichloridate, followed by removal of the blocking group with H$_2$:Pt, and C, the intramolecular transesterification of a triester. A phosphate monoester can be caused to cyclize to a suitably situated hydroxyl group in the molecule by treatment with dicyclohexylcarbodiimide

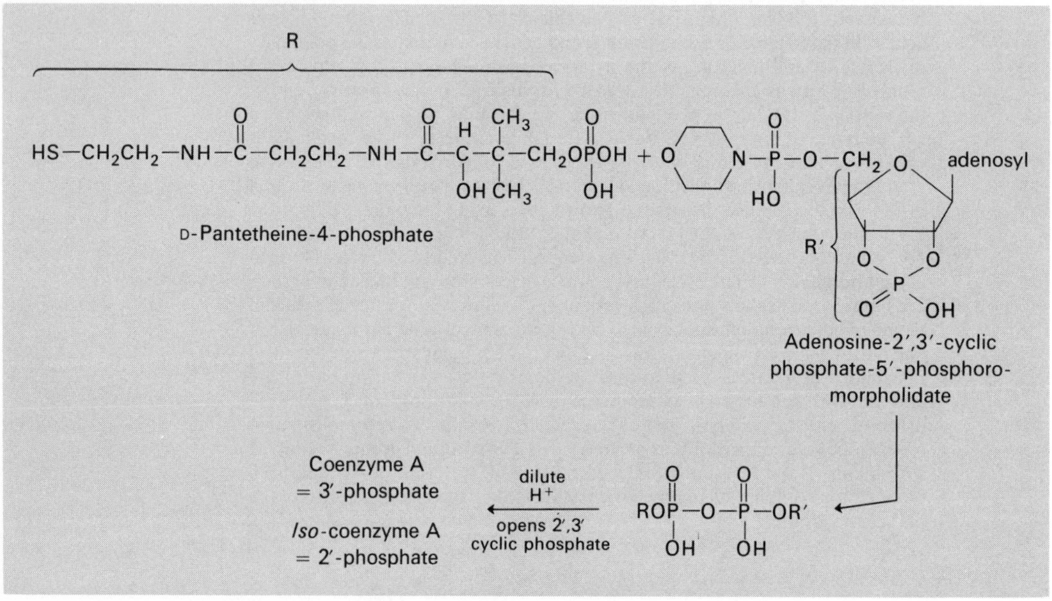

Figure 6.4   The synthesis of coenzyme A [J. G. Moffat and H. G. Khorana, *J. Am. Chem. Soc.*, **83**, 663 (1961)].

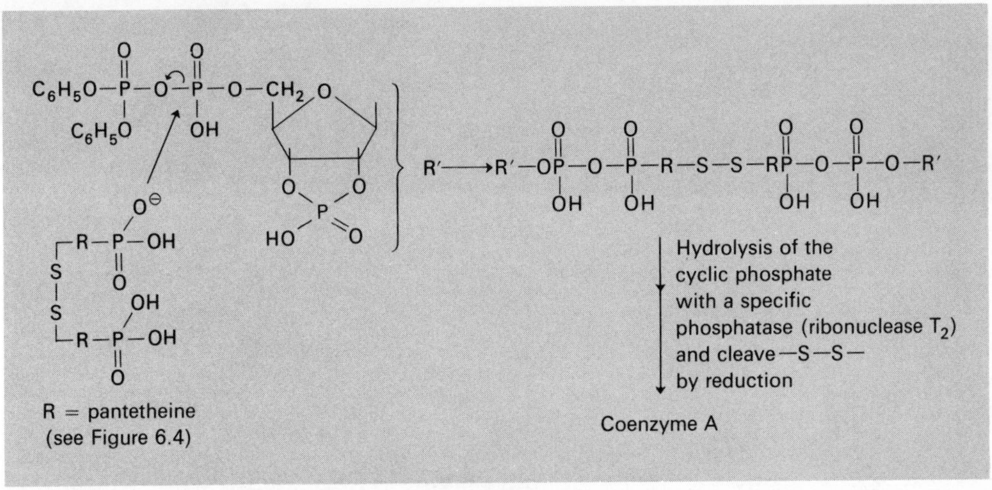
R = pantetheine
(see Figure 6.4)
Hydrolysis of the cyclic phosphate with a specific phosphatase (ribonuclease T2) and cleave —S—S— by reduction
Coenzyme A

Figure 6.5 The synthesis of coenzyme A by anionic displacement. (A. M. Michelson, *The Chemistry of Nucleosides and Nucleotides*, Academic Press, New York, 1963, p. 224.)

(DCC; $C_6H_{11}N{=}C{=}NC_6H_{11}$) (Route D). Finally, a suitably situated leaving group (such as Cl^-) can be displaced intramolecularly by the anion of a phosphate ester (Route E).

The six- and seven-membered cyclic phosphate rings are more stable than the five-membered rings. The latter can be prepared, however, from a monophosphate ester of a vicinal diol by reaction with DCC in the presence of ammonia or trialkylamines. Cyclic esters of ribofuranose are important in biochemical systems. The 3′,5′-cyclic phosphate (seven-membered ring) can be

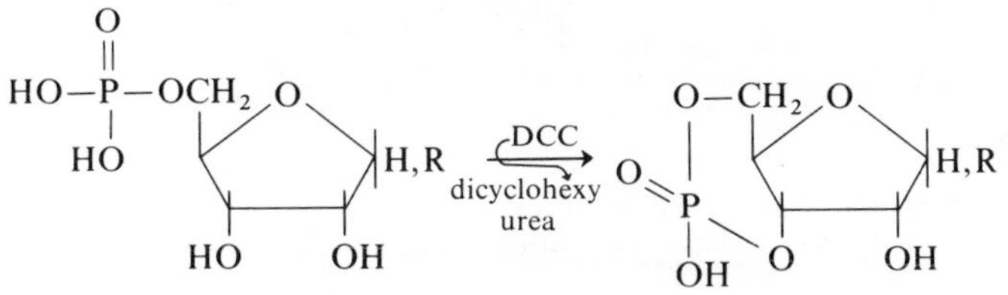
DCC
dicyclohexy
urea

formed from the 5′-monophosphate ester, and the 2′,3′-cyclic phosphate (five-membered ring) from either the 2′ or 3′ ester by the DCC method. The

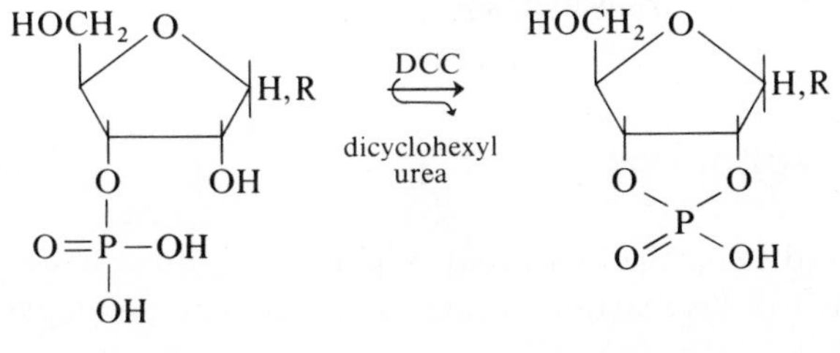
DCC
dicyclohexyl
urea

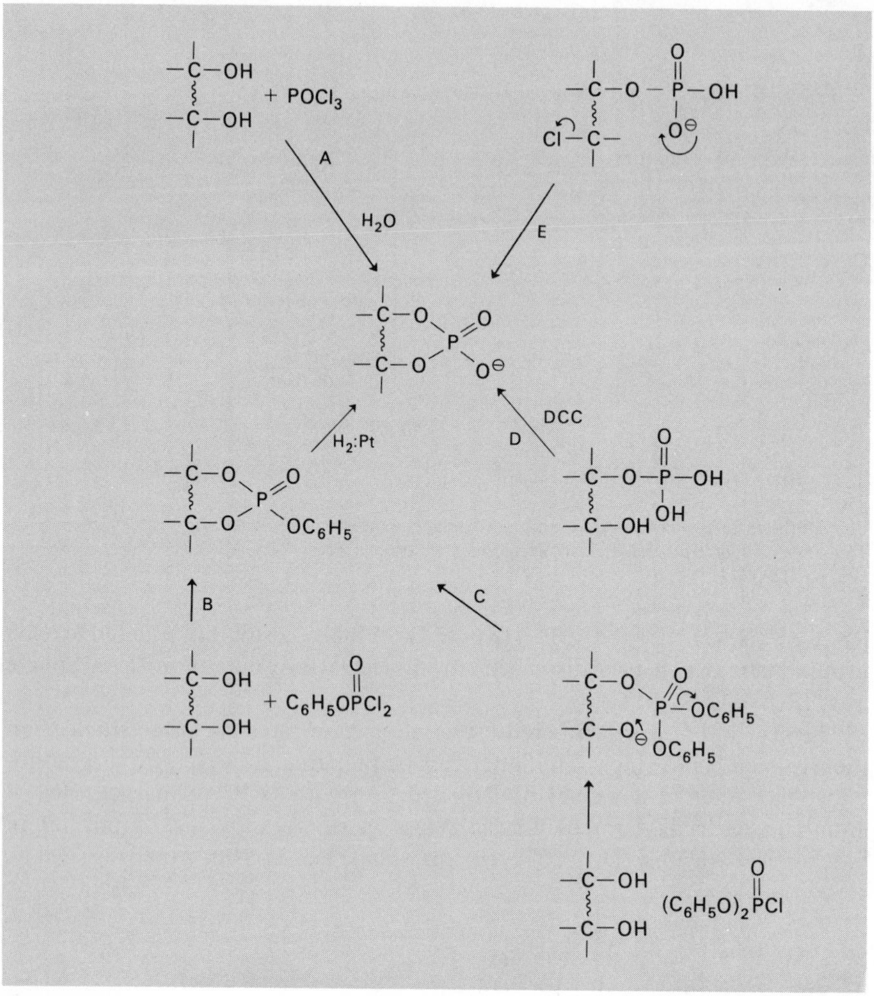

Figure 6.6 The synthesis of cyclic phosphates. Hydroxyls must be within 3 to 5 Å of each other but need not be on adjacent carbons. The various pathways are discussed in the text.

reaction can lead to pyrophosphate dimers starting from the 5'-phosphate, however, if high concentrations are used. Five-membered cyclic phosphates do not form between adjacent *trans* hydroxyl groups.

6.5   PHOSPHATE ESTER HYDROLYSIS

The importance of studies on the hydrolysis of phosphate esters lies not only in the fact that the hydrolysis reaction occurs as part of many biological processes (for example, muscle contraction) but in the fact that hydrolysis reactions serve as models for phosphate transfer reactions involving other acceptors. The process is frequently complex and many different mechanisms are operative. Much of the discussion to follow is of doubtful pertinence to biochemical systems, but understanding of phosphate transfer reactions in biochemical systems is fragmentary and achieving fuller understanding will depend largely on analogies drawn from model systems.

A. Monoesters

Phosphate esters can undergo hydrolysis by P—O bond cleavage or by C—O bond cleavage, depending on the site of attack by water at low pH values and depending on the presence of a proton for transfer to the alkoxyl group at higher pH values (Figure 6.7). For example, a diester cannot cleave by the

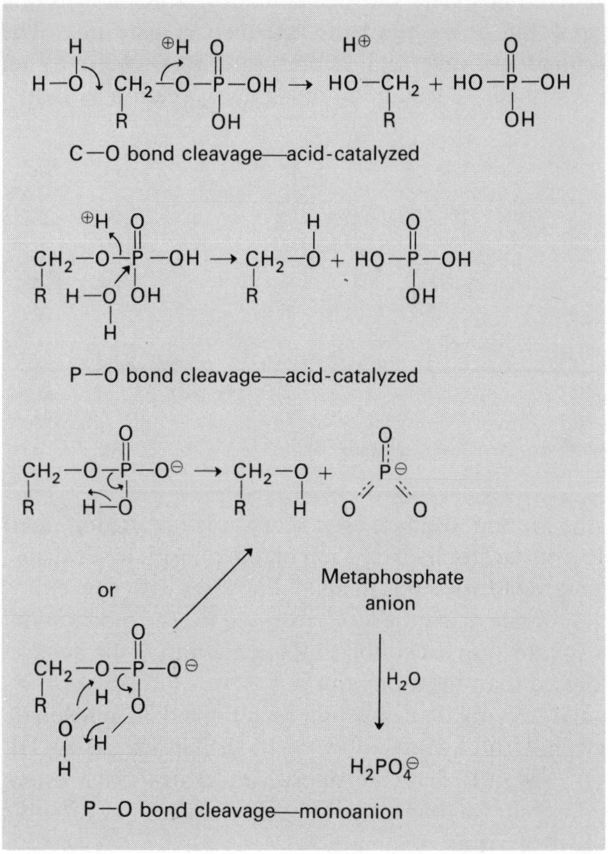

Figure 6.7

Mechanisms of phosphate ester hydrolysis.

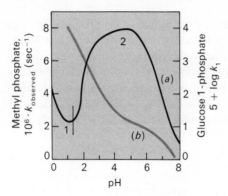

Figure 6.8

The effect of pH on monophosphate ester hydrolysis. (a) Methyl phosphate at 100°C. 1. Acid-catalyzed, C—O and P—O bond cleavage. 2. Monoanion, P—O bond cleavage. (b) Glucose 1-phosphate at 82°C. [C. A. Bunton, D. R. Llewellyn, K. G. Oldham, and C. A. Vernon, *J. Chem. Soc.*, 3,574 (1958), with permission of The Chemical Society.

monoanion mechanism. A typical example of the effect of pH on phosphate monoester hydrolysis is shown in Figure 6.8. This type of pH profile is observed with ordinary phosphate esters such as glycerol 1-phosphate and ethyl phosphate.

The hydrolysis of glucose 1-phosphate shows a similar dependence on rate over the range of pH 8 to 4, but below pH 4 the rate increases steadily. The reaction involves C—O bond cleavage and is presumed to proceed by an S_N1 mechanism with the intermediate formation of a glucosyl cation as shown below.

HOCH$_2$

OH

HO

(O—P—(OH)$_2$)

HO H$^{\oplus}$

α-D-glucopyranose 1-phosphate

→

HOCH$_2$

OH

HO

$\oplus$

OH

$+ \ H_3PO_4$

↓ H$_2$O

glucose

At pH values above 9 the monophosphate ester dianion is the predominant species. It is hydrolyzable by attack of hydroxyl ion on the phosphorus causing P—O bond cleavage. The rate of hydrolysis in alkali increases with concentration of hydroxide ion but does not reach the level observed for the monoanion. Apparently glucose 6-phosphate is an exception to this generality—the dianion being more rapidly hydrolyzed than the monoanion.

Another unusual reactivity in alkaline medium is observed in phosphate monoesters having electron-withdrawing substituents (such as —COOH, —SO$_3$H, —CN, —C=O, —SO$_3$H) β to the phosphate group. Such esters undergo facile β elimination as shown for serine phosphate. Below pH 7.0 some β elimination occurs but hydrolysis predominates.

$$
\begin{array}{c}
\text{COO}^{\ominus} \\
| \\
\text{H}_2\text{N}-\text{C}-\text{H} \quad \text{OH} \\
| \\
\text{O} \\
| \\
\text{H}_2\text{C}-\text{O}-\overset{\|}{\text{P}}-\text{O}^{\ominus} \\
| \\
\text{O}^{\ominus}
\end{array}
\xrightarrow[\text{PO}_4^{3\ominus}\, +\, \text{H}_2\text{O}]{\text{pH 7-14 +}}
\begin{array}{c}
\text{COO}^{\ominus} \\
| \\
\text{H}_2\text{N}-\text{C} \\
\| \\
\text{CH}_2
\end{array}
\longrightarrow
\begin{array}{c}
\text{COO}^{\ominus} \\
| \\
\text{C}=\text{NH} \\
| \\
\text{CH}_3
\end{array}
\xrightarrow[\text{NH}_3]{\text{H}_2\text{O}}
\begin{array}{c}
\text{COO}^{\ominus} \\
| \\
\text{C}=\text{O} \\
| \\
\text{CH}_3
\end{array}
$$

B. Diesters

Simple phosphate diesters are hydrolyzed by mechanisms similar to those discussed above with the notable exception that the monoanion is not readily cleaved since there is no intramolecular acid catalysis available. Several pathways are operative simultaneously and both C—O and P—O bond cleavage occurs.

C. Triesters

Triesters are labile in alkali. The reaction is a simple S_N2 displacement of alkoxide by hydroxide which leads to the diester by P—O bond cleavage. At lower pH values, C—O bond cleavage occurs via attack of water on carbon.

$$
\text{HO}^{\ominus} \quad
\begin{array}{c}
\text{O} \\
\| \\
\text{P}-\text{OR} \\
/ \quad \backslash \\
\text{RO} \quad \text{OR}
\end{array}
\longrightarrow
\begin{array}{c}
\text{O} \\
\| \\
\text{HO}-\text{P} \\
/ \quad \backslash \\
\text{RO} \quad \text{OR}
\end{array}
\quad + \quad \overset{\ominus}{\text{OR}}
$$

D. Cyclic Esters

As shown in Figure 6.9, the five-membered ring system of ethylene phosphate hydrolyzes by P—O bond cleavage 10^8 times more rapidly than dimethyl phosphate does. Only P—O bond fission occurs. The rate increase in the case of the five-membered ring is probably due to strain of the bond angles or to eclipsing of the hydrogen atoms or the unpaired electrons of oxygen. It is also possible that the noncyclic esters are stabilized because of partial double-bond character in the ester linkages. This would be more difficult to achieve in the cyclic esters, making them relatively less stable.

Cyclic esters having six- and seven-membered rings do not show such a large increase in hydrolysis rate relative to their acyclic counterparts. The rate at which they hydrolyze depends on the substituents of the ring, but in the case of

$$
\begin{array}{c}
\text{H}_2\text{C}-\!\!-\!\!-\text{O} \quad \text{O} \\
\backslash \quad \quad \backslash / \\
\text{H}_2\text{C} \quad \quad \text{P} \\
\backslash \quad \quad / \backslash \\
\text{H}_2\text{C}-\!\!-\!\!-\text{O} \quad \text{OH}
\end{array}
$$

trimethylene phosphate

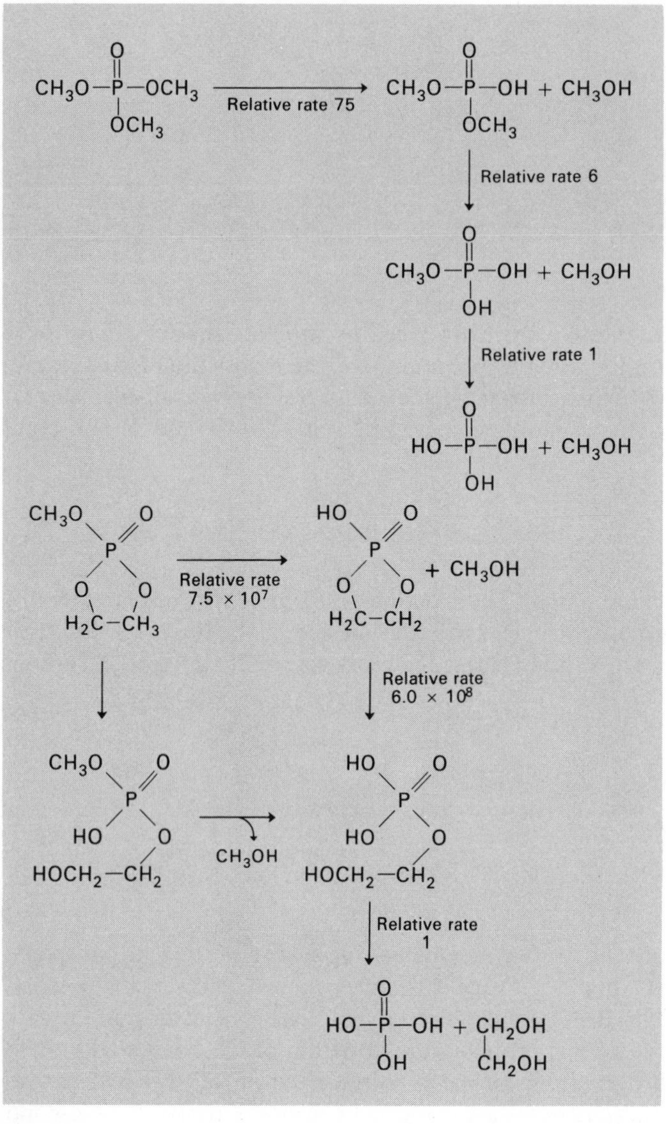

Figure 6.9

The relative rates of hydrolysis of phosphate esters. The rates are approximate values for the acid-catalyzed reaction. The mechanism (that is, C—O or P—O bond cleavage) has not been considered; however, the hydrolysis of the cyclic ester (ethylene phosphate) occurs exclusively by P—O bond cleavage.

trimethylene phosphate the hydrolysis rate is no more than 10 times that of the dimethyl ester.

Cyclic triesters having five-membered rings are also hydrolyzed 10^6 times more rapidly than acyclic triesters when the hydrolysis of the exocyclic ester linkage is considered. This has been explained as being due to the requirement that phosphate ester hydrolysis proceed through an intermediate in which the atoms surrounding phosphorus occupy the corners of a trigonal bipyramid and in which the oxygens of the ring occupy an equatorial and an apical position as shown in Figure 6.10. There are two conformations of the system possible which have approximately equal energy content and which are interconvertible by a process termed pseudo-rotation. Pseudo-rotation resembles the change from one chair form to another in the cyclohexane ring system in that, when it takes place, groups which were apical become equatorial. The presence of the ring favors the development of the trigonal-bipyramidal transition state, thus facilitating entry of the solvent at an apical position, followed by pseudo-rotation to bring the leaving alkoxyl group into the apical position to facilitate its departure (Figure 6.10).

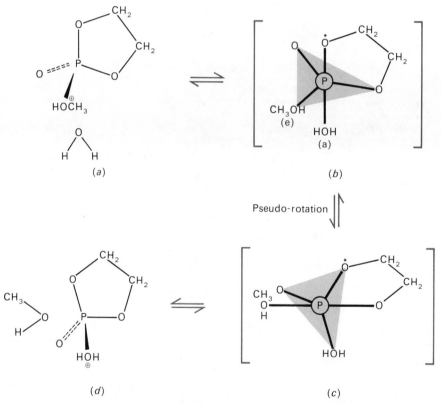

Figure 6.10 Pseudo-rotation in the hydrolysis of cyclic phosphates. The central plane of the trigonal bipyramid is emphasized to show the change in geometry of the intermediate during the process. The protonated triester (a) and the product (d) have a tetrahedral arrangement. Attack by a solvent molecule forces three of the substituents and the phosphorus into a single plane. One of the oxygen atoms is marked to show the change in the distribution of substituents during the process of pseudo-rotation. The change from (b) to (c) is produced by O* and HOH moving toward the right foreground as O and CH_3OH move to the left rear. [P. C. Haake and F. H. Westheimer, *J. Am. Chem. Soc.*, **83**, 1,102 (1961); also E. T. Kaiser, M. Panar, and F. H. Westheimer, *J. Am. Chem. Soc.*, **85**, 602 (1963).]

E. Acyl Phosphates

Acyl phosphates, such as acetyl phosphate, hydrolyze approximately 10^4 times faster than simple alkyl phosphates. The monoanion (a) or the dianion (b) in Figure 6.11 can react intramolecularly to give metaphosphate, which is rapidly

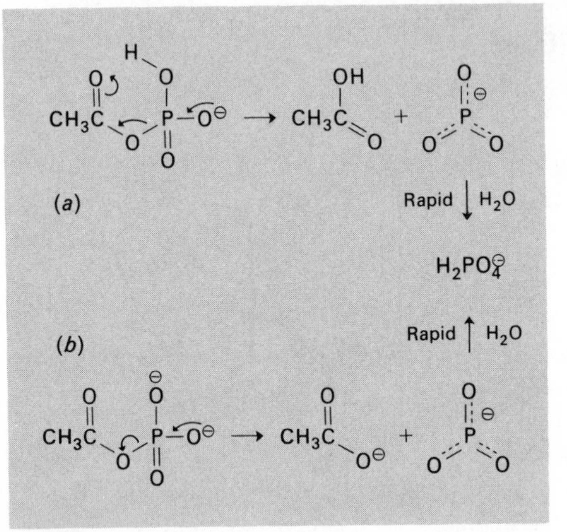

Figure 6.11

The hydrolysis of acyl phosphates. (a) Monoanion. (b) Dianion.

hydrated to give orthophosphate. A case of special interest is carbamoyl phosphate, which in biochemical systems functions as a donor of carbamoyl groups to amino groups (for example, in the formation of citrulline from

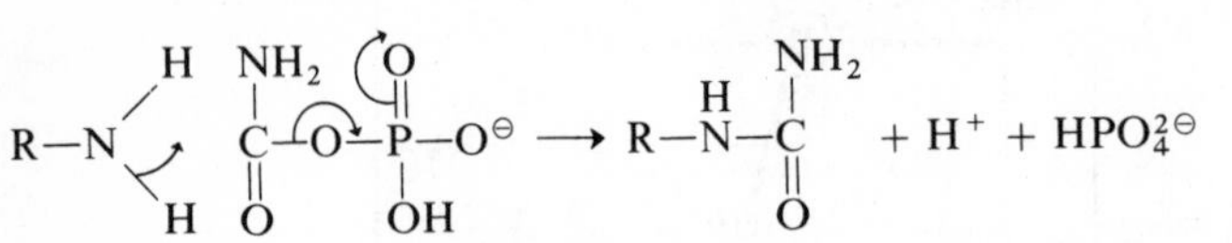

ornithine). The phosphate group is lost as orthophosphate by C—O bond cleavage. This reaction appears to occur only enzymatically. The hydrolysis of carbamoyl phosphate involves unimolecular elimination of cyanic acid (O=C=NH) assisted by participation of the relatively acidic proton of the amide group.

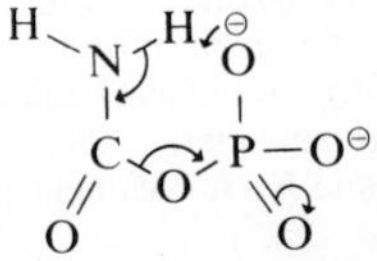

F. Intramolecular Displacements in Di- and Triesters

A suitably situated hydroxyl group (usually vicinal) can participate as a nucleophile in the displacement or migration of a phosphate ester (Figure 6.12).

The acid-catalyzed migration of phosphate groups can occur with mono- or diesters (A or A′). It may involve a pentacovalent intermediate (B) which can

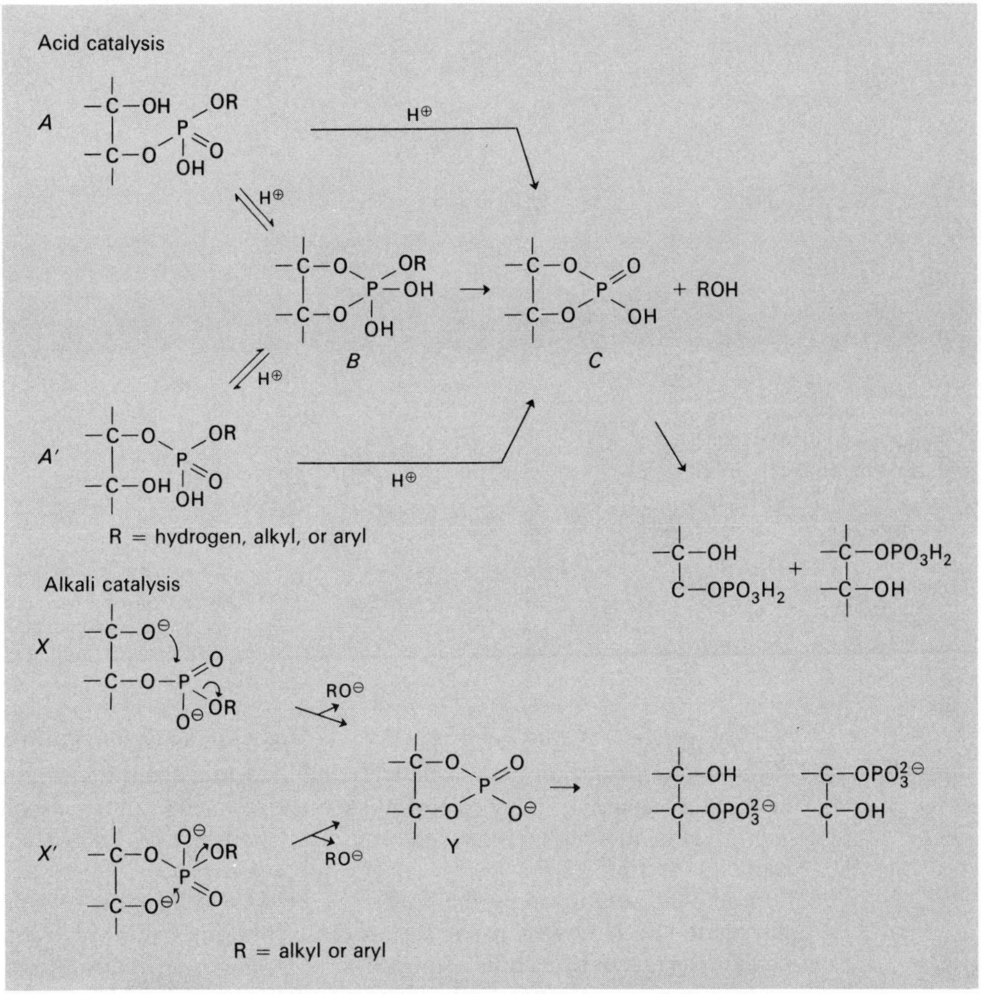

Figure 6.12 The formation of cyclic phosphate esters as intermediates in the hydrolysis of acyclic esters. A-A′ and X-X′ are pairs of compounds which differ only in the position of the phosphate ester group. If hydrolysis proceeds to the cyclic intermediates B and C or Y, then a mixture of products is produced and the original position of the ester linkage cannot be discerned.

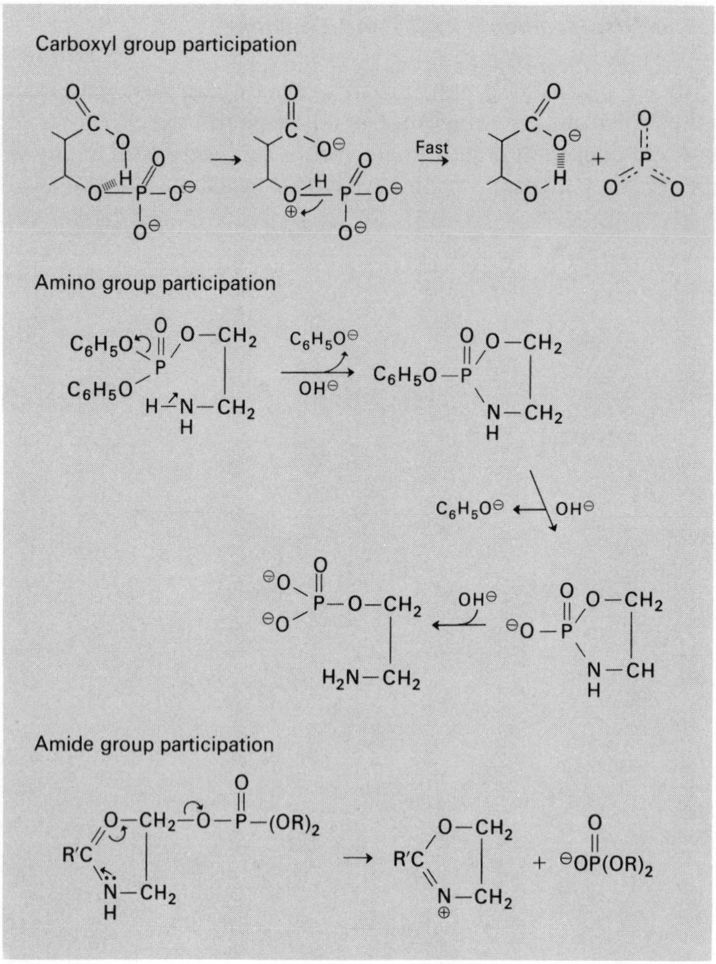

Carboxyl group participation

Amino group participation

Amide group participation

Figure 6.13
Participation by neighboring carboxyl, amine, and amide groups in phosphate ester hydrolysis.

decompose to give the migrated (A or A') or cyclized product (C). The occurrence of phosphate migration under relatively mildly acidic conditions led to difficulties in assigning structure to phosphate esters isolated from tissues following such treatment. The glycerol, glyceric acid, and inositol phosphates are examples (Section 7.5).

In alkaline solution only the diester (X or X', Figure 6.12) gives a cyclic phosphate (Y). However, participation of neighboring hydroxyl groups may increase the rate of hydrolysis of monoesters. Neighboring carboxyl and amino groups can also participate in hydrolysis and migration reactions and may be important in enzymatic processes (Figure 6.13).

The hydrolysis of cyclic esters is subject to general base catalysis. Pyridine bases and imidazole in the free base form increase the rate of hydrolysis of methyl ethylene phosphate by factors of 50 to 150. The differences in rate between the bases examined is at least partly explained by steric hindrance of the approach of the base to the phosphate ester. A possible role of base catalysis by the imidazole groups of ribonuclease in the hydrolysis of ribonucleic acids has been suggested (Figure 6.14); in this case the process can be considered as an intramolecular one involving the enzyme-substrate complex. It has been estimated that the rate enhancement produced by the enzyme may be as much as a factor of 10^9 even though the uncatalyzed reactions are relatively facile.

Monesters having an adjacent hydroxyl group are not hydrolyzed appreciably more rapidly in acid than ordinary monophosphate esters, but they are capable of forming cyclic intermediates and this results in migration of the phosphate group.

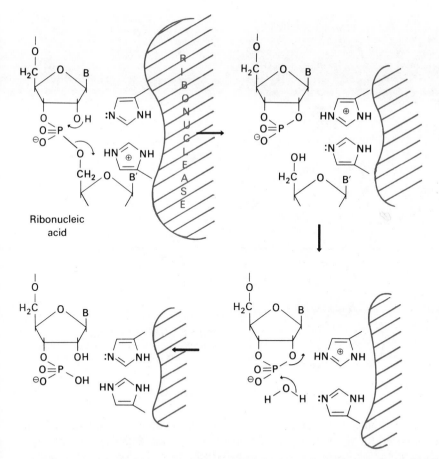

Figure 6.14   Participation by neighboring imidazole groups in the hydrolysis of RNA by ribonuclease. [J. P. Hummel and G. Kalnitsky, *Ann. Rev. Biochem.*, **33**, 15 (1963), with permission of Annual Reviews, Inc.]

G. Metal Ion Catalysis

Metal ion catalysis of phosphate hydrolysis is also of interest because enzymes which handle phosphate esters frequently require metal ions for activity. In most biochemical systems divalent ions such as $Mn^{2\oplus}$, $Mg^{2\oplus}$, and $Zn^{2\oplus}$ are encountered. In the case of salicyl phosphate, which in the absence of metal hydrolyzes by a mechanism involving carboxyl group participation, a seven-fold rate enhancement at pH 5.1 occurs in the presence of $Cu^{2\oplus}$. The metal is thought to induce the formation of a trianion which can undergo intramolecular phosphate migration to an acyl phosphate which hydrolyzes rapidly to give ortho- rather than metaphosphate [Figure 6.15(a)]. This

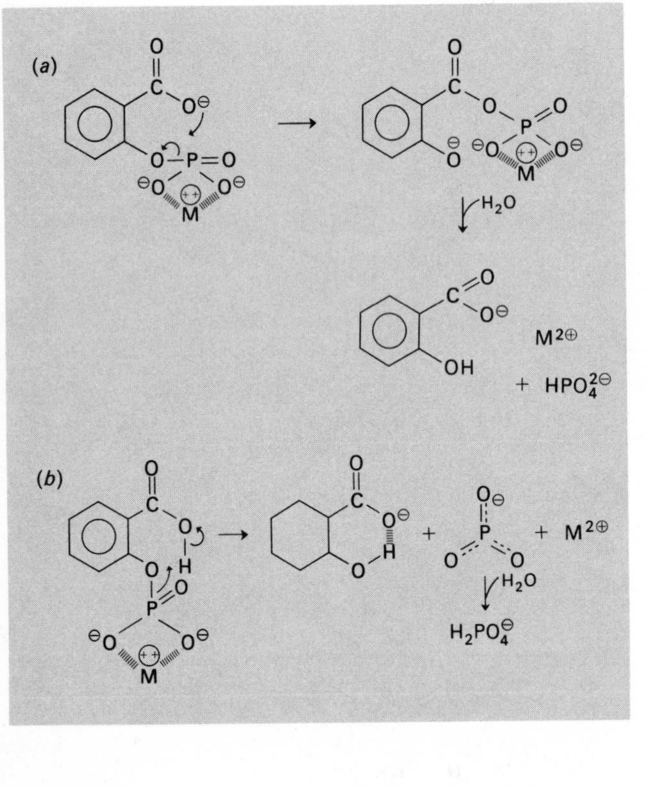

Figure 6.15

Metal catalysis of the hydrolysis of salicyl phosphate. (a) Trianion formation induced by metal ($M^{2\oplus}$). (b) Dianion stabilized by $M^{2\oplus}$.

mechanism should be considered to be tentative. It differs considerably from that proposed for the reaction in the absence of metal (Figure 6.13) although an entirely analogous mechanism can be written in which the metal stabilizes the phosphate dianion and increases the electronegativity of the ester oxygen [Figure 6.15(b)].

6.6 PYROPHOSPHATE HYDROLYSIS

The pyrophosphate bond is a typical anhydride type—it is quite susceptible to hydrolysis. This fact is of practical importance since much of the phosphate present in tissues is bound in pyrophosphate linkages and its amount can be estimated as the inorganic phosphate released by brief acid hydrolysis (7 min at 100°C in $1.0\ N\ H^{\oplus}$). Most, although not all, monophosphate esters are quite stable under these conditions. Triphosphates, such as adenosine triphosphate (ATP), are more easily hydrolyzed than diphosphates, such as adenosine diphosphate (ADP), although the difference in rate is not large (up to a factor of 10).

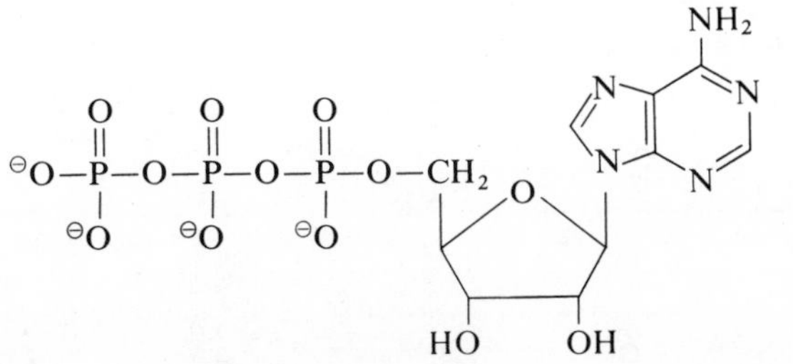

adenosine triphosphate
ATP

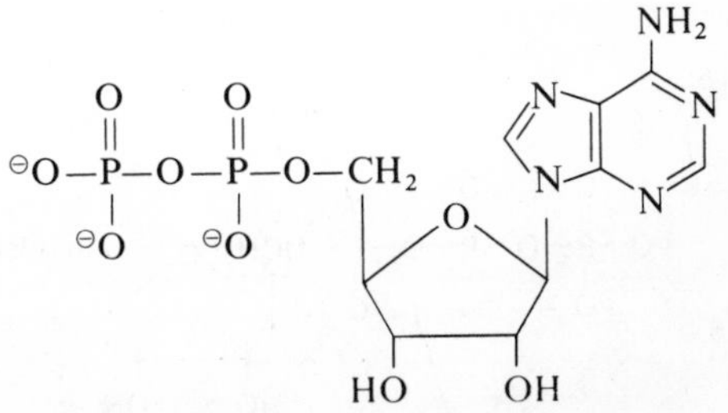

adenosine diphosphate
ADP

The effect of pH on the hydrolysis of pyrophosphate is similar to the effect on the hydrolysis of monomethyl phosphate [Figure 6.8(a)]. A maximum is observed at pH values where the monoanion predominates. There is no direct evidence, however, that the reaction involves the intermediate formation of metaphosphate which is the presumed intermediate in phosphate monoester, monoanion hydrolysis, and in acetyl phosphate hydrolysis (see Figures 6.7 and 6.11). The metaphosphate pathway is shown below for comparison.

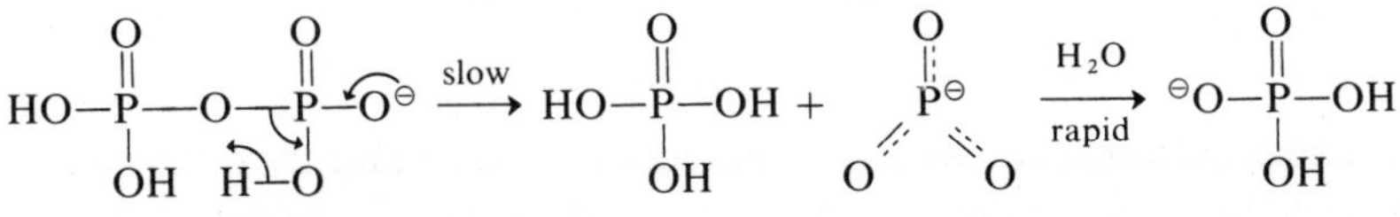

Hydrolysis is a special case of nucleophilic attack and the reactions of pyrophosphates with other nucleophiles are very important biochemically; many of the roles of ATP and similar compounds with high phosphate transfer potentials involve their ability to act as phosphorylating agents. As shown in Figure 6.16(a), the reaction can be written as the attack of the nucleophilic agent (glucose, acetate, and so forth) at the terminal phosphate with the displacement of ADP or at the central phosphorus with formation of a pyrophosphate and AMP. The other role of triphosphates involves reaction with a nucleophile to displace the terminal pyrophosphate residue or to displace all the phosphate

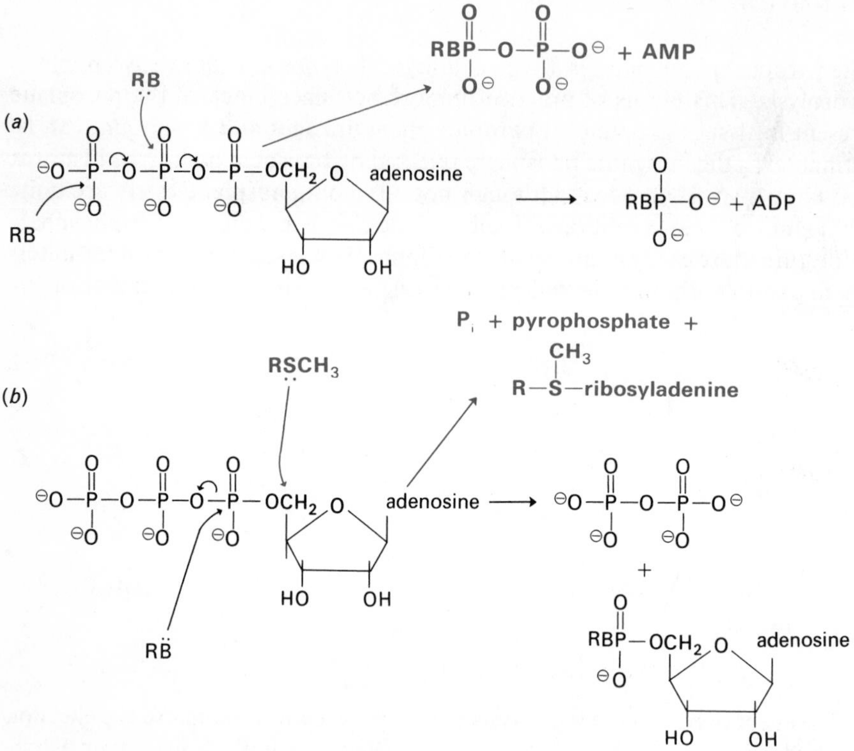

Figure 6.16 The roles of ATP in biological systems. (*a*) ATP acting as a phosphorylating or pyrophosphorylating agent. In living tissues, ADP is rephosphorylated by metabolic processes which generate groups with higher transfer potentials. (*b*) ATP acting in pyrophosphorolytic reactions. In most reactions of this type, pyrophosphate is subsequently hydrolyzed driving the reaction to the right.

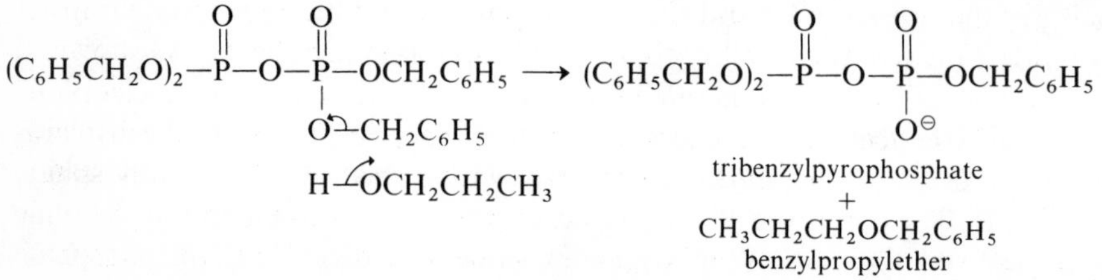

residues. The question arises as to the potential of ADP and pyrophosphate groups as leaving groups in this type of reaction. It has been demonstrated that tetrabenzylpyrophosphate is subject to solvolysis by 1-propanol and that the tribenzylpyrophosphate group is very efficient as a leaving group. The ionized pyrophosphate groups of ATP probably are not as effective as leaving groups, however, because of the presence of negative charges in the molecule. It is possible that in the enzyme-catalyzed reaction this negative charge is quenched by positive charges on the surface of the enzyme or by divalent metal ions which are commonly involved in the enzymatic process.

There have been attempts to elucidate the role of metal ions in the hydrolysis of ATP. Divalent ions such as $Ca^{2\oplus}$ and $Mn^{2\oplus}$ bind to ATP about 2×10^3 times more tightly than monovalent ions such as $K^\oplus$ and $Na^\oplus$. Some of the divalent ions appear to bind to the proximal [Figure 6.17(a)] and some to the distal pair of phosphate groups [Figure 6.17(b)]. Others have been shown to form complexes which involve both the phosphates and the adenine ring. In Figure 6.17 the different chelates of ATP are shown as they might function

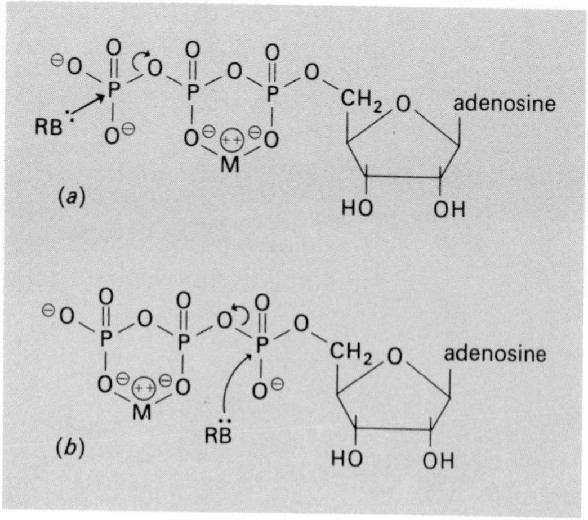

Figure 6.17

The role of metal ion in reactions of ATP. (a) In phosphorylation reactions. (b) In pyrophosphorolytic reactions.

in two of the reactions in which ATP participates. It is important to recognize that the mechanisms presented in Figure 6.17 are hypothetical—more so than most. The facts are that divalent metals catalyze the hydrolysis of ATP and ADP, are most effective in the pH range of 5 to 6, and form quite stable complexes with the nucleotides.

6.7 PHOSPHORAMIDATES

The most interesting phosphoramidates from a biochemical point of view are creatine phosphate and imidazole phosphate. The former is naturally occurring and can phosphorylate ADP to form ATP. The latter is of interest because of the possible involvement of the imidazole rings of histidines in transphosphory-

creatine phosphate

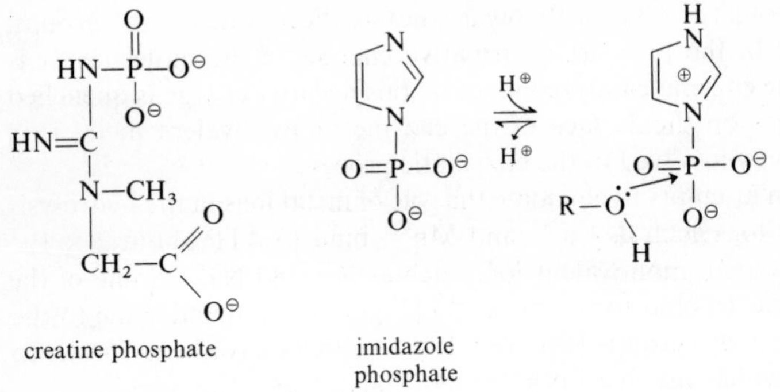

imidazole
phosphate

lation reactions. As shown above phosphoramidates exist near neutral pH as a zwitterion, which facilitates attack on phosphorus by nucleophylic agents leading directly to orthophosphate, ester, or anhydride, depending on whether the nucleophile is water, alcohol, or phosphate.

6.8   QUINONE PHOSPHATES

"Oxidative phosphorylation" is one of the processes of the cell by which the terminal pyrophosphate linkage of ATP is formed from ADP and ortho-

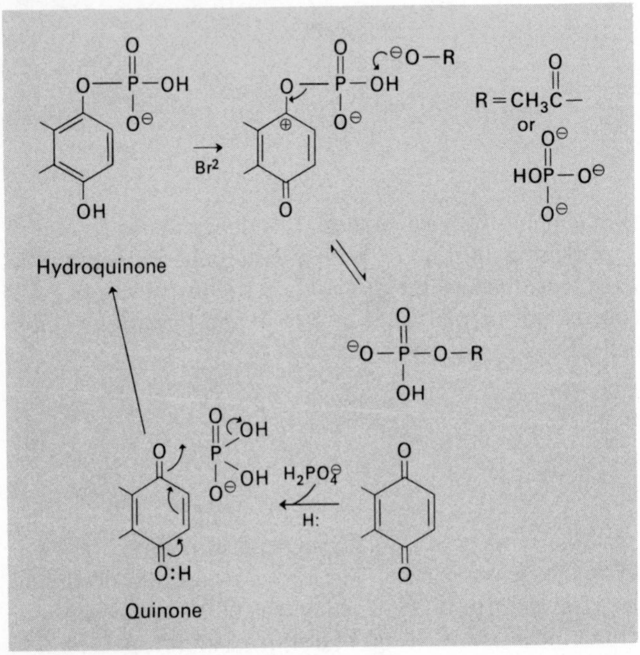

Figure 6.18

A model for oxidative phosphorylation. [V. M. Clark, D. W. Hutchinson, G. W. Kirby, and A. R. Todd, *J. Chem. Soc.*, 715 (1961).]

phosphate. It utilizes the energy released by the oxidation of a wide variety of substrates with the formation of water from oxygen. The sequence of events is very complex and is incompletely understood. It is possible, however, that oxidative phosphorylation involves phosphate transfer reactions which are more or less directly linked to oxidation. An interesting model system is shown in Figure 6.18.

In the presence of added orthophosphate or acetate, the oxidation of *A* leads to the formation of pyrophosphate and acetylphosphate. The formation of the ester bond occurs as the quinone is reduced to the hydroquinone state and the oxidation increases the phosphate transfer potential to the point that a pyrophosphate bond can be formed.

6.9 SUMMARY

This brief outline of phosphate ester formation and hydrolysis is meant only to illustrate the kinds of reaction in which phosphate can participate. Discussion of the role of inorganic phosphates in organized tissues such as bone and teeth has been omitted. Other phosphate esters with specific roles are described in other chapters and other volumes of this series.

REFERENCES

Bruice, T. C., and S. J. Benkovic, *Bioorganic Mechanisms*, Vol. II, W. A. Benjamin, Inc, 1966. Five of the chapters deal with the chemistry of phosphorus compounds of biological interest. There are many references.

Khorana, H. G., *Some Recent Developments in the Chemistry of Phosphate Esters of Biological Interest*, John Wiley and Sons, Inc., New York, 1961. While no longer very recent, this short monograph by the master of phosphate ester chemistry provides an excellent coverage of many of the synthetic techniques that have been developed.

SEVEN | LIPIDS

7.1 INTRODUCTION

It is more difficult to deal with the chemistry of lipids than with the chemistry of the other major constituents of living systems. The reason is simple: Lipids are more heterogeneous chemically than proteins, carbohydrates, or nucleic acids. The preceding statement implies that a class of compounds—the lipids— can be defined. For the purposes of writing this chapter, they are defined as those components of living systems which are soluble in organic solvents such as acetone, ether, chloroform, butanol, and so forth and which are essentially insoluble in water. They lack a common structural feature. Perhaps this definition conveys the impression that lipids are not functionally important— this is not the case. Although the difficulties in purifying and establishing the structures of lipids are enormous, it is clear that they are vitally important in energy metabolism and in the architecture of cell membranes and in the construction of subcellular units.

In this chapter, the lipids are subdivided somewhat arbitrarily into subclasses to facilitate their description. As becomes apparent, neither the definition of lipids as organically soluble materials nor the subdivision of them into classes for discussion are very precise. It should be kept in mind that such attempts to categorize and define are man's efforts to make his knowledge of nature more easily understood and transmitted. It is unfortunate that nature was not aware that we would pursue such an endeavor because not all things in nature can be readily fitted into classes. As the exploration of biological systems has developed to the present, however, it has become clear that there is a tendency for relatively simple precursors to be built up into very complex molecules, and it is possible to arrange natural products into categories on the basis of their biogenic origins. In this approach, all the compounds discussed in this chapter can be considered as derived from acetate. Some of them are polymers of acetate units (the fatty acids) and others are better considered as polymers of a five-carbon isopentenyl fragment. However, this in turn is derived from acetate. Unfortunately, acetate also serves as the carbon source for some aromatic and large ring compounds which are not easily classified as lipids. They are dealt with very briefly in the last chapter of this book.

7.2 FATTY ACIDS

The most abundant naturally occurring lipids are the fatty acids; some examples of the various kinds found are listed in Table 7.1. Most of them exist in nature

TABLE 7.1 A FEW EXAMPLES OF FATTY ACIDS FOUND IN NATURE

I. Saturated[a]   A Normal

Name[b]	Number of Carbons	Sources	Comment
Butyric	4	Milk fat	
Caproic Hexanoic	6	Milk fat, coconut	
Caprylic Octanoic	8	Milk fat Seeds (coconut, palm)	
Capric Decanoic	10	Milk fat Whale lipid	
Lauric Undecanoic	12	Milk fat Whale lipid	
Myristic Tetradecanoic	14 ⎫		
Palmitic			
Hexadecanoic	16 ⎬ Almost all fats, plant and animal	Account for the vast majority	
Stearic Octadecanoic	18 ⎭		
Long chain	20–32	Brain, waxes	

B. Branched

Iso series	10–28	Wool wax	Branched at penultimate carbon
			Formed from isobutyryl coenzyme A

TABLE 7.1   A FEW EXAMPLES OF FATTY ACIDS FOUND IN NATURE *(cont.)*

I. Saturated[a]   A Normal

Name[b]	Number of Carbons	Sources	Comment
Ante-iso series (optically active)	9–31	Wool wax	Branched at ante-penultimate carbon
			Formed from L-α-methyl butyryl coenzyme A; derived in turn from the amino acid isoleucine
Other		A wide variety of multiple-branched fatty acids have been isolated and characterized from bacterial sources.	

C. Carbocyclic[c]

Cyclopropyl
Lactobacillic acid
  cis-11,12-methylene
  octadecanoic acid

$$CH_3(CH_2)_5-\overset{\displaystyle H}{\underset{\diagdown}{C}}-\overset{\displaystyle H}{\underset{\diagup}{C}}-(CH_2)_9COOH$$
$$\overset{\displaystyle C}{\underset{\displaystyle H_2}{}}$$

Phospholipids of bacteria — Formed from unsaturated precursors by enzymes utilizing *S*-adenosyl methionine

II. Unsaturated[d]   A Normal

Monoethenoid = 1 double bond

Name	Number of Carbons	Sources	Comment
Myristoleic 9-tetradecenoic	14	Whale oil Dolphin oil	
Palmitoleic 9-hexadecenoic	16	Whale oil Milk fat	
Oleic 9-octadecenoic	18	Pork fat	Most abundant
Gadoleic 9-eicosenoic	20	Cod-liver oil	Many other unsaturated fatty acids have been isolated from plants and

Polyethenoid[e]

Name	Number of Carbons	Sources	Comment
Linoleic *cis*, *cis*-9, 12-octa-decadienoic	18	Linseed oil Most seeds	Essential dietary constituents for man
Linolenic *cis*-9,12,15-octa-decatrienoic	18	Soybean oil Most seeds	
Arachidonic 5,8,11,14-eicosatetraenioic	20	Liver Lecithins	

B. Branched Chain[f]

Name	Number of Carbons	Sources	Comment
Mycolipenic 2,4,6-trimethyl-2-tetraeicosenoic	27	Tubercle bacillus	Optically active with 4(R), 6(R)

TABLE 7.1 A FEW EXAMPLES OF FATTY ACIDS FOUND IN NATURE *(cont.)*

II. Unsaturated   A Normal

Name[b]	Number of Carbons	Sources	Comment
		C. Carbocylic	
Cyclopentenyl e.g., chaulmoogric acid	8–18	Seeds of the family Flacourtiaceae	Used in the treatment of leprosy
Cyclopropenyl e.g., sterculic acid		Seeds of *Sterculia foetida*	

III. Hydroxyl containing[gh]

β-Hydroxybutyric	4	Many tissues as a coenzyme A ester As a polymer in some bacteria	An important intermediate in the utilization of fatty acids
Mevalonic acid 3,5-dihydroxy-3-methyl-pentanoic acid	6	Most tissues as a phosphate ester	A precursor of sterols and terpenes
Ricinoleic acid	18	$CH_3(CH_2)_5CHOHCH_2CH=CH(CH_2)_7COOH$ As a glyceride in castor oil	
Mycolic	88	*Mycobacterium* species	A very complex family of fatty acids having hydroxyl and methoxyl groups
Corynomycolenic 2-*n*-tetradecyl-3-hydroxy-cis-11-octadecenoic	32	$CH_3-(CH_2)_5-\overset{H}{C}=\overset{H}{C}-(CH_2)_7-CHOH-\underset{\underset{C_{14}H_{29}}{|}}{CH}COOH$	
		Corynebacterium diphtheriae	A branched, unsaturated hydroxy acid

[a] Saturated fatty acids having odd numbers of carbon atoms are known. They occur mainly in marine species (25 percent of the fatty acids in the fat of the mullet have been found to be odd numbered C-15 to 17). However, the depot fats of many animals may contain traces of them. They arise from biosynthesis starting with proprionyl coenzyme A, from the diet, or in brain tissue by α oxidation of even chain fatty acids.

[b] The names used in this table are the trivial names and are followed by the IUPAC names for some of the compounds.

[c] This is not the only naturally occurring cyclopropyl compound. Hypoglycin A, found in the seeds of *Blighia sapida*, has the structure $CH_2=$ ◁ $CH_2CH(NH_2)COOH$ and possess marked ability to lower blood glucose.

[d] The position of a double bond is indicated by a single number. The carbon chain is numbered from the carbonyl group and the number used is that of the lowest numbered carbon participating in the bond. The position of the double bond can also be designated by the symbol Δ. For example, myristoleic acid has a double bond between carbon 9 and 10. It is named 9-tetradecenoic acid or Δ9-tetradecenoic acid.

[e] Note that the double bonds are not conjugated.

[f] Many other branched-chain unsaturated fatty acids are known. The structures of relatively few are firmly established.

[g] In addition to the hydroxy acids, the corresponding *keto* acids can be expected to be found since the biosynthesis of hydroxy acids involves the reduction of the *keto* acid.

Naturally occurring epoxides (for example, 9,10-epoxystearic acid) have been isolated.

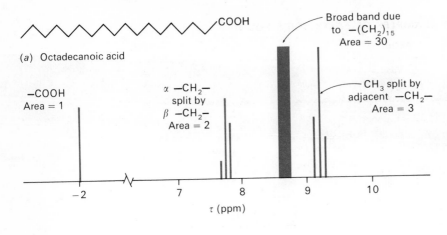

(a) Octadecanoic acid

Broad band due
to −(CH$_2$)$_{15}$
Area = 30

−COOH
Area = 1

α −CH$_2$−
split by
β −CH$_2$−
Area = 2

CH$_3$ split by
adjacent −CH$_2$−
Area = 3

τ (ppm)

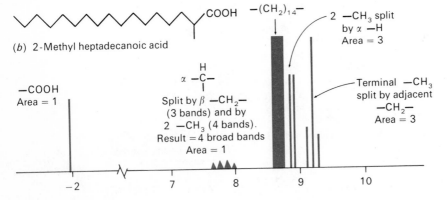

(b) 2-Methyl heptadecanoic acid

−(CH$_2$)$_{14}$−

2 −CH$_3$ split
by α −H
Area = 3

−COOH
Area = 1

H
α −C−
|
Split by β −CH$_2$−
(3 bands) and by
2 −CH$_3$ (4 bands).
Result = 4 broad bands
Area = 1

Terminal −CH$_3$
split by adjacent
−CH$_2$−
Area = 3

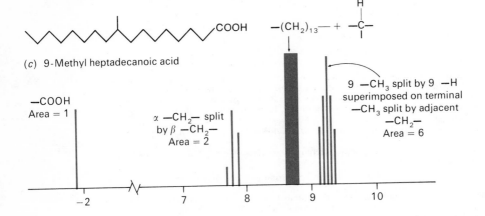

(c) 9-Methyl heptadecanoic acid

−(CH$_2$)$_{13}$− +

H
|
−C−
|

9 −CH$_3$ split by 9 −H
superimposed on terminal
−CH$_3$ split by adjacent
−CH$_2$−
Area = 6

−COOH
Area = 1

α −CH$_2$− split
by β −CH$_2$−
Area = 2

as esters of glycerol, sphingosine, long-chain alcohols, or coenzyme A (Figure 6.4) with only small amounts ever present in the free form. Even when "free," the fatty acids tend to associate with other components of biological systems such as proteins. Consider for a moment palmitic acid. It is the most abundant fatty acid and has 16 carbon atoms and a saturated straight-chain structure. Given this number of carbon atoms, it is possible to write 328,092 different structures having various branching arrangements and stereochemistries. Many of these structures may not be capable of existence because of crowding of atoms in highly branched structures, but a large number of isomers could still exist. It is remarkable that so few of these branched isomers do occur and that in terms of abundance (but perhaps not function) they are relatively unimportant. As discussed later, the fact that relatively few isomers of fatty acids are naturally occurring may make the characterization of lipids more difficult, but it must also be realized that distinguishing between geometric isomers, particularly those with long branches, would be difficult.

Nmr spectrometry is useful in assigning structures to some isomers. For example, the spectra of the isomeric octadecanoic acid having a straight chain, a 2-methyl, a 9-methyl, and a 15-methyl group are shown in Figure 7.1. The absorbances due to —CH_3 groups, to the α hydrogens, and to the intrachain —CH_2— groups are readily assigned, and the structural changes clearly affect the areas and splittings observed. It is possible to distinguish 2-methyl substitution from a methyl branch at other positions, but it is doubtful that other isomers can be clearly distinguished with most common spectrometers. In Table 7.2 values for chemical shifts of most of the common structures in the fatty acids are presented. From these data it can be concluded that the presence of some features such as cyclopropene rings is easily established.

The analytical mass spectrometer offers another approach to structural assignments. In Figure 7.2 are the mass spectra of an 18-carbon straight-chain fatty acid methyl ester (*A*) and its 10-methyl derivative (*b*). It is quite easy to identify the peaks due to the molecular ions (indicated by M) giving the molecular

Figure 7.1

Nmr spectra of some isomeric fatty acids. The spectra are presented diagrammatically. (*a*) The signal due to the carboxyl hydrogen is easily identified because it is so far downfield. The absorption due to the methyl hydrogens at C-18 and the hydrogen at C-2 (the α-CH_2) are identifiable and are split by interactions with adjacent methylene protons. (*b*) The signal due to the second set of methyl protons (at C-2) is downfield from the protons of the terminal methyl group because of their proximity to the carboxyl group. They are split by the single proton remaining at C-2. The signal due to the single proton at C-2 is, in turn, split by the three protons of this methyl group. (*c*) The protons of the methyl group at C-9 absorb at the same position as those of the terminal methyl group but are split into a doublet by the single proton at C-9. This proton is not distinguishable since it absorbs in the region of the methylene groups. The protons at C-2 can be distinguished. [C. Y. Hopkins, in *Progress in the Chemistry of Fats and Other Lipids*, **VII: 2**, 215 (1965), with permission of Pergamon Publishing Company.]

TABLE 7.2 CHEMICAL SHIFT OF PROTONS IN LONG-CHAIN FATTY ACIDS AND ESTERS (EXPRESSED AS τ = ppm RELATIVE TO TETRAMETHYLSILANE = 10.00)^a (cont.)

τ	Group Formula	Description	Type of band
3.0	CH=C.CO	β-Proton of $\alpha\beta$-unsaturated acid (trans)	Multiplet
4.0	(CH=CH)$_3$	Conjugated triene group near center of chain	Multiplet
4.2	C=CHCO	α Proton of $\alpha\beta$-unsaturated acid	Doublet
4.2	(CH=CH)$_2$	Conjugated diene group near center of chain	Multiplet
4.3	CH$_2$=CH—	Methine of terminal vinyl	Multiplet
4.3	$\begin{array}{c} \text{C—C—C} \\ \mid \quad \mid \\ \text{CH = CH} \end{array}$	Olefine group of cyclopentene ring	Singlet
4.4	CH=CH.C.CO	Protons of double bond in the 3 position	Multiplet
4.5	CH=CH	Protons of double bond in the 4 position	Multiplet
4.7	CH=CH	Isolated double bond protons (not conjugated)	Multiplet
4.9	CH—O—C	CH of glycerol esterified in the 2 position	Quintuplet
5.1	CH$_2$=	Terminal vinyl	{ Apparent triplet
5.7	FCH$_2$—	Terminal methylene in ω-fluoro acid	Multiplet
5.8	CH$_2$—O—C	CH$_2$ of glycerol esterified in the 1 or 3 position	—
5.9–6.1	CH—OH	CH of glyceryl ester not esterified in the 2 position	Multiplet
6.3	CH_2—OH	CH$_2$ of glyceryl mino- or diester	Apparent doublet
6.3	CH$_3$—O—CO	CH$_3$ of methyl ester	Singlet
6.3	HO.CH_2	Terminal methylene in ω-hydroxy acid	—
6.4	CH—OH	CH of CHOH in chain	Unresolved multiplet
6.7	CH$_3$OC	Methoxyl	Singlet
6.9	=CCH$_2$CO	αCH$_3$ group in Δ^3 unsaturated acid	Apparent doublet
7.2	$\overset{\displaystyle\ulcorner\text{O}\urcorner}{\text{CH—CH}}$	Epoxy ring protons	—
7.2	=CH.CH$_2$.CH=	Diallyic methylene	Triplet
7.2	$\equiv$C.CH$_2$.CH=	Methylene-interrupted enzyme group	Apparent doublet
7.5	CH $\equiv$	Terminal acetylene	—
7.6	=C.CH$_2$CH$_2$CO	2- and 3-methylene groups in Δ^4 unsaturated acid	Apparent singlet

TABLE 7.2 CHEMICAL SHIFT OF PROTONS IN LONG-CHAIN FATTY ACIDS AND ESTERS (EXPRESSED AS τ = ppm RELATIVE TO TETRAMETHYLSILANE = 10.00)^a (cont.)

τ	Group Formula	Description	Type of band
7.75	CH_2COO	Protons on 2-carbon in acid or ester	Unsymm. triplet
7.9	CH_3COOR	CH_3 of acetoxy group	Singlet
8.0	$=CHCH_2$	CH_2 adjacent to isolated olefine group	Apparent doublet
8.3	$CH_2 . C . C=$	CH_2 in β position to olefine carbon	—
8.4	$CH_3CH=$	CH_3 attached to olefinic carbon	Doublet
8.7	$CH_2CH_2CH_2$	CH_2 of carbon chain	Broad band
8.9	$CH_3 . \overset{\displaystyle}{\underset{OH}{C}}$	Terminal methyl adjacent to CHOH	—
9.0	CH_3	Terminal methyl in β position to olefinic carbon	Triplet
9.1	CH_3	Terminal methyl remote from double bonds and substituents	Unsymm. triplet
9.2	$\underset{C \ = \ C}{\overset{\lceil CH_2 \rceil}{}}$	CH_2 of cyclopropene ring	Singlet
9.4 } 10.3 }	$\underset{CH \ - \ CH}{\overset{\lceil CH_2 \rceil}{}}$	Protons of cyclopropane ring near center of chain	—
Variable	COOH	Carboxyl (usually below 0)	Singlet
Variable	C—OH	Hydroxyl (usually 3 to 8)	Singlet
Variable	$C—NH_2$	Amino and amido	—

^{a} From C. Y. Hopkins, *Progress in the Chemistry of Fats and Other Lipids*, vol. VIII, Part 2, page 215 (1965). With permission of Pergamon Publishing Company.

weight, and in the branched-chain acid the peaks due to rupture of the chain on either side of the methyl group (171 and 199 or M-113). As with other forms of spectrometry, the interpretation of the spectra from an unknown substance depends on the accumulation and interpretation of observations on known compounds. The application of this technique to studies of lipids has been very fruitful, particularly when the sample is introduced to the spectrometer directly from a gas chromatograph so that separation and identification are achieved virtually simultaneously.

The average fatty acid has a pK_a value close to 4.8, and at pH 7.0 it exists as a salt. Depending on the cation, the salts have very different properties. The divalent cations ($Ca^{2\oplus}$, $Ba^{2\oplus}$, and so forth) form salts which are appreciably less soluble than those of the monovalent cations ($Na^{\oplus}$, $K^{\oplus}$, and so forth). The latter are the common *soaps*. Some comments about the solubility of soaps and unionized fatty acids are in order. The sodium salts of fatty acids are almost completely miscible with water. Except in very dilute solution, however, these

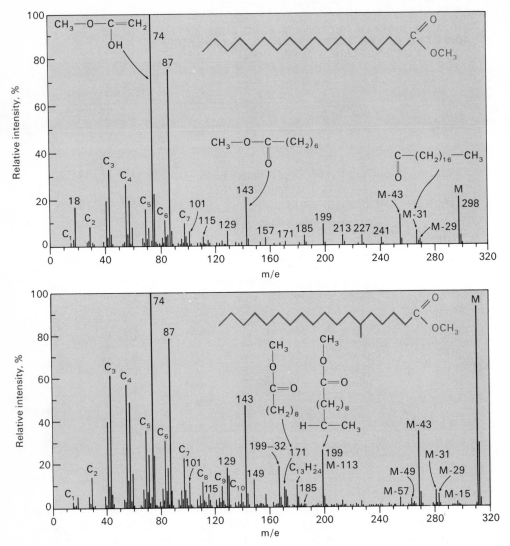

Figure 7.2   The mass spectrum of (a) methyl octadecanoate; (b) methyl 10D-methyl-octadecanoate. The spectrum of (b) shows peaks at m/e 171 and 199 derived from rupture of the carbon chain on either side of the methyl branch. [R. Ryhage and E. Stenhagen, *Journal of Lipid Research*, **1**, 361 (1960), with permission of Lipid Research, Inc.]

salts exist as micelles (Figure 7.3). These micelles are frequently spherical and of such a size that all the molecules have their carboxyl groups at the surface, and the surface consists of a tightly packed array of carboxylate anions. An electrical double layer is formed with the counter ion, and the structure of water is perturbed at the surface. The micelles being negatively charged, tend to repel each other. The exchange of fatty acid between the dispersed and micellar

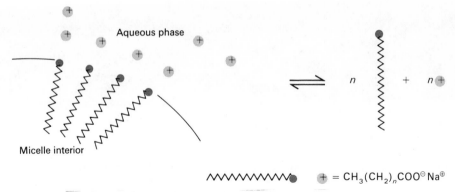

Figure 7.3   A micelle of fatty acid salt in equilibrium with the monodisperse form.

situations is quite rapid. The dispersed form is, in a sense, a hybrid of two quite different kinds of compound, the strongly polar (hydrophilic, water-soluble) carboxylate group and the nonpolar (hydrophobic, organic-soluble) alkyl chain. To bring the latter into solution in water requires that a cavity of suitable dimensions be created in the water. As discussed in Chapter 2, such cavities are thermodynamically difficult to maintain, and the solubility of long alkyl residues in water is very limited. The apparent solubility of fatty acid salts in water is therefore a measure of their tendency toward micelle formation rather than a true solubility.

The conversion of a fatty acid salt to the unionized acid completely alters its solubility in water. The undissociated acid tends to undergo dimerization through hydrogen bonding of the carboxyl group as shown below. Since the

carboxyl group is the only hydrophilic portion of the molecule, and its hydrogen-bonding tendency (and hence hydrophilic character) is completely satisfied by the formation of dimers, the water solubility of the longer-chain fatty acids is essentially zero. Again, water solubility is a function of the ease with which a cavity of the required size can be formed. Incidentally, it is not necessary that an alkyl residue exist in an extended conformation when in solution in a polar solvent. It can be folded upon itself to form a compact mass which has intrachain hydrophobic interactions and which can fit into a water cage most conveniently. In nonpolar solvents, the chains tend to be fully extended and are engaged in interchain hydrophobic interactions (Figure 7.4).

The unionized fatty acids are easily obtained from the salts by acidification and extraction of the aqueous phase with an organic solvent. This provides an easy method for obtaining the fatty acids from a tissue. Whether present as

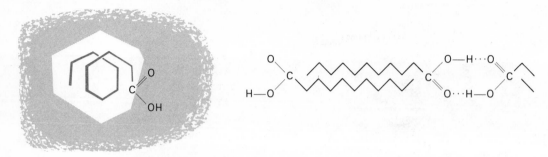

Figure 7.4　　Inter- and intrachain hydrophobic interactions of fatty acids. (*a*) Fatty acid folded (intramolecular hydrophobic interactions) to facilitate fitting into a water cavity—diagrammatic. (*b*) Fatty acids extended to facilitate intermolecular interactions.

esters or as the "free" acids, fatty acids can be converted to the salt form by treatment with alkali (saponification—the making of a soap). The fatty acids can be separated from other lipid components by extraction of the alkaline solution with an organic solvent to remove nonsaponifiable materials followed by acidification and extraction with an organic solvent to obtain a solution which contains essentially only the fatty acids. The composition of the fatty acid mixture so obtained is characteristic of the tissue examined and is of some commercial importance. since the fatty acids are important industrially; their value varies with degree of unsaturation, and so forth.

The carboxyl group. This group of the fatty acids can be chemically modified in many ways. Of significance is the ease with which it can be converted to the methyl ester by treatment with diazomethane or methanol-hydrochloric acid. The methyl esters of fatty acids can be separated and the composition of mixtures determined with precision by gas-liquid chromatography (Figure 7.5). The technique can also be used to prepare quantities of pure fatty acids.

In the synthesis of lipids containing fatty acid esters, the ester bond is usually formed by treating the alcohol (or amine or sulfhydryl) group with the acid chloride. The acid chlorides are easily prepared by treatment of the free acid with thionylchloride or oxalylchloride. Some of the conversions in which the acid chlorides have been utilized are shown in Figure 7.6 along with some conversions of the acid itself. The utility of the reactions in the study of fatty acids is also indicated.

The alkyl chain. The reactivity of the saturated alkyl chain is dependent on the presence of the carboxyl group as a point of activation. This is most apparent in the α bromination of fatty acid chlorides which proceeds through the intermediate enol (Figure 7.6).

The unsaturated alkyl chain. This chain is much more reactive. The addition of halogens to the *cis*-olefinic bond produces a *threo*-dihalide (Figure 7.7). The addition of halide (IBr is used) to double bonds in lipids (fatty acids or their esters) is the basis for a method of estimating the proportion of unsaturated

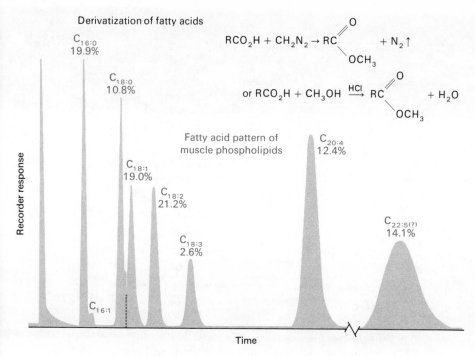

Derivatization of fatty acids

$$RCO_2H + CH_2N_2 \rightarrow RC\overset{\displaystyle O}{\underset{\displaystyle OCH_3}{\diagup}} + N_2\uparrow$$

$$\text{or } RCO_2H + CH_3OH \xrightarrow{\text{HCl}} RC\overset{\displaystyle O}{\underset{\displaystyle OCH_3}{\diagup}} + H_2O$$

Recorder response

$C_{16:0}$
19.9%

$C_{18:0}$
10.8%

$C_{18:1}$
19.0%

$C_{18:2}$
21.2%

$C_{18:3}$
2.6%

$C_{16:1}$

Fatty acid pattern of
muscle phospholipids

$C_{20:4}$
12.4%

$C_{22:5(?)}$
14.1%

Time

Figure 7.5 The separation and quantification of the fatty acids in muscle phospholipids. Conditions—5 percent ethylene glycol adipate on 80–100 mesh on ABS chromosorb. Isothermal 185°C, hydrogen detector with N_2 carrier. (Courtesy of Dr. R. L. Dryer, Department of Biochemistry, University of Iowa.)

fatty acids in a sample. The amount of halide reacted is estimated by titration and calculated in terms of grams of iodine reacted per 100 g of lipid (the iodine number).

The addition of HBr and HI to the olefinic bonds of the unsaturated fatty acids occurs readily and in a normal fashion. If the double bond is polarized by being adjacent to either terminus of the fatty acid, addition of HX is directed by the polarization (Markownikoff) [Figure 7.8(a)]. In the absence of these directing influences, addition is random. The reaction is sensitive to the presence of peroxides which produce atomic bromine which in turn can initiate hydrohalogenation through a radical mechanism [Figure 7.8(b)]. In this case the addition is anti-Markownikoff. The unsaturated fatty acids are readily converted to saturated fatty acids by the usual catalytic procedures (hydrogen in the presence of Raney nickel, platinum, palladium, and so forth). The reaction is not very selective but can be controlled to some extent by varying the catalyst, temperature, and pressure.

The oxidation of unsaturated acids is important in many of the commercial applications of these materials such as the food and paint industries. Autoxidation involving molecular oxygen is initiated by formation of a radical α to a double bond (most readily between two double bonds). The radical reacts with

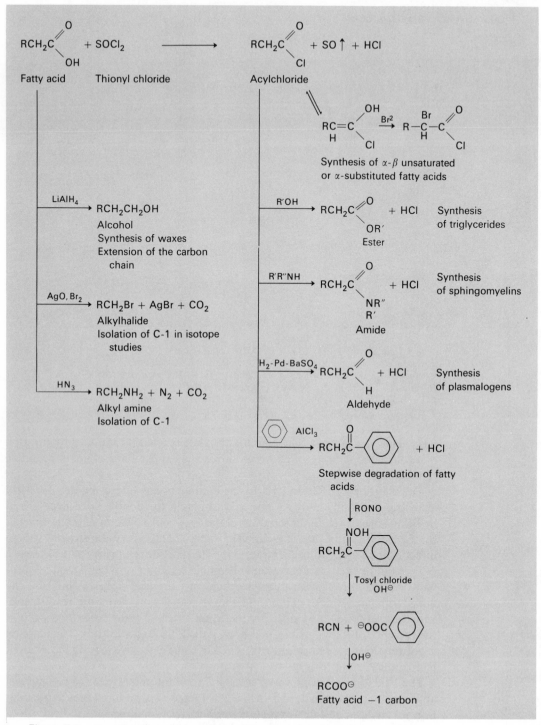

Figure 7.6 Some reactions of fatty acids involving the carboxyl function.

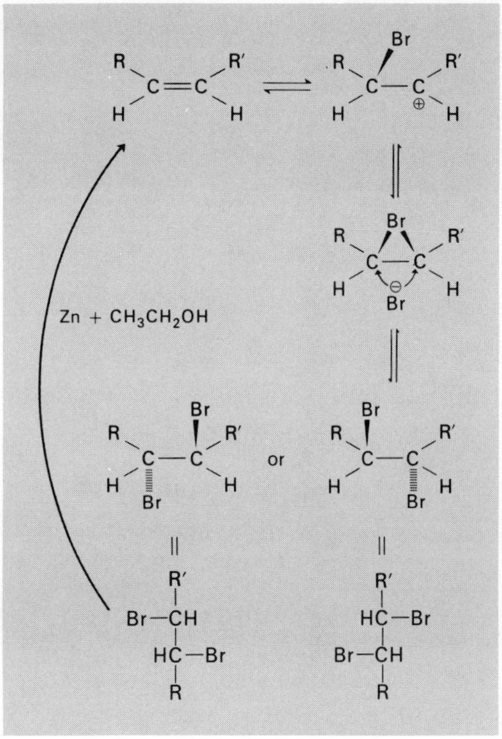

Figure 7.7

The bromination of a *cis*-olefin with the production of *threo*-dibromide.

oxygen to form a peroxide radical which abstracts a proton and forms a daughter radical and a hydroperoxide. The latter can then decompose to form an alkoxyl and a hydroxyl radical and initiate two more chain reactions (Figure 7.9). This cascade produces a marked autocatalytic effect. The radical chain reactions are terminated in a variety of ways resulting in dimers and polymers.

The controlled oxidation of unsaturated fatty acids is very important as a method for locating the position of double bonds. If the chain can be cleaved at the double bonds and the resulting fragments characterized, the position of the double bond can be located. Methods which have proved useful are shown in Figure 7.10.

The potential of the double bonds of unsaturated fatty acids to isomerize (*cis* ⇌ *trans*) is of great importance in assigning structures to biologically active compounds. Invariably, enzymes are capable only of acting on one isomer and the other is not metabolized. The majority of naturally occurring unsaturated fatty acids are *cis* and the most common of these is oleic acid (Table 7.1). It can be isomerized by heating to an equilibrium mixture consisting of 67 percent of the *trans* isomer, elaidic acid. *cis-trans*-Isomerization is also catalyzed by a variety of agents such as nitrous acid, nitrates, sulfurous acid, sulfur, phosphorus, and selenium. The reaction is proposed to involve a π complex between the catalyst and the olefin which isomerizes and dissociates.

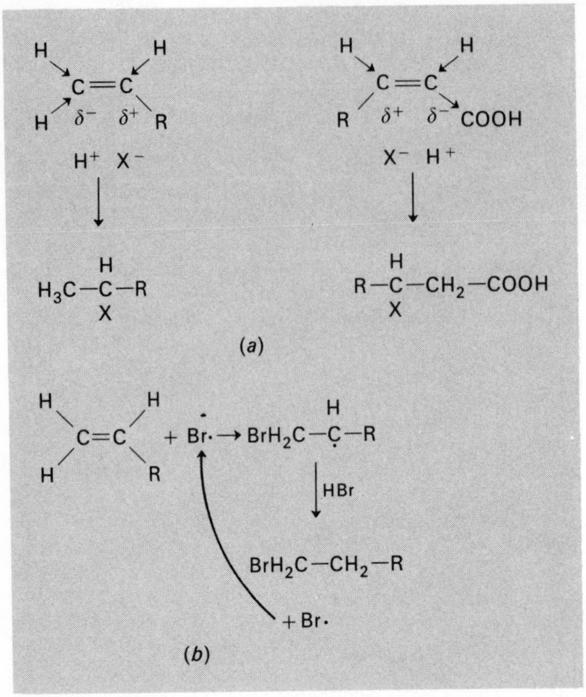

Figure 7.8

The addition of halogen acids to double bonds. (*a*) Markownikoff addition directed by polarization of the double bond. (*b*) Anti-Markownikoff addition by a radical mechanism.

Double bonds of unsaturated fatty acids can be caused to migrate in concentrated alkaline medium. The reaction is initiated by abstraction of a proton from an α-methylene group and rearrangement of the anion. Since the carboxyl group is electron-withdrawing, the equilibrium position of the double bond is in conjugation with the carbonyl group (Figure 7.11). Using milder conditions, compounds containing more than one double bond rearrange to give conjugated systems which have enhanced ultraviolet absorbance. This rearrangement has been used to evaluate the polyolefin content of lipids. The rearrangement and migration of double bonds in almost all cases leads to a mixture which has 67 percent *trans* and 33 percent *cis* isomers.

The hydroxyl groups. This group of the fatty acids show almost normal reactivity toward esterifying and oxidizing agents. They are usually somewhat less reactive than those of the simpler alcohols. The phenyl and naphthyl urethans (esters of *N*-aryl carbamic acids) are commonly used to prepare characteristic derivatives of the alcohols.

$$RR'CHOH + \phi NCO \longrightarrow RR'CHO-\overset{\overset{\displaystyle O}{\displaystyle \|}}{C}HN\phi$$

aryl isocyanate   urethan

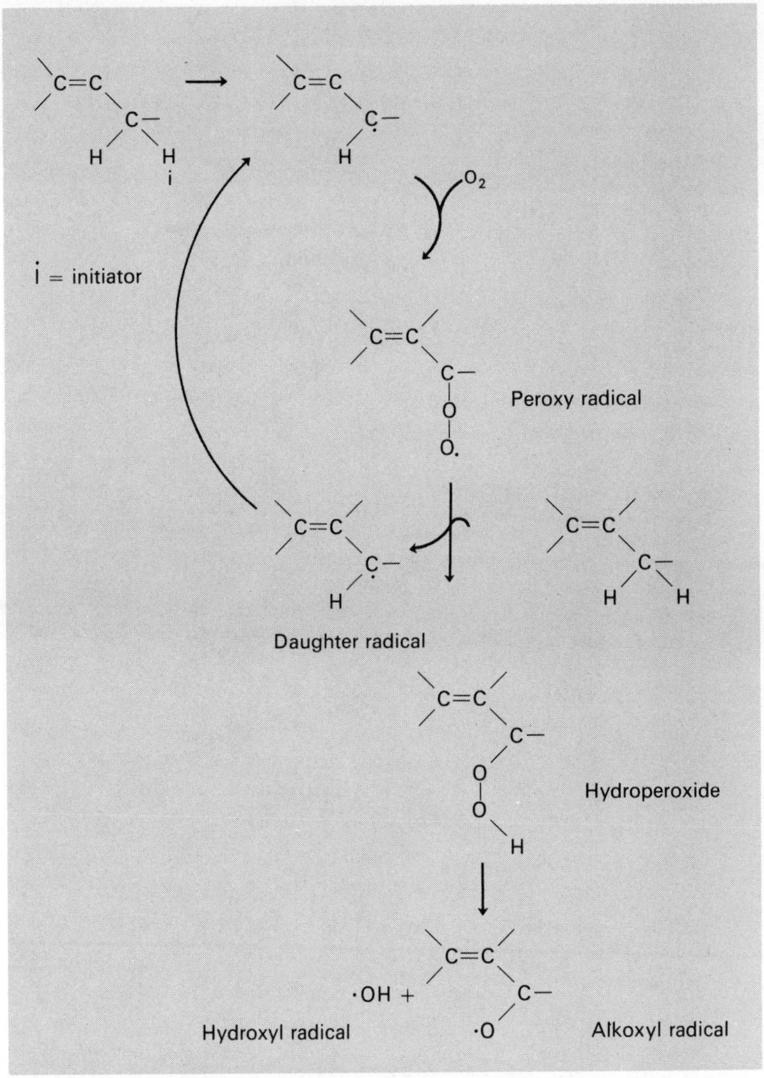

Figure 7.9   The autoxidation of unsaturated fatty acids.

7.3   NEUTRAL FATS

Many of the naturally occurring lipids contain no charged groups. They are literally neutral. It will become clear later in the chapter that many different structural types might be found in a neutral lipid fraction; however, among the

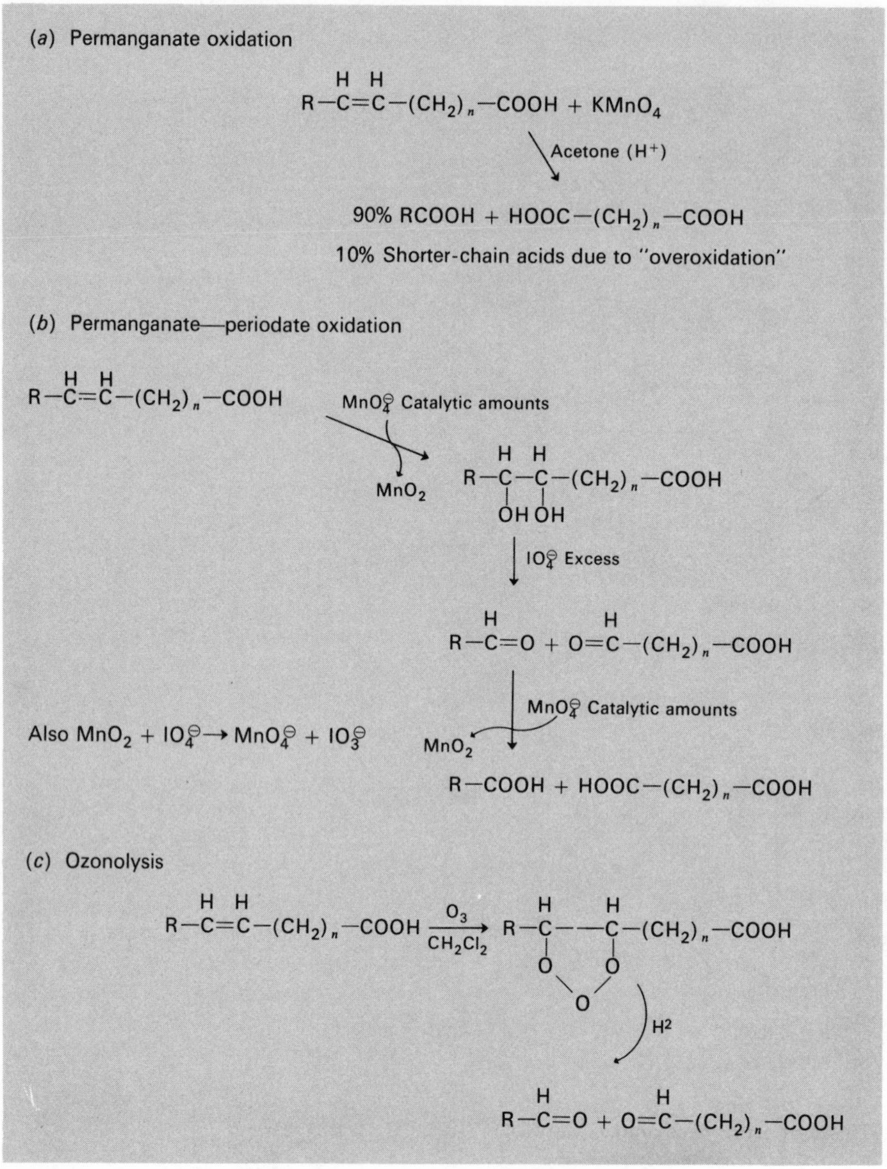

(a) Permanganate oxidation

$$R-\overset{H}{\underset{}{C}}=\overset{H}{\underset{}{C}}-(CH_2)_n-COOH + KMnO_4$$

Acetone (H$^+$)

90% RCOOH + HOOC$-(CH_2)_n-$COOH

10% Shorter-chain acids due to "overoxidation"

(b) Permanganate—periodate oxidation

$$R-\overset{H}{\underset{}{C}}=\overset{H}{\underset{}{C}}-(CH_2)_n-COOH$$

MnO$_4^{\ominus}$ Catalytic amounts

MnO$_2$

$$R-\overset{H}{\underset{OH}{C}}-\overset{H}{\underset{OH}{C}}-(CH_2)_n-COOH$$

IO$_4^{\ominus}$ Excess

$$R-\overset{H}{\underset{}{C}}=O + O=\overset{H}{\underset{}{C}}-(CH_2)_n-COOH$$

MnO$_4^{\ominus}$ Catalytic amounts

MnO$_2$

Also MnO$_2$ + IO$_4^{\ominus} \rightarrow$ MnO$_4^{\ominus}$ + IO$_3^{\ominus}$

R$-$COOH + HOOC$-(CH_2)_n-$COOH

(c) Ozonolysis

$$R-\overset{H}{\underset{}{C}}=\overset{H}{\underset{}{C}}-(CH_2)_n-COOH \xrightarrow[CH_2Cl_2]{O_3}$$

$$R-\overset{H}{\underset{O}{C}}---\overset{H}{\underset{O}{C}}-(CH_2)_n-COOH$$

H^2

$$R-\overset{H}{\underset{}{C}}=O + O=\overset{H}{\underset{}{C}}-(CH_2)_n-COOH$$

Figure 7.10   The controlled oxidation of unsaturated fatty acids. (a) D. E. Dolby, L. C. A. Nunn, and I. Smedley-MacLean, *J. Biochem.*, **34**, 1,422 (1940). (b) E. von Rudloff, *J. Am. Oil Chemists*, **33**, 126 (1956). (c) E. Klenk and W. Bognard, *J. Physiol. Chem.*, Hoppe-Seyler's, **291**, 104 (1952).

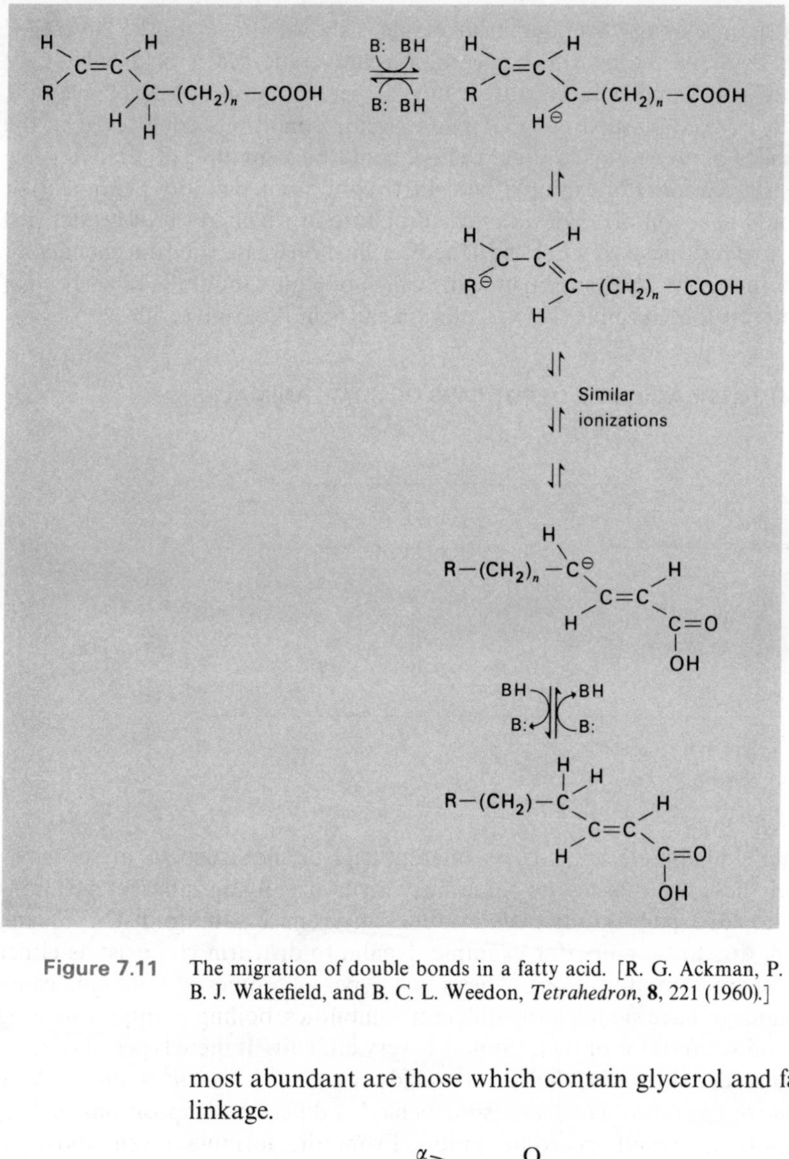

Figure 7.11   The migration of double bonds in a fatty acid. [R. G. Ackman, P. Linstead, B. J. Wakefield, and B. C. L. Weedon, *Tetrahedron*, **8**, 221 (1960).]

most abundant are those which contain glycerol and fatty acids joined in ester linkage.

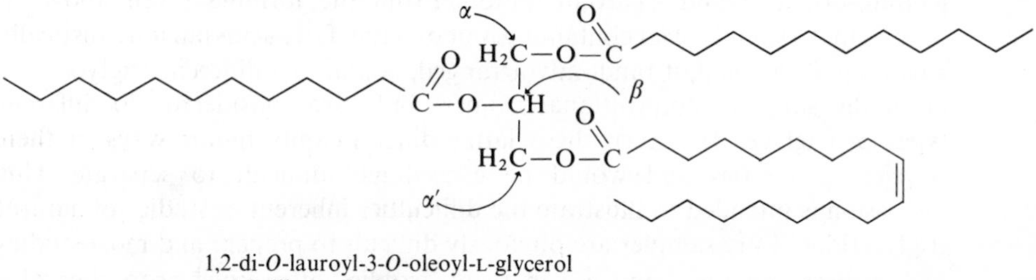

1,2-di-*O*-lauroyl-3-*O*-oleoyl-L-glycerol
or 1,2-dilauro-3-olein
or 1-*O*-oleoyl-2,3-dilauroyl-D-glycerol

The structure of an "average" triglyceride is shown above. In the "average" biological system, the neutral fats contain relatively few fatty acids which are usually about 50 percent saturated and 50 percent unsaturated (Table 7.3). It should be readily apparent that two extreme conditions could exist in the triglycerides of an average animal. There could be a mixture of relatively few specific triglycerides (for example, triolein, tripalmitin, 1-palmito-2,3-diolein) or there could be a completely randomized mixture in which glycerol is esterified by fatty acid residues as it would be if the esterification was carried out chemically. Since there are few biochemical systems which operate randomly, it has been of some interest to determine which situation exists in the triglycerides.

TABLE 7.3   COMPONENT FATTY ACIDS OF DEPOT FATS OF SOME ANIMALS[a]

	Saturated			*Unsaturated[b]*				
	14	16	18	14	16	18	20	22
Hen	1	26	6	—	7	60	—	—
Horse	3	26	5	—	5	39	—	—
Pig	1	28	15	—	4	50	—	—
Tiger	1	22	25	—	7	43	—	—
Seal	3	10	2	2	15	32	17	17
Porpoise	12	5	—	5	27	17	10	7

[a] Expressed as percentage by weight of total acids.
[b] Includes mono- and polyethenoic.

Within n different fatty acids, n^3 different triglyceride structures are possible. Some of these are due to the same fatty acids occupying different positions (for example, 1-palmito-distearin is different from 2-palmito-distearin) and some are due to DL pairs (for example, 1-palmito-distearin can exist as either a D or as an L modification). These two types of isomerisms do not cause compounds to have significantly different solubilities, boiling points, or melting points and separations of such isomers is very difficult. If these types of isomers are disregarded, however, then n fatty acids can give rise to $(n^3 + 3n^2 + 2n)/6$ different triglycerides. These are isomers having different compositions such as palmito-distearin and stearodipalmitin. From the formula given above, it follows that a sample of fat containing four different fatty acids (an unrealistically small number) could, if randomly arranged, contain 64 different triglycerides, or, if the simpler compositional isomers only are considered, 20 different types of triglycerides. Even these latter differ in only minor ways in their physical properties and would be exceedingly difficult to separate. This discussion is intended to illustrate the difficulties inherent in studies of natural triglycerides. Pure samples are obviously difficult to prepare and most studies have utilized mixtures and are in reality studies of average properties of a heterogeneous sample. Some of the methods used are given in Table 7.4.

TABLE 7.4 METHODS USED TO CHARACTERIZE AVERAGE PROPERTIES OF FATS (TRIGLYCERIDES)

Method	Chemical Reaction	Property Described
Saponification number is mg of KOH required to saponify 1 g of lipid (1,000 mg)	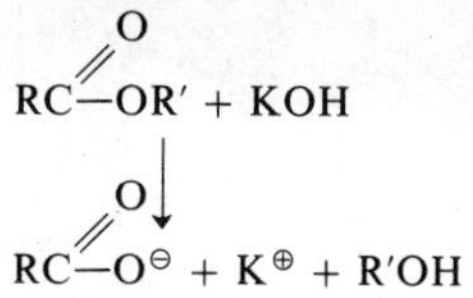 $$\underset{\text{O}}{RC}\!-\!OR' + KOH$$ $$\downarrow$$ $$RC\!-\!O^{\ominus} + K^{\oplus} + R'OH$$	Average molecular weight of the lipid. 1 mmole of ester bonds utilizes 1 mmole of KOH; therefore 1 mmole of triglyceride $\equiv$ 3 mmole (168 mg) of KOH. It follows that $168/MW =$ saponification number/1,000.
Iodine number is number of g of I_2 taken up per 100 g of fat	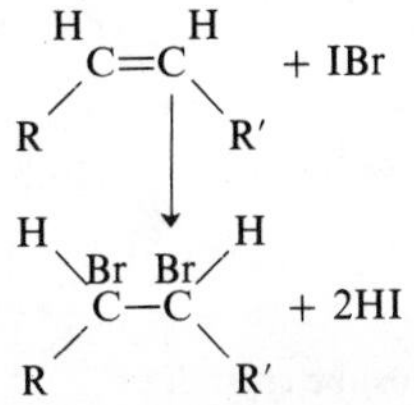 $$\underset{R}{\overset{H}{C}}\!=\!\underset{R'}{\overset{H}{C}} + IBr$$ $$\downarrow$$ $$\underset{R}{\overset{H}{\underset{Br}{C}}}\!-\!\underset{R'}{\overset{H}{\underset{Br}{C}}} + 2HI$$	Degree of unsaturation; 1 mmole of olefin consumes 2 equivalents of halogen $\equiv$ 256 mg of I_2.
Oxidation KMnO$_4$	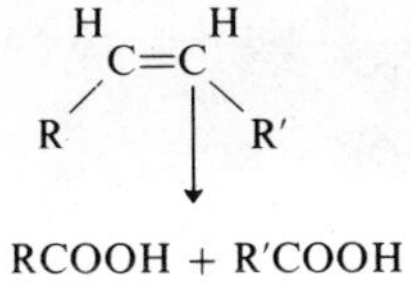 $$\underset{R}{\overset{H}{C}}\!=\!\underset{R'}{\overset{H}{C}}$$ $$\downarrow$$ $$RCOOH + R'COOH$$	Degree of unsaturation and distribution of fatty acids. (a) Completely saturated triglycerides are not oxidized. (b) Triglycerides containing unsaturated fatty acids give a triglyceride containing an equivalent number of carboxyl groups.

Triglyceride

G = glycerol
S = saturated FA
U = unsaturated FA
A = dibasic FA produced by oxidation of U
FA = fatty acid

Products

$GS_3 \longrightarrow GS_3$
$GS_2U \longrightarrow GS_2A + 1\,FA$
$GSU_2 \longrightarrow GSA_2 + 2\,FA$
$GU_3 \longrightarrow GA_3 + 3\,FA$

Method	Chemical Reaction	Property Described
Ozonolysis and reduction	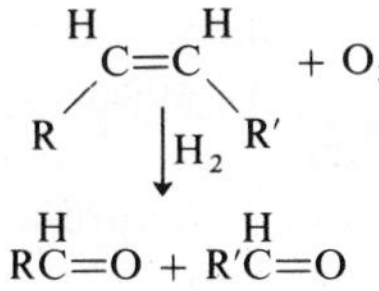 $$\underset{R}{\overset{H}{C}}\!=\!\underset{R'}{\overset{H}{C}} + O_3$$ $$\downarrow H_2$$ $$\overset{H}{RC}\!=\!O + \overset{H}{R'C}\!=\!O$$	Degree of unsaturation and distribution of fatty acids. Analysis similar to that shown above except that aldehydes are involved.
Spectroscopy (a) Ultraviolet	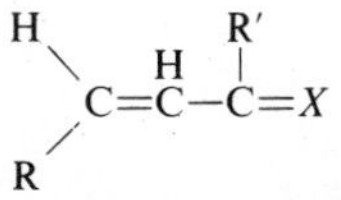 $$\underset{R}{\overset{H}{C}}\!=\!\overset{H}{\underset{}{C}}\!-\!\overset{R'}{\underset{}{C}}\!=\!X$$ X may be C or O; λ_{max} varies with structure and increases as the number of substituents increases.	Proportion of conjugated double bonds, before and after isomerization with alkali (see text)

TABLE 7.4   METHODS USED TO CHARACTERIZE AVERAGE PROPERTIES OF FATS (TRIGLYCERIDES) *(cont.)*

Method	Chemical Reaction	Property Described
(b) Infrared	$2.75–13.9\mu$ ($3,640–720$ cm^{-1})	Various functional groups such as

Additional detail for the infrared row:

		cm^{-1}
	—OH	$3,250–3,700$
	H H $\,$C=C	$3,010–3,040$ and $1,620–1,680$
	C=O	$1,735–1,750$ esters $1,700–1,725$ acids $1,705–1,725$ ketones
	—C≡C—	$2,190–2,260$
	—C—O—C—	$1,060–1,150$

Method	Chemical Reaction	Property Described
X-Ray diffraction	Sample must be crystalline	Chain lengths of fatty acids
Nmr	See Figure 7.1 and text	
Gas-liquid chromatography (GLC) See Figure 7.5		The fatty acid composition of a lipid.

$$
\begin{array}{c}
\underset{\text{O}}{RC}{-}OR' + KOH \\
\downarrow \\
\underset{\text{O}}{RC}{-}O^{\ominus} + K^{\oplus} + R'OH \\
\downarrow H^{\oplus} \\
\underset{\text{O}}{RC}\diagdown_{OH} \xrightarrow{CH_3N_2} \underset{\text{O}}{RC}\diagdown_{OCH_3} \\
\text{separation by GLC}
\end{array}
$$

Let us return to the question raised concerning the randomness, or otherwise, of natural triglycerides. The most successful approach to the solution of this problem involves the use of enzymes as highly specific and very gentle reagents which allow the exploration of structure in very elegant fashion. The way in which enzyme systems of great purity of function have been applied is shown in Figure 7.12.

7.4   WAXES

A wax is a plastic (pliable) material which is easily molded when warm, becomes hard when cold, and is insoluble in water. The composition of several natural waxes is presented in Table 7.5. The chemistry of these compounds will not be

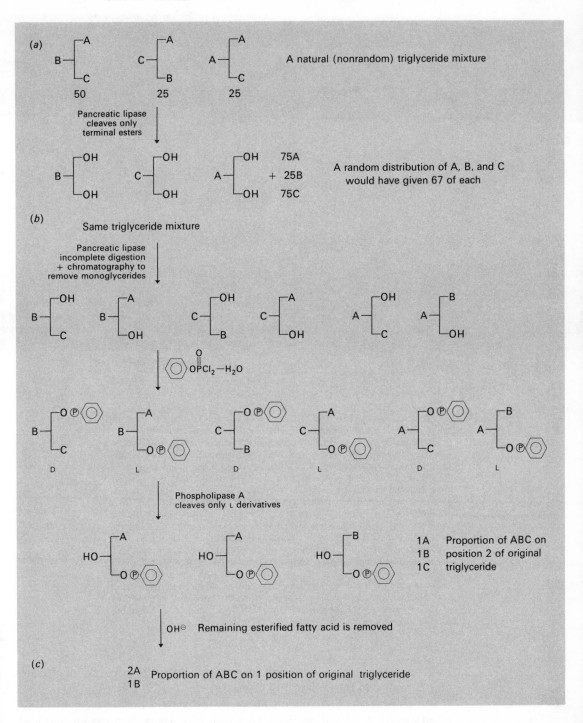

Figure 7.12 (continued on next page)

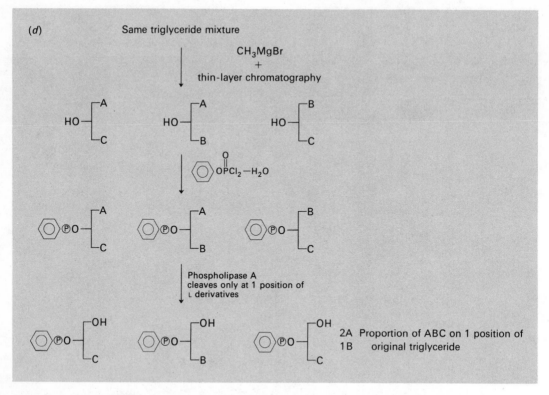

Figure 7.12   The determination of fatty acid distribution in a sample of mixed triglyceride using highly purified enzymes. (*a*) The sample contains only three fatty acids in equal amounts and they are present in the form of the three triglycerides shown. This experiment demonstrates that the triglycerides do not have the fatty acids randomly distributed. [H. Brockerhoff, *J. Lipid Res.*, **6**, 10 (1965).] (*b*) The same triglyceride mixture is treated with pancreatic lipase for a shorter time and diglycerides are isolated. After making the phenylphosphate derivatives, the enzyme phospholipase A is used to remove the fatty acids from position 2 of the L isomers. Since the pancreatic lipase acts randomly, the same amount of D isomer has the same fatty acid on position 2, and the distribution of fatty acids at this position is determined. [H. Brockerhoff, *J. Lipid Res.*, **8**, 167 (1967).] (*c*) The proportions of the fatty acids in position 1 are established by isolating the monoacyl compounds and treating them with alkali. (*d*) Treatment of the triglyceride mixture with methylmagnesium bromide gives the α,α′ diesters which can be phosphorylated and then treated with phospholipase A to demonstrate the identity of fatty acids at position 1 of the original triglyceride mixture. [M. Yurkowski, and H. Brockerhoff, *Biochim. Biophys. Acta*, **125**, 55 (1966).]

discussed. Their characterization has depended on the application of techniques similar to those shown in Table 7.4.

Some of the alcohol components of waxes contain two adjacent hydroxyl groups which when not esterified react with such agents as periodate and lead tetraacetate. In addition, they can form cyclic acetals and ketals. In other

TABLE 7.5   WAXES

A. Wax Esters

Alcohol Components		Number of Carbons	
Primary alcohols			
Normal	RCH$_2$OH	Saturated	8–36 even
		Monounsaturated	14–22 even
Iso-series	CH$_3$ $\backslash$ HC(CH$_2$)$_n$CH$_2$OH $/$ CH$_3$		18–26 even
Ante-iso-series	CH$_3$CH$_2$ $\backslash$ HC—(CH$_2$)$_n$—CH$_2$OH $/$ CH$_3$		17–27 odd only
Secondary normal	R $\backslash$ CHOH $/$ R′		18, 20, and 29
1,2 diols			
Normal	RCHOHCH$_2$OH		16
	or		
Iso	CH$_3$ $\backslash$ HC—(CH$_2$)$_n$—CHOH—CH$_2$OH $/$ CH$_3$		18–24 even
α,ω-diols	HOCH$_2$—(CH$_2$)$_n$—(CH$_2$OH		20–32 even
Sterols and triterpenes			27–30

Lanosterol and many other ring modifications

Acid Components	
Normal acids	18–30
Iso and ante-iso-acids	10–26
Hydroxy acids	14–20
Aromatic acids	10

Ferulic acid from Douglas fir

TABLE 7.5　WAXES *(cont.)*

B. Paraffins

	Alkyl Components	*Number of Carbons*
Normal	$CH_3-(CH_2)_{\overline{n}}-CH_3$	19–37 odd and even
Iso	$\begin{array}{c} CH_3 \\ \diagdown \\ HC-(CH_2)_n-CH_3 \\ \diagup \\ CH_3 \end{array}$	29–35 odd
Ante-iso	$\begin{array}{c} CH_3CH_2 \\ \diagdown \\ HC-(CH_2)_{\overline{n}}-CH_3 \\ \diagup \\ CH_3 \end{array}$	28–34 even

instances ethers of glycerol and long-chain alcohols have been identified. The compounds are optically active, clearly indicating that the linkage is to the α position of the glycerol. In addition, these compounds have all the characteristics of vicinal diols (1,2-glycols).

7.5　PHOSPHOLIPIDS

The diversity of structure, and presumably of function, of this class of lipids is very great. They are all phosphate esters and fall into two main groups, depending on whether they contain glycerol or sphingosine. These two classes can be further subdivided depending on the other constituents present. The main concerns of this section are (1) to present the structures of the classes of phospholipids, which is done in Table 7.6, (2) to discuss some of the methods of establishing structure, and (3) to describe their properties.

Table 7.6 is not an exhaustive listing; other compounds are known with other structural features; however, these serve to illustrate the complexity of this group of compounds.

The phospholipids are located in the cell usually in combination with proteins. The complex is presumably the active or functional unit and the structures described here are in fact only portions of such units. The complexes formed by phospholipids are due in part to the ability of these compounds to occupy interfaces between organic and aqueous environments. This affinity for interfaces depends on the presence of hydrophobic and hydrophilic portions of the molecule. The phospholipids are in fact natural soaps or detergents (Figure 7.13).

A word or two about nomenclature. Phosphatidyl (a prefix) is diacylglycerophosphoryl, derived from phosphatidic acid which is diacylglycerophosphatidic

TABLE 7.6   PHOSPHOLIPIDS

Glycerol Phosphate Derivatives

$$\begin{array}{l} H_2COR \\ R'OCH \qquad O \\ \quad\;\; H_2C-O-\overset{\displaystyle\|}{P}-OR'' \\ \qquad\qquad\;\; \underset{\displaystyle O^{\ominus}}{|} \end{array}$$

Lecithins	R is saturated fatty acid R′ is unsaturated fatty acid R″ is derived from choline; $HOCH_2CH_2\overset{\oplus}{N}(CH_3)_3$
Phosphatidyl ethanolamine (cephalins)	R and R′ as for lecithins R″ is derived from ethanolamine; $HOCH_2CH_2NH_2$
Phosphatidyl serine	R and R′ as for lecithins R″ is derived from serine; $HOCH_2CHNH_3^{\oplus}COO^{\ominus}$
Plasmalogens	R is $-\overset{\displaystyle H}{C}=\overset{\displaystyle H}{C}-R'''$ R′ is unsaturated fatty acid R″ is derived from ethanolamine
Glyceryl ethers	R is CH_2R R′ is unsaturated fatty acid
Phosphatidyl inositol	R and R′ are fatty acids R″ is derived from *myo*-inositol
Phosphatidic acid	R and R′ are fatty acids R″ is H or $\ominus$
Phosphatidyl glycerol	R and R′ are fatty acids R″ is derived from glycerol; $HOCH_2-CHOH-CH_2OH$
Diphosphatidyl glycerol (cardiolipin)	

point of attachment

$$\begin{array}{c} OH \\ \qquad OH \\ OH \quad HO \\ HO \\ \qquad OH \end{array}$$

$$\begin{array}{l} \qquad\qquad\qquad O \\ CH_2OR \qquad H_2C-O-\overset{\displaystyle\|}{P}-O-CH_2 \\ R'OCH \quad\; O \;\; HOCH \quad O^{\ominus} \quad HCOR' \\ \;\, H_2C-O-\overset{\displaystyle\|}{P}-O-CH_2 \qquad\quad H_2COR \\ \qquad\qquad \underset{\displaystyle O^{\ominus}}{|} \end{array}$$

R and R′ are fatty acids

Lysoglycerophosphatides	R is fatty acid R′ is H R″ is derived from ethanolamine or choline

TABLE 7.6 PHOSPHOLIPIDS *(cont.)*

Sphingosine Phosphate Derivatives

D configuration

OH

O‖P
O—P—OR″
O⊖

NH
R′

trans

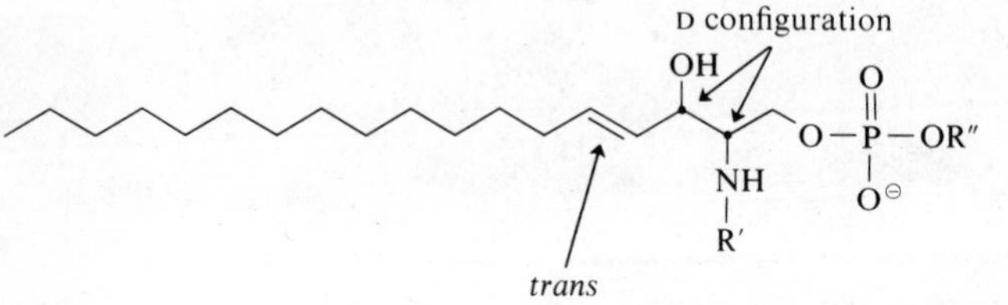

Sphingomyelins	R′ is fatty acid Commonly lignoceric C, 24: saturated cerebronic C, 24: 2-OH nervonic C, 24: Δ15

R″ is derived from choline; $HOCH_2CH_2\overset{\oplus}{N}(CH_3)_3$

Ceramides R′ as for sphingomyelin

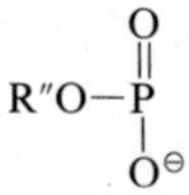

$$R''O-\overset{\overset{\displaystyle O}{\|}}{\underset{\underset{\displaystyle O^{\ominus}}{|}}{P}}$$ replaced by H

[*Note:* Compounds containing dihydrosphingosine have been isolated (double-bond saturated)]

Phytosphingosine Phosphate Derivatives

OH

O‖P
O—P—OR″
O⊖

HO   NH
R′

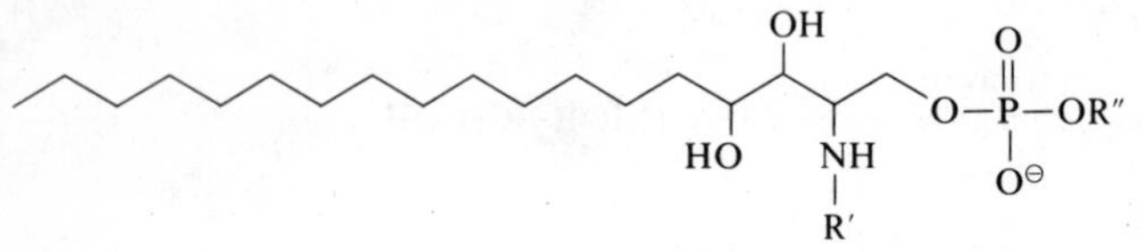

R′ is fatty acid
R″ is derived from inositol

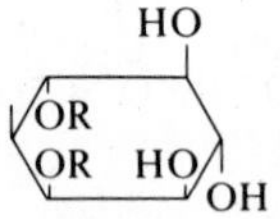

HO

OR
OR HO
OH

R is glucuronic acid
glucosamine
mannose
galactose
arabinose
fucose

acid. All the compounds in nature appear to be derivatives of L-glycerol 3-phosphate. Since the strict application of numbering preferences requires that substituents have the lowest numbers, some confusion can develop. For example, L-glycerol 3-phosphate is also D-glycerol 1-phosphate. However, the natural lipids are derived from dihydroxyacetone phosphate which is reduced to the "L-glycerol isomer;" whereas D-glyceraldehyde-3-phosphate,

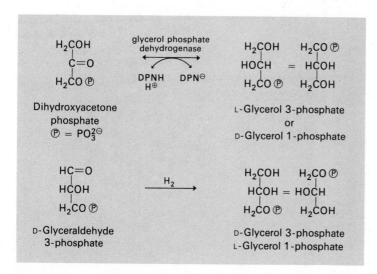

2-*O*-Oleoyl-1-*O*-stearoyl-3-*O*-phosphorylcholine glycerol

Figure 7.13   A phospholipid. (*a*) The hydrophobic part which will interact with lipids or
the hydrophobic regions of proteins. (*b*) The hydrophilic region which can
interact with water, carbohydrates, or the hydrophilic parts of proteins.

which is derivable from the standard reference compound D-glyceraldehyde
simply by phosphorylation at position 3, on reduction gives the "D isomer."
The confusion in this area seems to be resolvable by acceptance that the compounds are L when the phosphate is in position 3 (Figure 7.14).

The chemical stabilities of the glycerophospholipids are those of simple
oxygen esters and phosphate esters (Section 6.5, A and B). Both linkages can be
cleaved by acid or alkali and the components recovered. The position of the
phosphate group in the glycerol ester recovered from this treatment is open to
question because of phosphate migration (Section 6.5F, Figure 6.12). Some
chemical and biochemical approaches to the elucidation of the structure of a
lecithin are shown in Figure 7.15. These methods illustrate the kinds of
approaches which have been taken to the determination of structure. There is
no intention of presenting a complete description of the methods used. They are
constantly being modified as new and improved chemical and biochemical

The relationship between dihydroxyacetone phosphate D-glyceraldehyde 3-phosphate and the
glycerol phosphates.

(a) Lecithin $\xrightarrow{CH_3O^\ominus}$ $RC\begin{smallmatrix}O\\\\\\OCH_3\end{smallmatrix}$ + L-Glycerol 3-phosphorylcholine

2 Fatty acid
methylesters

$\downarrow$ $\begin{smallmatrix}H^\oplus \text{ or } OH^\ominus\\H_2O\end{smallmatrix}$

L-Glycerol 3-phosphate
+
Glycerol 2-phosphate
+
choline

(b)

$$
\begin{matrix}H_2COH\\HOCH\\H_2CO\,\textcircled{P}\end{matrix} + IO_4^\ominus \longrightarrow \begin{matrix}HC{=}O\\+\\HC{=}O\\H_2CO\,\textcircled{P}\end{matrix} + IO_3^\ominus
$$

L-Glycerol
3-phosphate

Glycolaldehyde
phosphate

$\textcircled{P_i}$ $\xleftarrow{H_2O}$

$$
\begin{matrix}H_2COH\\HCOH\\H_2COH\end{matrix} + 2IO_4^\ominus \longrightarrow \begin{matrix}2HC{=}O\\+\\1HCOOH\end{matrix} + 2IO_3^\ominus
$$

(c) Lecithin + NH_2OH $\xrightarrow{OH^\ominus}$ $RC\begin{smallmatrix}O\\\\\\NHOH\end{smallmatrix}$ + L-Glycerol 3-phosphorylcholine

2 Fatty acid
oximes

$\searrow$ LiAlH₄

RCH_2OH + L-Glycerol 3-phosphorylcholine

2 Fatty alcohols

(d) The biochemical examination of lecithin structure

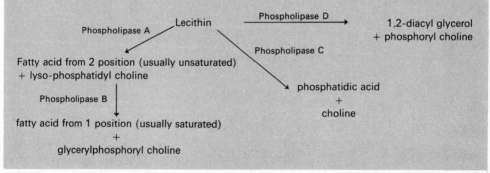

Phospholipase A

Lecithin $\xrightarrow{\text{Phospholipase D}}$ 1,2-diacyl glycerol
+ phosphoryl choline

Phospholipase C

Fatty acid from 2 position (usually unsaturated)
+ lyso-phosphatidyl choline

phosphatidic acid
+
choline

Phospholipase B $\downarrow$

fatty acid from 1 position (usually saturated)
+
glycerylphosphoryl choline

approaches are developed. Similar comments can be made about the synthesis of phospholipids.

Two routes to the synthesis of a specific lecithin, in which two different fatty acids occupy the 1 and 2 positions, are shown in Figure 7.16. The first route shows a complete *de novo* synthesis from mannitol; the positioning of the fatty acids relies on the fact that the primary hydroxyl group of iodoglycerol is more readily esterified than the secondary group. The second method starts from a natural, mixed lecithin with various fatty acids at position 1 and 2; these are removed by *trans*esterification and both hydroxyls are reacted with the fatty acid which is finally to occupy position 1. Use is then made of the highly specific enzyme phospholipase A to open the 2 position and re-esterify it with a second specific fatty acid.

The inositol containing glycerol phosphates presented an interesting problem of structure elucidation. There are two questions to be answered. First, what is the position of attachment to the inositol ring? Second, what is the position of attachment to the glycerol? The ways in which these questions were answered are shown in Figure 7.17. The phosphatidyl inositol was hydrolyzed in alkali and the inositol phosphates recovered from the reaction mixture. First, the inositol phosphates themselves are stable to alkali; therefore, they are the products of hydrolysis and not of a subsequent rearrangement. Myoinositol has a plane of symmetry through carbons 2 and 5 and is not optically active. If it is substituted at either 1, 3, 4, or 6, however, the derivative is optically active and the 1/3 and 4/6 pairs are enantiomeric (mirror images). Of the phosphate monoesters isolated, one was optically active (the major product) and the other was not (minor product). This type of product distribution is probable only if the linkage in the starting material involved positions 1(3) or 4(6) and if hydrolysis proceeded through a cyclic intermediate. If the original linkage had been to the 2 position, there would be equal likelihood of cyclization involving either the 1,2- or 2,3-hydroxyls and of opening to give 1- and 3-phosphates in equal amounts and the 2-phosphate, none of which are optically active. A similar result would be expected if the original linkage had involved the 5 position.

Figure 7.15

The chemical examination of lecithin structures. (*a*) The structures of the products are established by comparison with standard substances. The fatty acids are examined by gas-liquid chromatography. [G. Hubscher, J. N. Hawthorne, and P. Kemp, *J. Lipid Res.*, **1**, 433 (1960).] (*b*) The glycerol phosphate can be examined by chromatography, by periodate oxidation before (and after) removal of the phosphate, or by oxidation to dihydroxy acetone phosphate by the enzyme L-α-glycerol phosphate dehydrogenase. The structure of choline can be established after isolation by chromatography by comparison with authentic material as a phosphomolybdate derivative. (D. J. Hanahan and H. Brockerhoff, *Comprehensive Biochemistry* (M. Florkin and E. H. Slotz, ed.), **6**, Elsevier Publishing Co., New York, p. 83, 1965. (*c*) Alternate methods of examining the fatty acids in lecithins which allow for isolation of the glycerol phosphate. (*d*) The determination of lecithin structure using enzymes. The phospholipases are specific for the reactions shown. The products can be examined by chromatographic procedures.

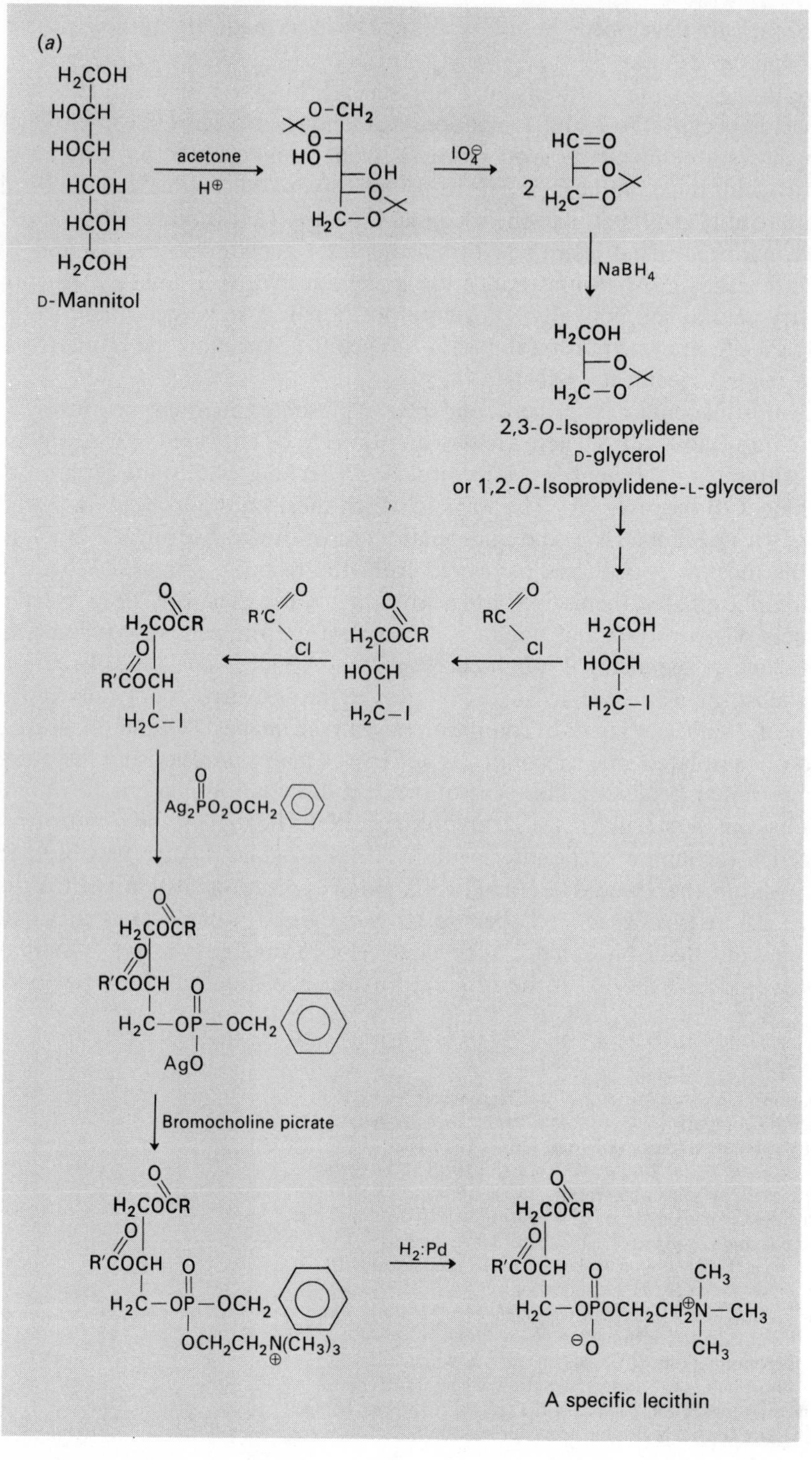

(a)

D-Mannitol

acetone
H⊕

IO₄⊖

2,3-O-Isopropylidene
D-glycerol

or 1,2-O-Isopropylidene-L-glycerol

NaBH₄

Ag₂PO₂OCH₂⟨⟩

Bromocholine picrate

H₂:Pd

A specific lecithin

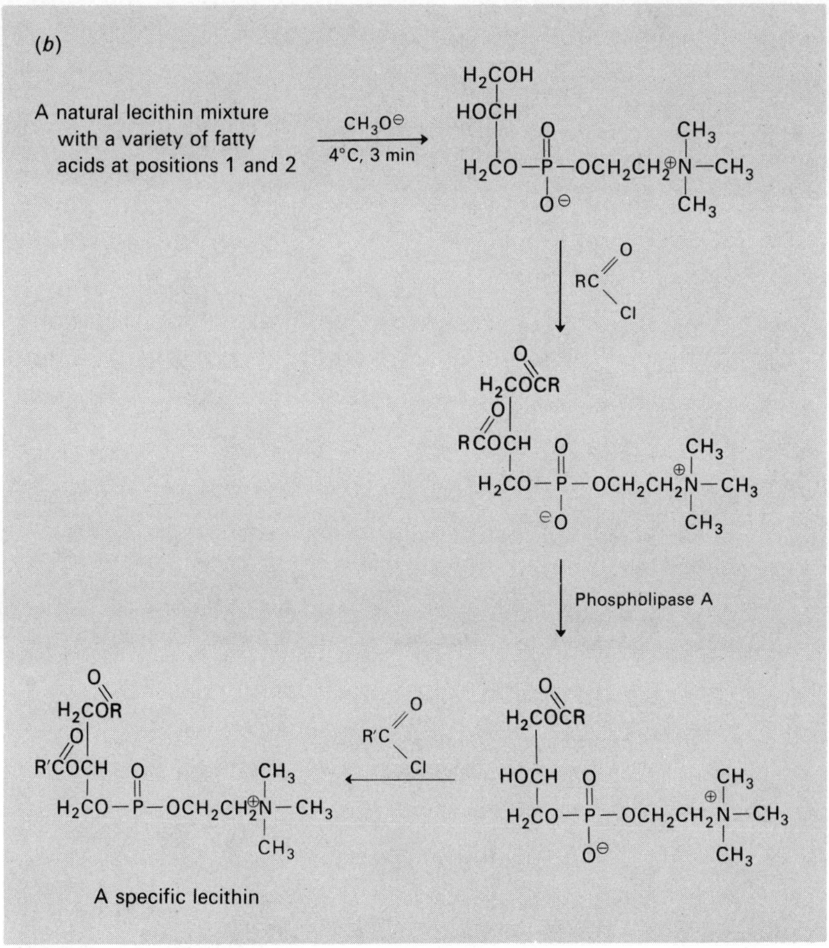

Figure 7.16   The synthesis of a lecithin. (a) Total chemical synthesis. [G. H. de Haas and L. L. M. van Deenen, *Rec. Trav. Chim.*, **82**, 1,163 (1963).] (b) Starting from a natural lecithin. [D. J. Hanahan and H. Brockerhoff, *Arch. Biochem. Biophysics*, **91**, 326 (1960).]

The presence of optical activity in the major product clearly rules out the possibility of the original linkage being either the 2 or 5 positions. However, the optically inactive product was shown to be the 2-phosphate by comparison with the authentic material, indicating that the cyclic intermediate had involved either the 1,2- or 2,3-hydroxyls and that the product was not the 4(6) derivative. The demonstration that the linkage involved the 1-hydroxyl and not the 3 depended on the synthesis of the 3 compound and the demonstration that it was enantiomeric to the isolated material. The position of attachment to the glycerol residue was shown by hydrolyzing the fatty acids with dilute alkali and treating the product with 1 molar equivalent of periodate. The product

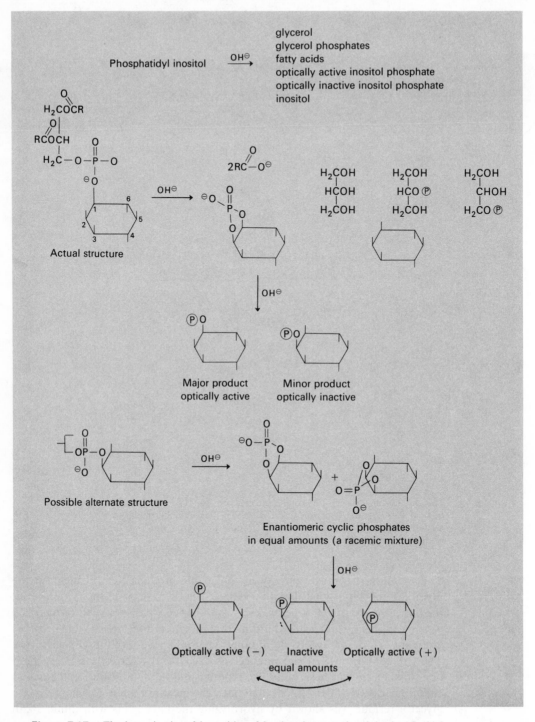

Figure 7.17   The determination of the position of the phosphate ester bonds in phosphatidyl inositol. [C. E. Ballou and L. I. Pizer, *J. Am. Chem. Soc.*, **82**, 3,333 (1960).]

was shown to have the intact inositol ring, indicating that the glycerol had been oxidized and that the usual 3-phosphate linkage was present. The final products also had optical activity, indicating that the position of attachment to the inositol was to the 1(3) or 4(6) positions.

The three cases discussed in this section serve to illustrate the methods commonly used to establish the structures of phospholipids. Each different chemical type requires a different approach but, as with many other natural products, the enzymes which degrade them are proving to be the most powerful reagents for their study.

7.6   GLYCOLIPIDS

This group of compounds is similar in some respects to the phospholipids. They have hydrophobic and hydrophilic parts, but in this case the latter are carbohydrate moieties. They can also be classified in a similar fashion to the phospholipids on the basis of their content of glycerol and sphingosine (Table 7.7). An additional subgroup, however, the complex glycosides of steroids, should perhaps be included. That they are not is due to the fact that they represent a very diverse group of compounds which have been studied primarily from the point of view of their steroid content. An example is listed in Table 7.10.

TABLE 7.7   GLYCOLIPIDS

I Glycerol Derivatives

$$
\begin{array}{l}
\text{H}_2\text{COR}\\
\text{R'OCH}\\
\text{H}_2\text{COR''}
\end{array}
$$

A. Galactosyl Diglycerides

R and R' are fatty acyls
R" is β-D-galactopyranosyl
   or 6(α-D-galactopyranosyl)-β-D-galactopyranosyl
   (shown)

B. Sulfolipid

R and R' are fatty acyls
R" is 6-deoxy-6-sulfonic acid-α-D-glucopyranosyl

TABLE 7.7 GLYCOLIPIDS *(cont.)*

II Sphingosine Derivatives

A. Cerebrosides

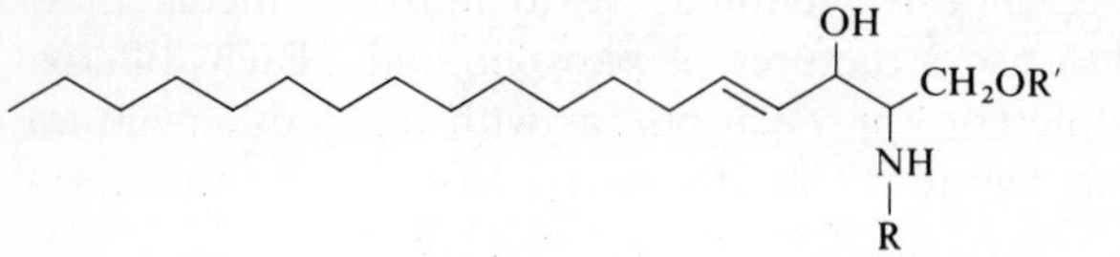

OH

CH$_2$OR'

NH

R

R is fatty acyl—lignoceric cerebronic . . . , etc.
R' is carbohydrate, usually galactose, occasionally glucose

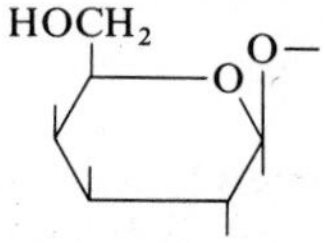

HOCH$_2$

O—

β-D-linkage

B. Cerebroside Sulfates—similar to cerebrosides but galactose has 3-*O*-sulfate group.

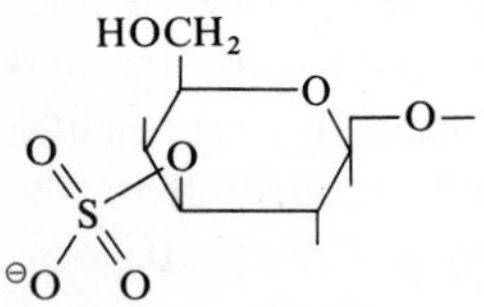

HOCH$_2$

O—

C. Gangliosides

e.g. Monosialoganglioside

HO

CH$_2$—O

NH

C

O

HOCH$_2$

HOCH$_2$

HOCH$_2$

HOCH$_2$

HOCH$_2$

HOCH$_2$

HOCH$_2$

—CH

HOCH

CH$_3$C—N

O

COO$^{\ominus}$

R is stearic acid
R' is complex carbohydrate containing neuraminic acid

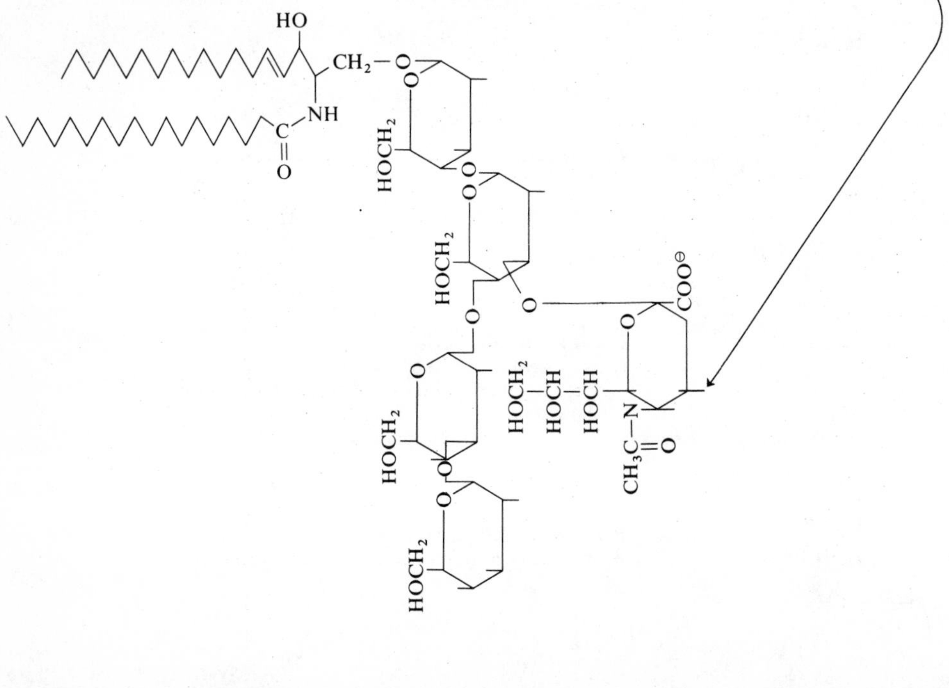

The complexity of the glycolipids is due mainly to their carbohydrate content. The general principles of structure elucidation for carbohydrates have been described in Section 5.17 and there is no need to repeat them here.

7.7 LIPIDS DERIVED FROM Δ³-ISOPENTENYL PYROPHOSPHATE

A very large number of lipids have been shown to be derived from the polymerization of Δ³-*iso*pentenyl pyrophosphate which, in turn, is formed from acetate. It seems useful to distinguish these from those lipids which are formed from acetate without the intermediary of the 5-carbon fragment. The biosynthetic sequences resulting in this relationship between diverse lipids is shown in Figure 7.18. An enormous chemical literature has developed pertaining to each of the end products shown in Figure 7.18. For the most part, little can be said here concerning the chemistry of these systems. A few structures are shown to illustrate the diversity which has been uncovered and in some instances a little coverage of some aspects of the chemistry is given. The reader can apply the rule of thumb that any compound behaves as the sum of all its parts. Thus the chemical and physical properties of the compounds shown will depend on the numbers and kinds of all the groups present and can in part be inferred from the structures shown.

A. Terpenes

The name *terpene* was originally applied to the ethereal oils which could be obtained from turpentine (an extract from pine). The ethereal oils are those materials which are steam distillable. Most of those first characterized had the formula $C_{10}H_{15}$ and it was soon recognized that along with a large number of other natural substances they could be considered as deriving from the condensation of the 5-carbon isoprene units (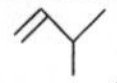). This formalism was useful in establishing relationships, and the discovery that the biosynthesis of the compounds in fact involved a compound related to isoprene (Δ³-isopropenylpyrophosphate) vindicated this approach to establishing relationships between compounds. The various classes of terpenes are designated mono- (10 carbons), sesqui- (15 carbons), di- (20 carbons), tri- (30 carbons, tetra- (40 carbons, and penta- (50 carbons). The polyisoprenoid compounds rubber and gutta-percha can be considered as the limiting members of the class. In Table 7.8 a few structures of terpenes are given that are selected almost at random to illustrate the types of compounds that have been observed.

Two classes of terpenes require more comment. They are the triterpenes, specifically those which give rise to the steroids, and the tetraterpenes, which are the cartenoids.

The biosynthesis of isoprenoid substances

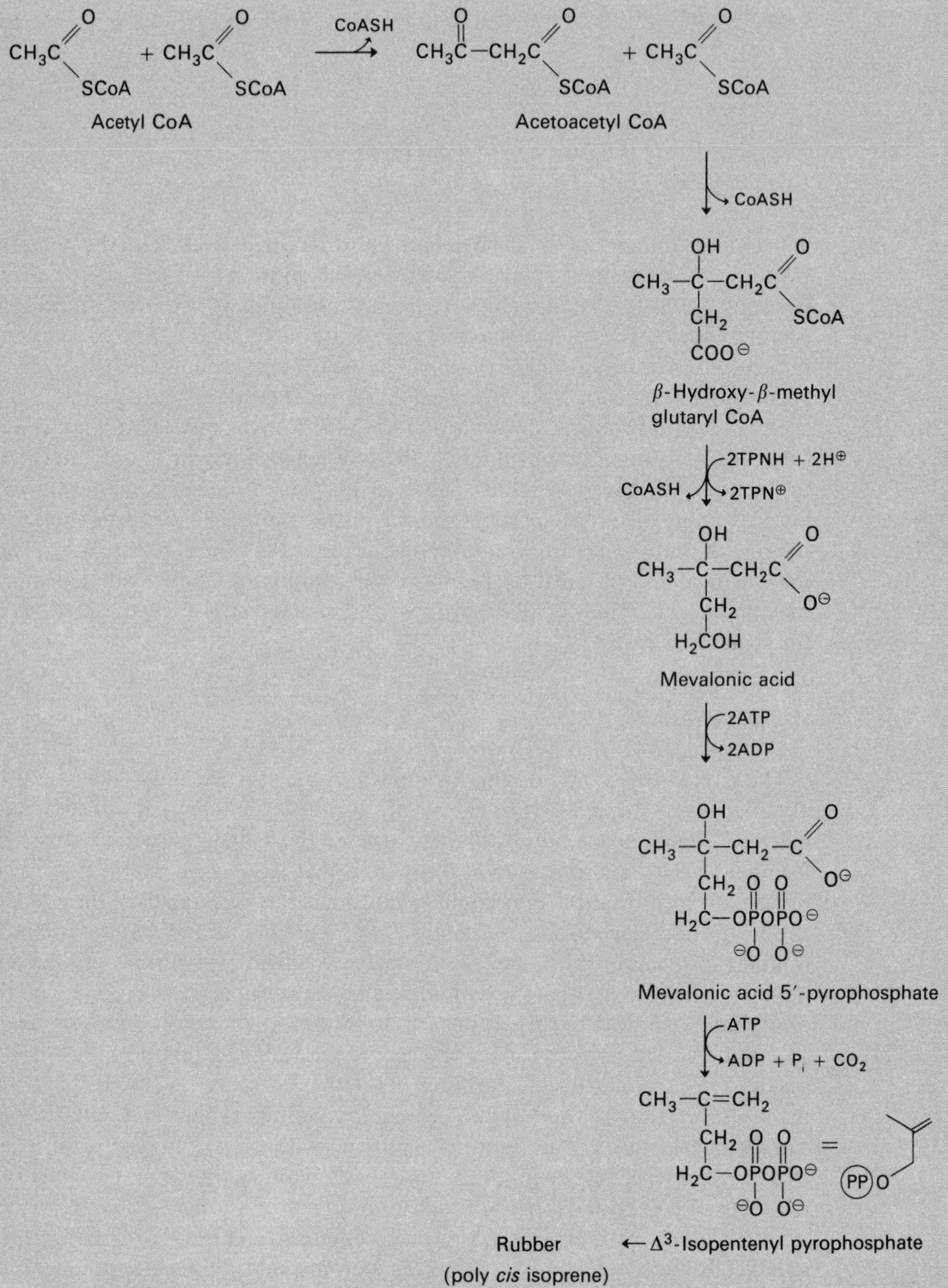

$$CH_3C\overset{O}{\underset{SCoA}{}} + CH_3C\overset{O}{\underset{SCoA}{}} \xrightarrow{\text{CoASH}} CH_3C\overset{O}{}-CH_2C\overset{O}{\underset{SCoA}{}} + CH_3C\overset{O}{\underset{SCoA}{}}$$

Acetyl CoA   Acetoacetyl CoA

CoASH

$$CH_3-\overset{OH}{\underset{\underset{COO^{\ominus}}{CH_2}}{C}}-CH_2C\overset{O}{\underset{SCoA}{}}$$

β-Hydroxy-β-methyl glutaryl CoA

CoASH $\quad$ 2TPNH + 2H$^\oplus$ / 2TPN$^\oplus$

$$CH_3-\overset{OH}{\underset{\underset{H_2COH}{CH_2}}{C}}-CH_2C\overset{O}{\underset{O^{\ominus}}{}}$$

Mevalonic acid

2ATP / 2ADP

$$CH_3-\overset{OH}{\underset{\underset{H_2C-OPOPO^{\ominus}}{\overset{CH_2}{}}}{C}}-CH_2-C\overset{O}{\underset{O^{\ominus}}{}}$$

Mevalonic acid 5′-pyrophosphate

ATP / ADP + P$_i$ + CO$_2$

$$CH_3-\overset{}{\underset{\underset{H_2C-OPOPO^{\ominus}}{CH_2}}{C}}=CH_2 \qquad = \text{(PP)O}$$

Rubber $\quad \leftarrow \Delta^3$-Isopentenyl pyrophosphate
(poly cis isoprene)

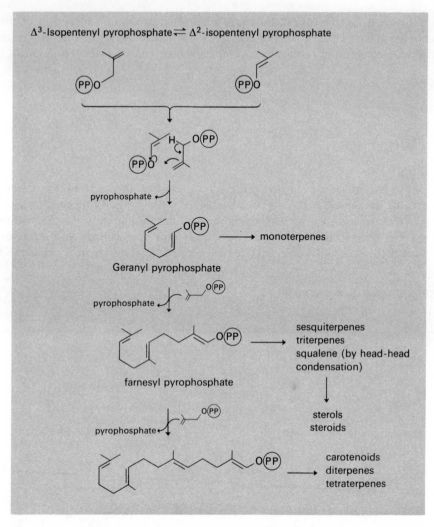

Δ³-Isopentenyl pyrophosphate ⇌ Δ²-isopentenyl pyrophosphate

pyrophosphate

Geranyl pyrophosphate → monoterpenes

pyrophosphate

farnesyl pyrophosphate → sesquiterpenes
triterpenes
squalene (by head-head condensation)

pyrophosphate → sterols
steroids

carotenoids
diterpenes
tetraterpenes

Figure 7.18 The biosynthesis of lipids derived from Δ³-isopentenyl pyrophosphate. Each step is catalyzed by a specific enzyme (or set of enzymes). The cofactors ATP, CoASH, TPNH, and so forth are described elsewhere in the text.

B. Steroids

Steroid means sterol-like. *Sterols* (Greek, stereos is solid, plus "ol," the generic ending for alcohols) are solid alcohols having from 27 to 29 carbons. The larger designation *steroid* covers those compounds containing the "parent nucleus" consisting of the *A, B, C,* and *D* rings shown in Figure 7.23.

The triterpene squalene is the precursor of the steroids. It displays the common feature of all the terpenes—willingness to cyclize. Cyclization of olefins is a specific case of the more general electrophilic addition to double bonds. It has been well-established that these additions require that the groups added be *trans* and in the same plane as shown below. If this requirement is

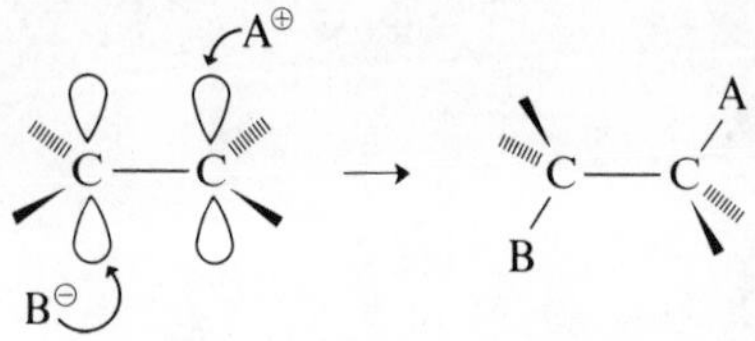

met in the intramolecular cyclization of polyolefins, then a complex molecule such as squalene can give rise to a limited number of fused ring systems which have fixed stereochemistry reflecting the conformation of the open chain which was involved in the rearrangement. An attempt has been made to represent one of the cases in Table 7.8. In this case the carbon chain of squalene is

TABLE 7.8   TERPENES AND TERPENOIDS

A. Monoterpenes

I. Acyclic

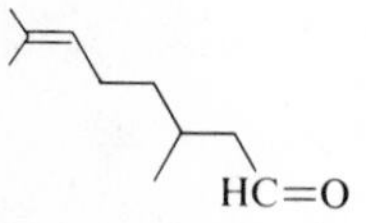

HC=O	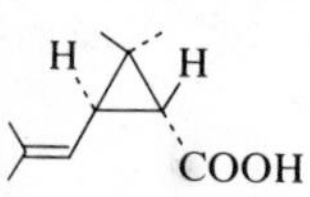 H⟍ ⟋H ⟍COOH	O ‖ O−CCH₂CH₂CH₃
Citronellal (eucalyptus)	Chrysanthemum carboxylic acid involves 1,3-condensation of (isopentenyl residues)	2,2-Dimethyl-3-isopropylidene cyclopropyl propionate (sex attractant from female cockroaches)

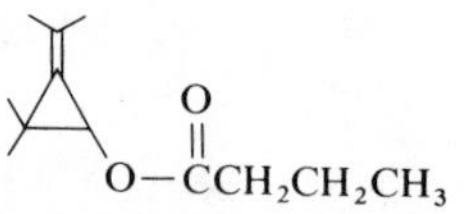

II. Cyclic

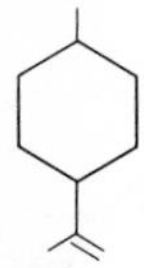

		HOOC
Limonene (lemon)	Cineole (eucalyptus)	Thujic acid (fir)
α-Pinene (pine)	Camphor	Nepatolactone (catnip)

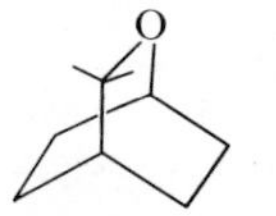

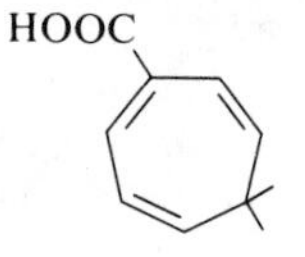

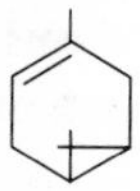

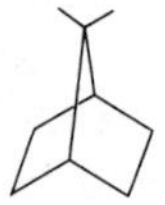

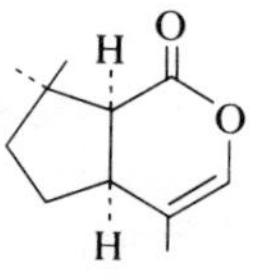

TABLE 7.8   TERPENES AND TERPENOIDS (cont.)

B. Sesquiterpenes   II. Cyclic

Ipomeamarone
(fungus)

Bisabolene

Longifolene
(pine)

C. Diterpenes

HOOC

Abietic acid

OH

Geranyllinalool
(jasmine oil)

OH

Vitamin A
(carrots)

D. Triterpenes

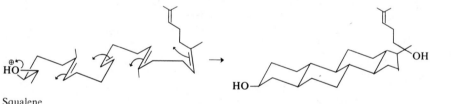

Squalene

folded so as to yield dammarenediol

Dammarenediol

E. Tetraterpenes—see Figure 7.24

F. Higher terpenes

CH_3O

CH_3O

CH_3

Ubiquinones
coenzyme Q, $n = 10$ (mitochondria)

arranged *chair-chair-chair-boat-unfolded* prior to ring closure and the product has a *chair-chair-chair-pentano-unfolded* arrangement—the boat in the precursor giving rise to a five-membered ring. A second example is shown in Figure 7.19 and a number of other possibilities in Table 7.9. In the case of lanosterol formation (Figure 7.19) the folding is such that the product has the *chair-boat-*

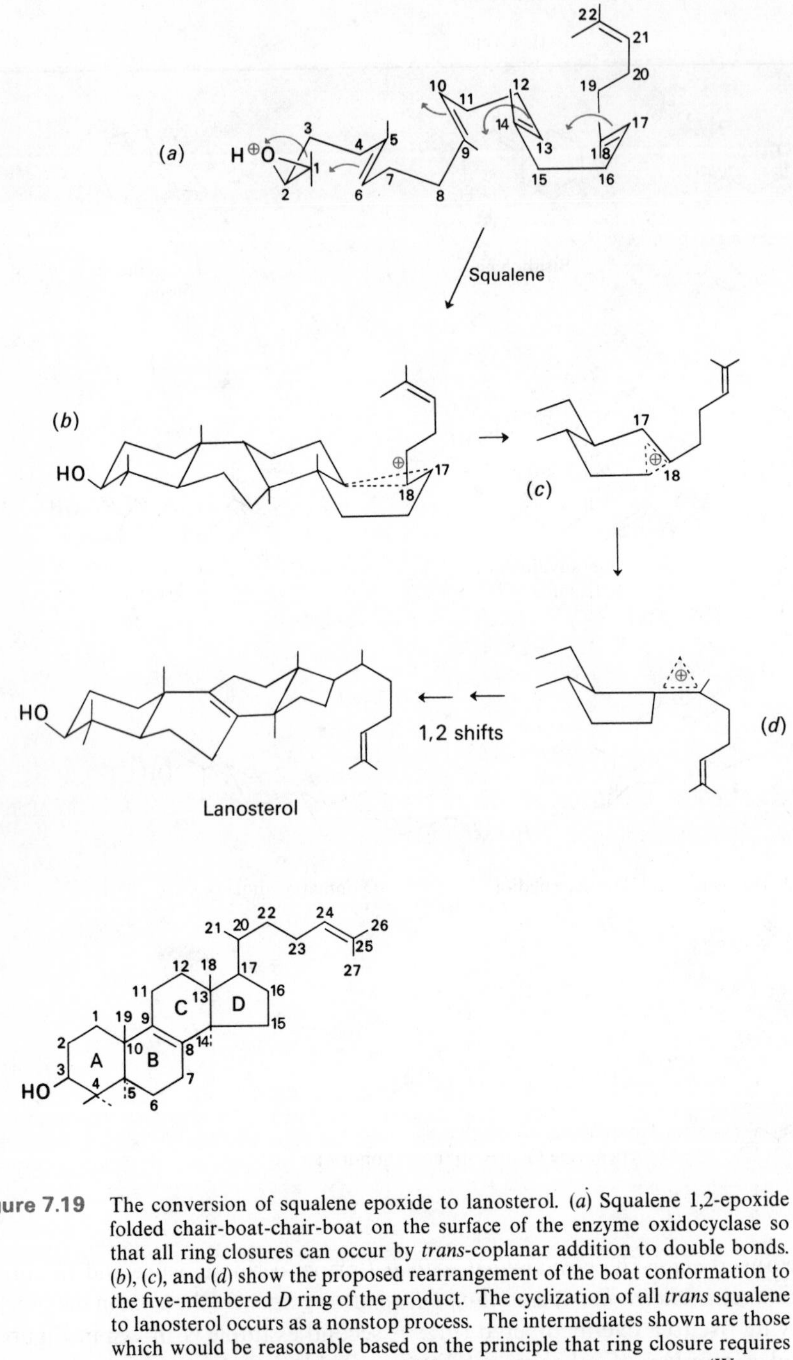

Figure 7.19   The conversion of squalene epoxide to lanosterol. (*a*) Squalene 1,2-epoxide folded chair-boat-chair-boat on the surface of the enzyme oxidocyclase so that all ring closures can occur by *trans*-coplanar addition to double bonds. (*b*), (*c*), and (*d*) show the proposed rearrangement of the boat conformation to the five-membered *D* ring of the product. The cyclization of all *trans* squalene to lanosterol occurs as a nonstop process. The intermediates shown are those which would be reasonable based on the principle that ring closure requires the *trans*-coplanar arrangement and that subsequent rearrangements (Wagner-Meerwein type) require similar conformations.

TABLE 7.9   STRUCTURES THAT CAN BE DERIVED FROM SQUALENE BY CONCERTED RING-CLOSURE REACTIONS

Folding of Squalene

Chair-chair-unfolded-unfolded-chair

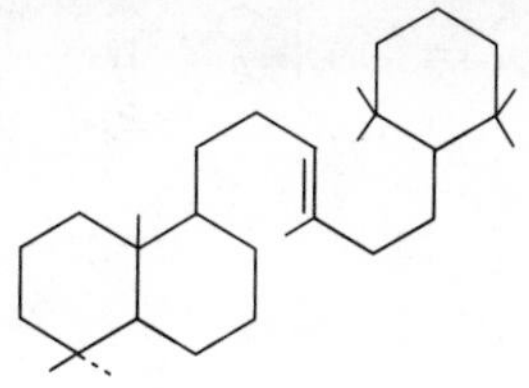

Ambrein

Chair-chair-unfolded-chair-chair

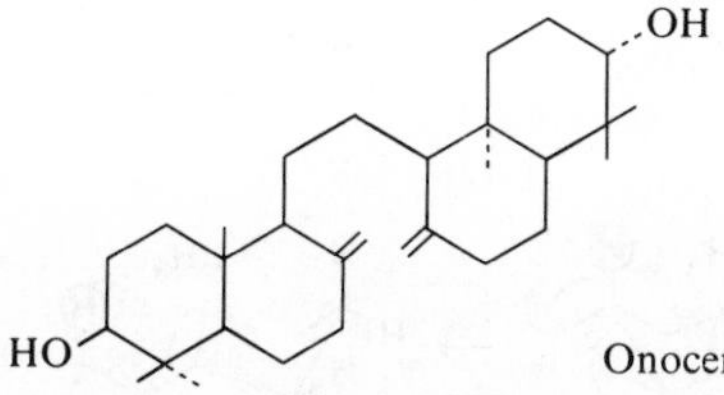

Onocerin

Chair-chair-chair-boat-boat

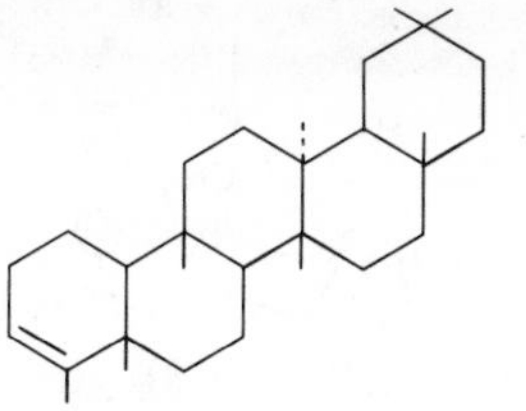

Friedelin

Chair-chair-chair-chair-chair

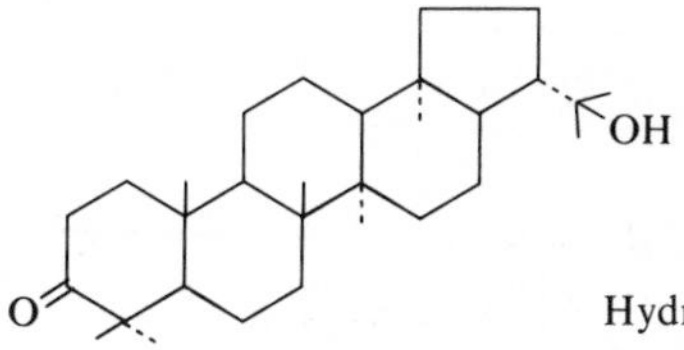

Hydroxyphenone

chair-pentano-unfolded arrangement. Lanosterol is further converted to cholesterol, whose structure is shown in Figure 7.20. This process requires the loss of the three methyl groups, hydrogenation of the side-chain double bond and migration of the Δ^8 double bond across the *B* ring to the 5 position.

The migration of groups across the steroid nucleus can be achieved under relatively mild conditions and also appears to take place during the enzymatic conversion of squalene to lanosterol. The rearrangements are 1,2 shifts (Wagner-Meerwin) of H_2C: or H: which require a *trans*-coplanar arrangement of the groups involved similar to that for ring closure. An illustration of this rearrangement occurring in a cyclohexane ring is shown in Figure 7.21. The process

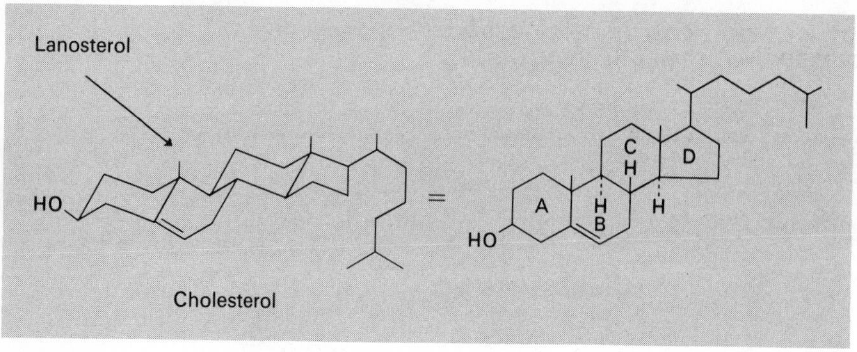

Figure 7.20   The structure of cholesterol.

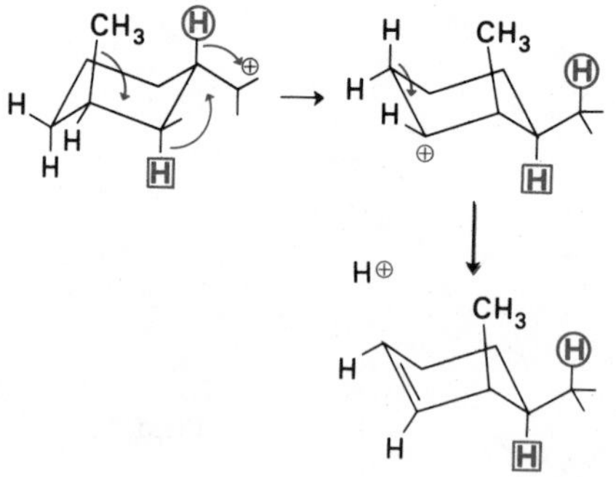

Figure 7.21

The spatial arrangement necessary for migration of methyl or hydrogen in cyclic systems.

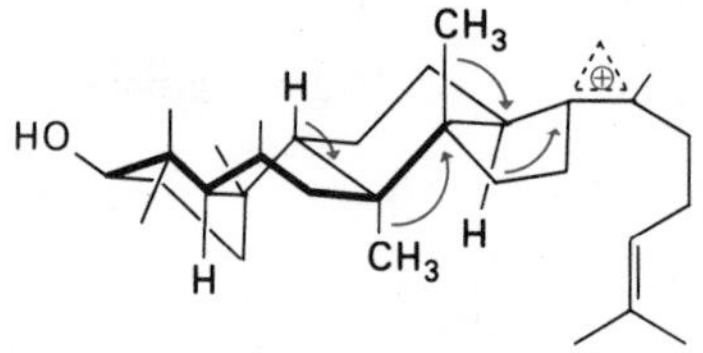

Figure 7.22

1,2 Shifts occurring in the formation of lanosterol.

thought to be involved in the conversion of squalene to lanosterol is shown in Figure 7.22.

There is an unfortunate difference in the numbering used for squalene and for the steroids derived from it (Figure 7.19). The groups attached to steroids above the plane of the ring are indicated by a solid line and are designated as β-oriented. Those which project below the ring are represented by dashed or dotted lines and are α-oriented.

There are many biologically interesting steroids and sterols. A selection of them is presented in Table 7.10.

TABLE 7.10   SOME BIOLOGICALLY IMPORTANT STEROIDS

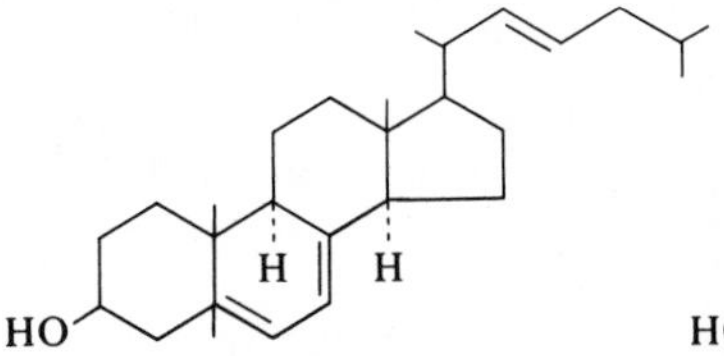

Ergosterol
(yeast)

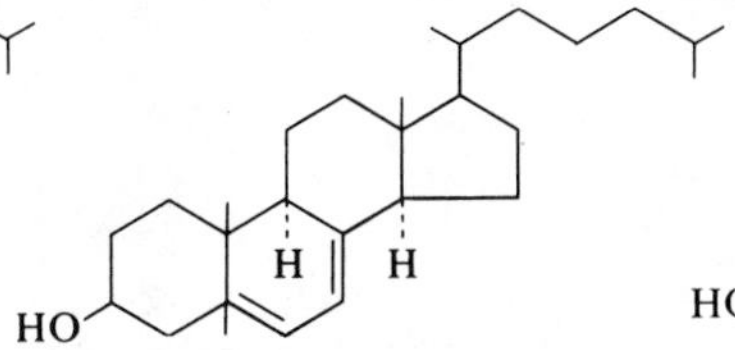

7-Dehydrocholesterol = provitamin D
(skin)

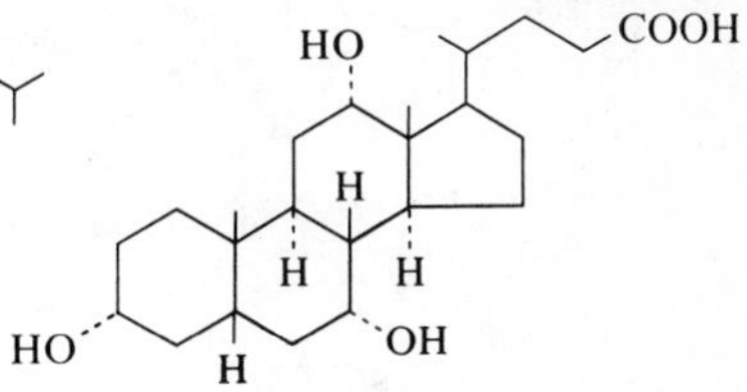

Cholic acid
(bile)

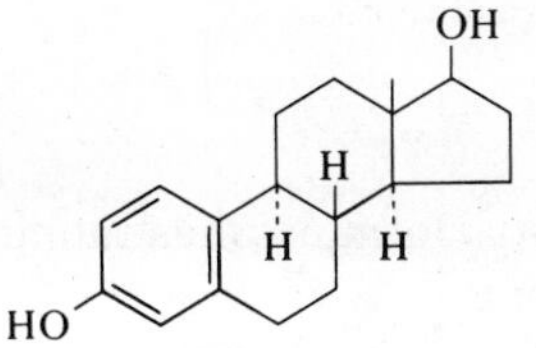

Estradiol
(women)

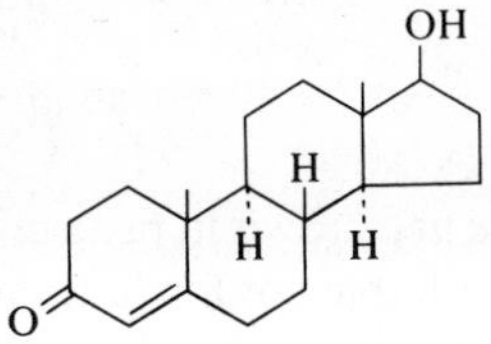

Testosterone
(men)

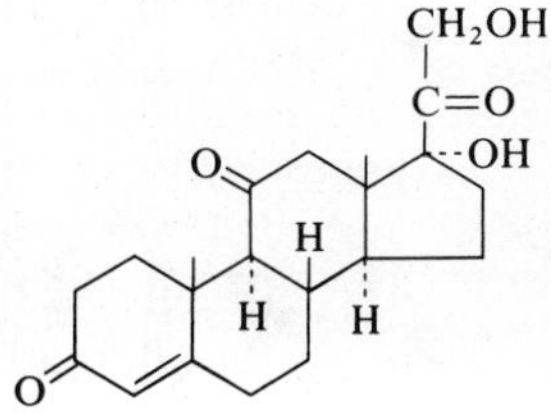

Cortisone
(mammals)

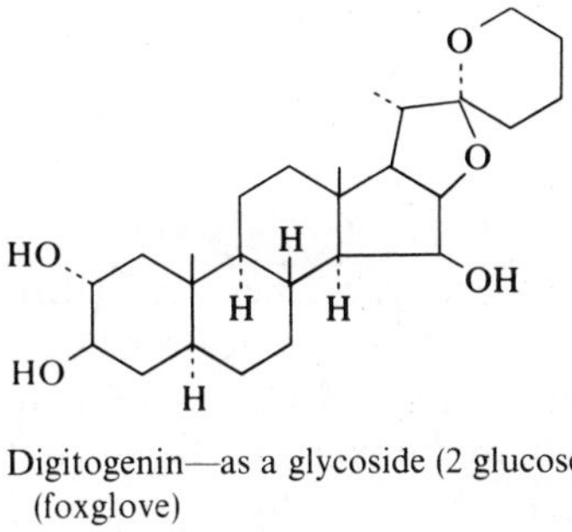

Digitogenin—as a glycoside (2 glucose,2 galactose,1 xylose)
(foxglove)

It should be noted that it is necessary to designate the position (α or β) of groups at positions 5, 8, 9, 10, 13, and 14 of the saturated A, B, C, D fused ring systems since these designations determine the kinds of ring fusion (*cis* or *trans*) and thus the overall geometry of the molecule. There are two main classes of isomers. The cholestane group related to the common saturated sterols has 5-α (A and B rings fused *trans*)—the coprostanol group of saturated sterols differs only in that the A/B ring fusion is *cis* due to a β arrangement at position 5

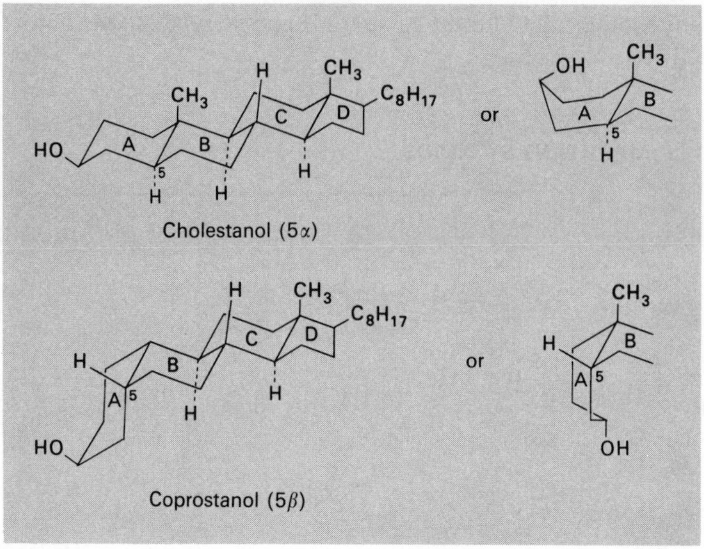

Figure 7.23

The differing geometries of cholestanol and coprostanol.

(Figure 7.23). The two are shown in projection with partial formulas illustrating the alternative, less probable conformations of the *A* ring.

The chemistry of the steroids has been the subject of so many investigations and has played such a significant role in the development of organic chemistry that it is most reasonable at this point to refer the reader to authoritative monographs on that subject.

C. Carotenes and Carotenoids

Many of the yellow, red, or purple colors observed in biological systems are due to the presence of carotenes and carotenoids. The carotenes are hydrocarbons and the carotenoids are hydroxylated compounds derivable from them. These groups of compounds can be considered as derivatives of lycopene (Figure 7.24). Naturally occurring compounds (about 150 are known) can be derived from lycopene by one or more of the following processes: hydrogenation, dehydrogenation, cyclization, methyl migration, aromatization, oxygenation, and cleavage. There is little opportunity here to deal with much of the chemistry of these substances although much has been done. To illustrate the kinds of reactions utilized, the total chemical synthesis of β-carotene is shown in Figure 7.25 and the biosynthetic route in Figure 7.18. The processes involved in carotene biosynthesis serve as a beautiful illustration of the economy of biosynthetic systems. It involves repeated addition of the same precursor Δ^3-isopentyl pyrophosphate (itself derived from acetyl CoA) to form a series of polymers of increasing size which in turn are cyclized, condensed, and oxidized to give a wide variety of products.

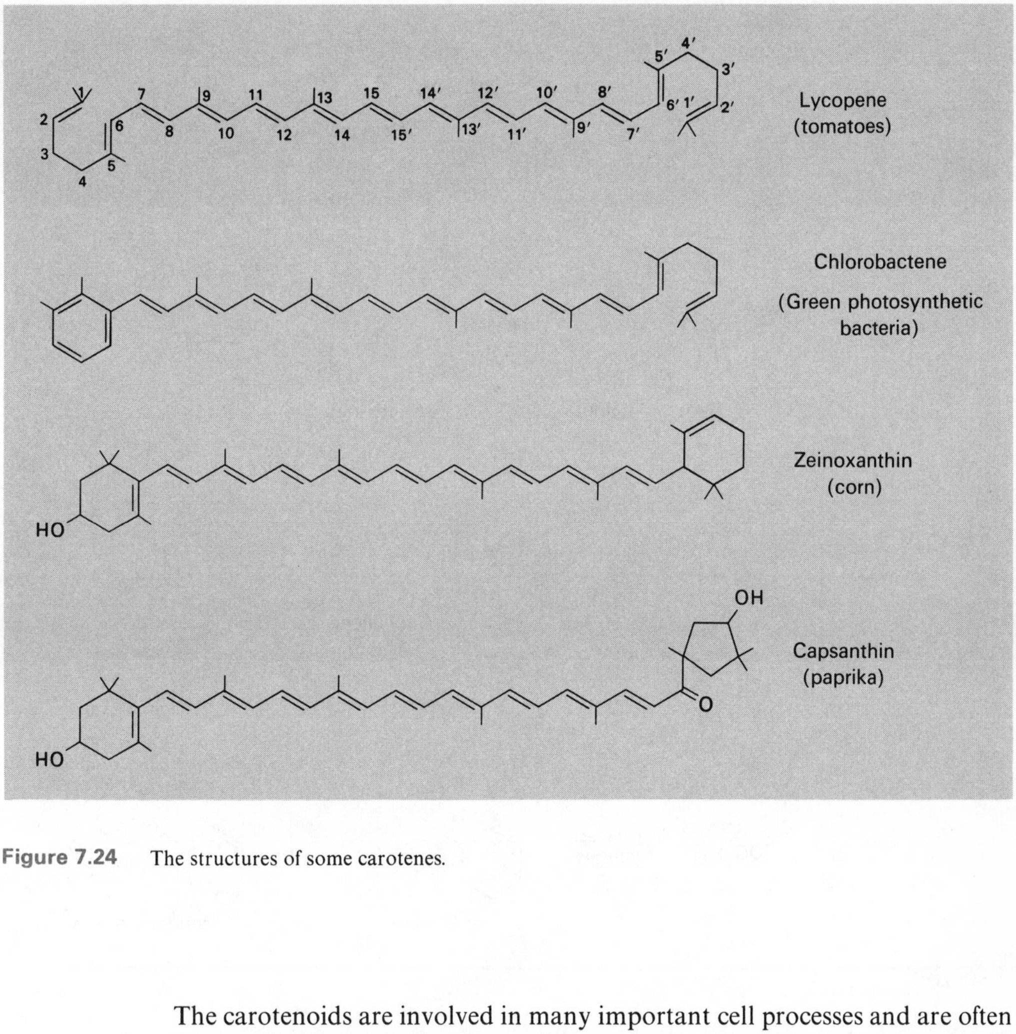

Figure 7.24   The structures of some carotenes.

The carotenoids are involved in many important cell processes and are often concerned with the interactions of the system with light. For example, all photosynthetic organisms synthesize carotenoids and they are incorporated into the chloroplasts along with the chlorophylls. They can absorb light and transmit the energy to the photosynthetic apparatus for the production of chemical energy. They also play a role in phototaxis in plants (movement toward or away from light). In photosynthetic bacteria, they may protect against destruction by light. In animals, β-carotene is converted to vitamin A which in turn is necessary for visual activity.

The structures of a few carotenoids are given in Figure 7.24. Almost all carotenoids have 40 carbons and an all *trans* arrangement of double bonds. Compounds with *cis* double bonds have been described and the *cis-trans* interconversion at the 11 position of retinene is involved in the response of the eye to light.

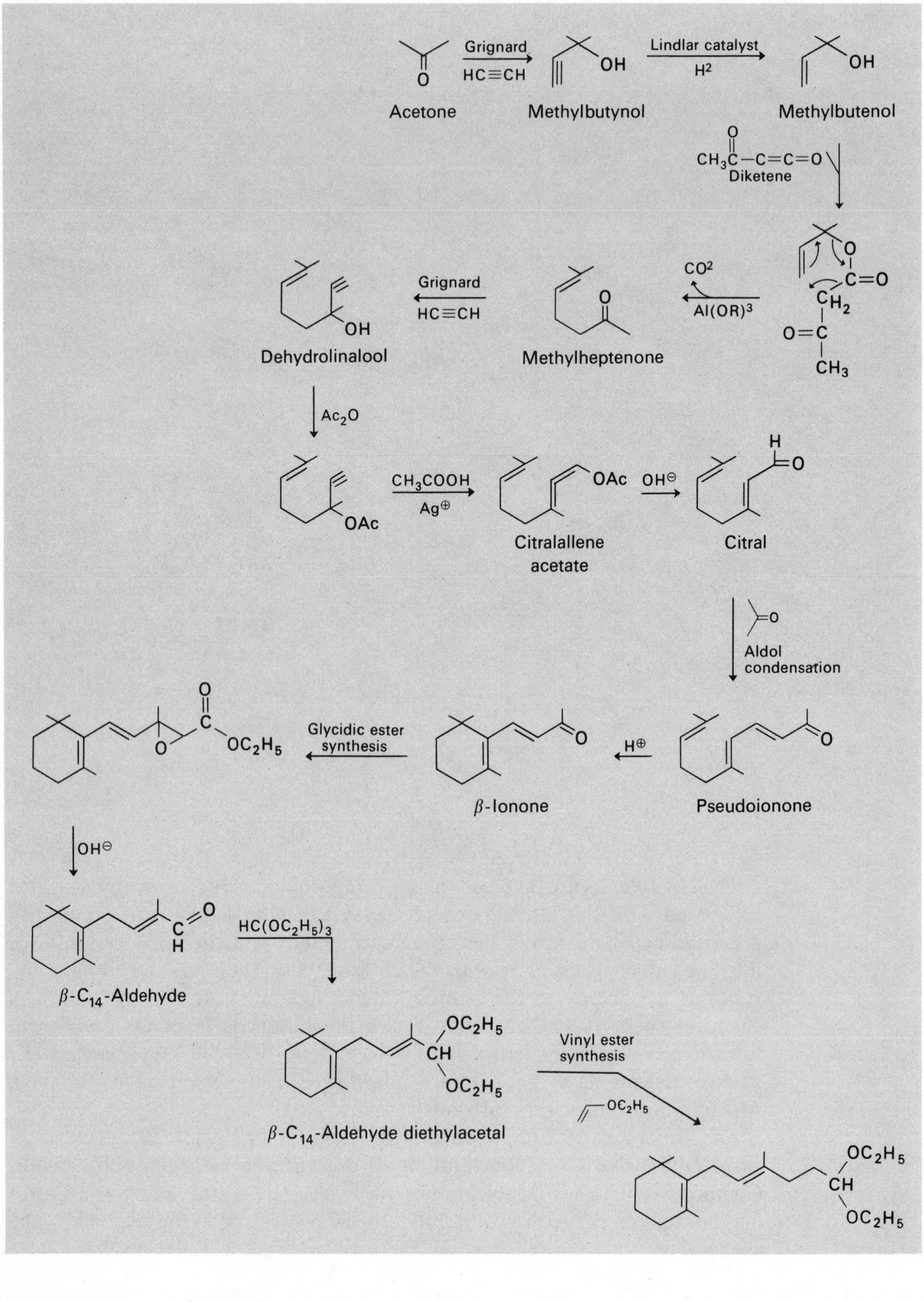

Acetone   Methylbutynol   Methylbutenol

Grignard / HC≡CH

Lindlar catalyst / H2

$CH_3C-C=C=O$
Diketene

CH_2
$O=C$
CH_3

CO^2
$Al(OR)^3$

Dehydrolinalool   Methylheptenone

Grignard / HC≡CH

Ac_2O

CH_3COOH / $Ag^\oplus$

$OH^\ominus$

Citralallene acetate   Citral

Aldol condensation

Glycidic ester synthesis

β-Ionone   Pseudoionone

$H^\oplus$

$OH^\ominus$

β-C_{14}-Aldehyde

$HC(OC_2H_5)_3$

β-C_{14}-Aldehyde diethylacetal

Vinyl ester synthesis

$-OC_2H_5$

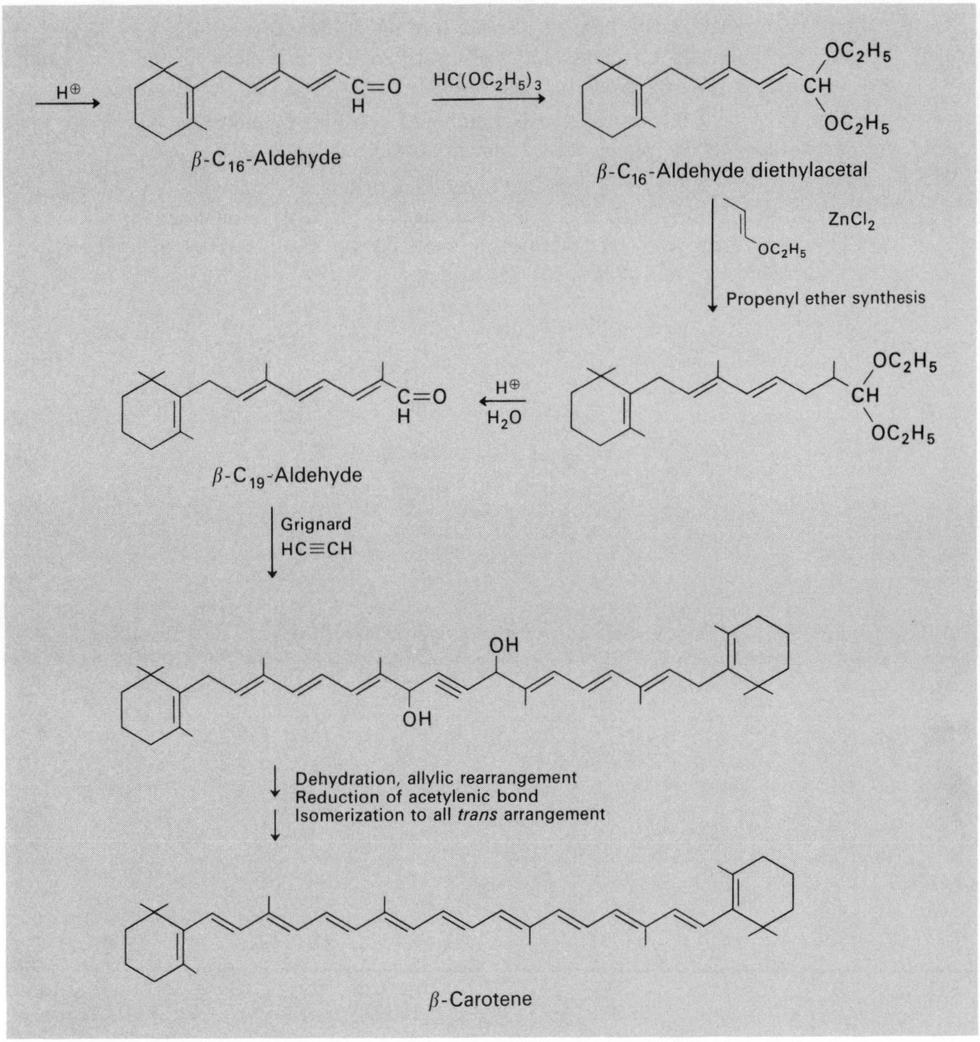

Figure 7.25 The total chemical synthesis of β-carotene.

REFERENCES

Advances in Lipid Research (Rodolfo Paoletti and David Kritchevsky, eds.), Vol. 1, Academic Press, New York, 1963. The annual publication containing contributions from experts in lipid chemistry and biochemistry with emphasis on the latter.

Deuel, Jr., H. J., *The Lipids I and II*, Interscience Publishers, Inc., New York, 1951. A comprehensive treatise on the chemistry and biochemistry of the lipids, which is out-of-date with respect to biochemistry but it contains a useful compilation of the chemistry of the lipids.

Fatty Acids (Klare S. Markley, ed.), Vols. 1 through 5, Interscience Publishers, New York. A very complete treatment of the chemistry of the fatty acids including reactions, separations, physical properties, and so forth.

Fieser, L. M., and M. Fieser, *Steroids*, Reinhold Publishing Corporation, New York, 1959. An authoritative coverage of the chemistry of steroids up to 1959.

Progress in the Chemistry of Fats and Other Lipids (R. T. Holman, ed.), Vol. 1 Pergamon Press, New York, 1958. This series is published in parts throughout the year. Each part contains several articles covering specialized areas of interest in lipid research. Both chemistry and biochemistry are treated.

EIGHT | PURINES, PYRIMIDINES, NUCLEOSIDES, NUCLEOTIDES, AND NUCLEIC ACIDS

Although environment can influence the appearance and behavior of biological systems, it is obvious that mechanisms exist for the transmission of the information which serves as the basic design for a biological system from generation to generation. The chemical compounds which store the message are the nucleic acids. These compounds are polymers of nucleotides, which in turn are formed from phosphate, monosaccharide, and a purine or pyrimidine base. The average cell contains from 2 to 5 percent of its dry weight as nucleic acids and nucleotides. Nucleosides or purine and pyrimidine bases are usually not present.

In this chapter the chemistry of the purine and pyrimidine bases and of the nucleosides, nucleotides, and nucleic acids is presented. Included in the nucleotide section are a number of compounds which are not found in nucleic acids but which are present in almost all cells and play roles in the metabolic processes of the cell. They are involved in the production and utilization of energy.

The bases are discussed first; they constitute the unique feature of this group of compounds and are responsible for the information storing and transmission capabilities of the nucleic acids.

8.1   PURINES AND PYRIMIDINES

In this section a few of the properties of the purine and pyrimidine bases which illustrate their behavior are presented and also methods which have been used to synthesize them and their analogs. Further properties will be covered in subsequent sections.

The structures of the *common* naturally occurring bases are given in Figure 8.1.

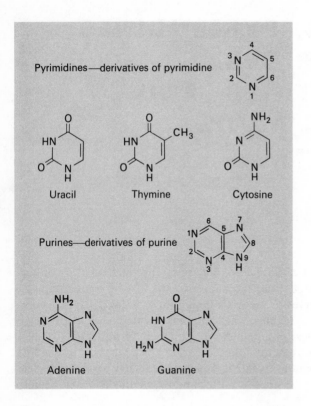

Figure 8.1

The common naturally occurring pyrimidine and purine bases.

Pyrimidines

The pyrimidine ring is aromatic but the *meta*-positioned nitrogen atoms produce an electronic distribution which is markedly deficient at positions

2, 4, and 6. Position 5 on the other hand is *relatively* electron rich. The pyrimi-

dine ring is subject, therefore, to nucleophilic substitution at positions 2, 4, and 6 and to electrophilic substitution at position 5 when electron-donating groups are present in other positions of the ring. Some of the conversions which can be achieved with pyrimidines are shown in Figures 8.2 and 8.10.

The synthesis of pyrimidines has most frequently been accomplished by condensing a three-carbon compound with a guanidine, thiourea, alkyl thiourea, amidine, or urea. Typical examples are shown in Figure 8.3.

Purines

The purine ring system is shown in Figure 8.1 along with the abundant naturally occurring derivatives. The nitrogen atoms are electron-withdrawing,

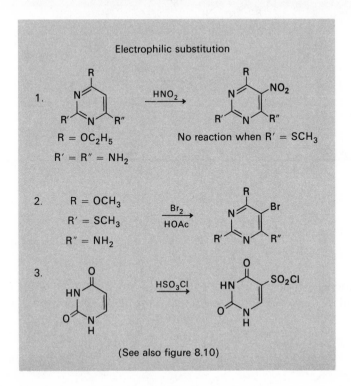

Electrophilic substitution

1. $R = OC_2H_5$
    $R' = R'' = NH_2$
    HNO$_2$ →
    No reaction when $R' = SCH_3$

2. $R = OCH_3$
    $R' = SCH_3$
    $R'' = NH_2$
    $\xrightarrow[\text{HOAc}]{Br_2}$

3. $\xrightarrow{HSO_3Cl}$

(See also figure 8.10)

Figure 8.2   Reactions of pyrimidines (*continued on next page*).

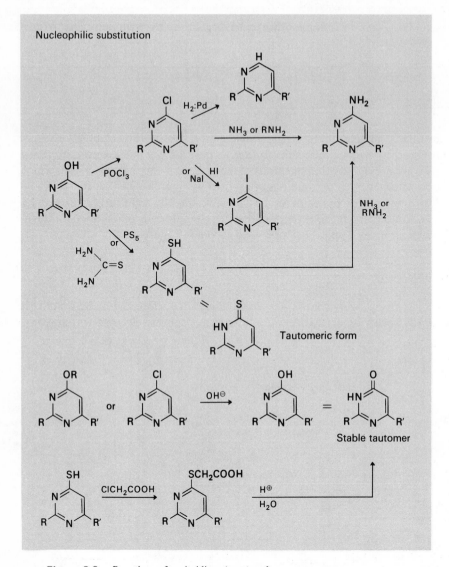

Figure 8.2 Reactions of pyrimidines (*continued*).

producing a deficiency of electrons at positions 2 and 6 of the pyrimidine ring and to a lesser extent at position 8 in the imidazole portion of the ring.

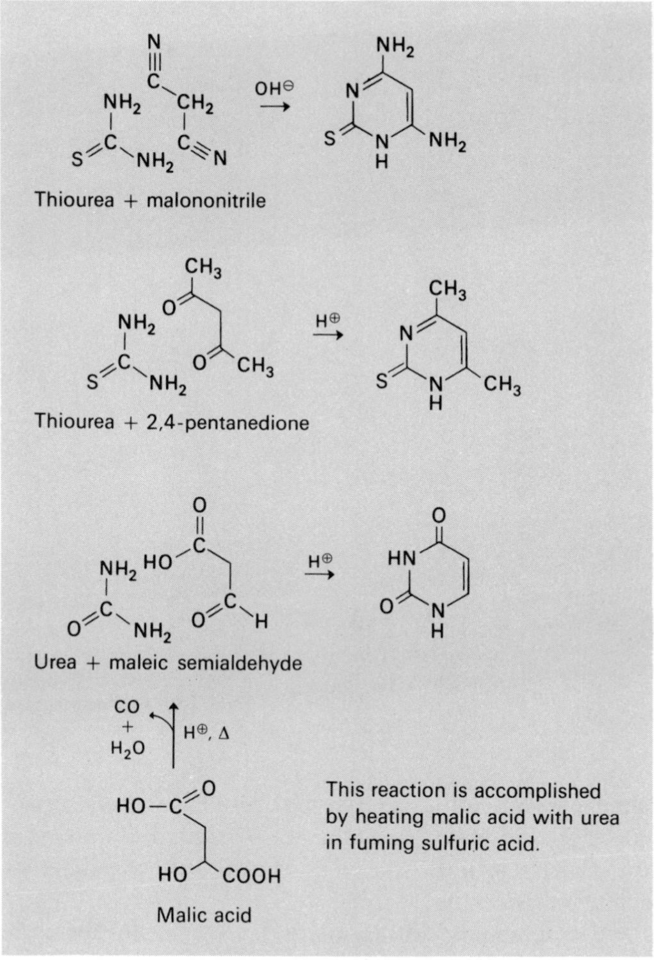

Thiourea + malononitrile

Thiourea + 2,4-pentanedione

Urea + maleic semialdehyde

This reaction is accomplished by heating malic acid with urea in fuming sulfuric acid.

Malic acid

Figure 8.3

Synthetic routes to pyrimidines. [A. Bendich, *The Nucleic Acids* (E. Chargaff and J. N. Davidson, ed.), Academic Press, Inc., New York 1955, p. 81].

The purine ring system has no centers susceptible to attack by electrophilic reagents but it can react in a similar fashion to the pyrimidine ring toward nucleophilic agents (Figure 8.2).

The purine ring system is most readily synthesized from 5,6-diaminopyrimidines which can be synthesized from the 6-amino compound (Figure 8.4).

The tautomeric forms of the naturally occurring purine and pyrimidine bases are of interest primarily because of the role that the tautomers of the nucleosides play in the functioning of the genetic code. The forms shown in Figure 8.1 are considered to be the major forms present in the bases and these assignments were established by comparison of the properties of the free bases with those of substituted bases of fixed tautomeric structure (Figure 8.5). For example, uracil and its 1- and 3-methyl-substituted derivatives have essentially the same ultraviolet spectrum—since the methyl compounds must have the lactam

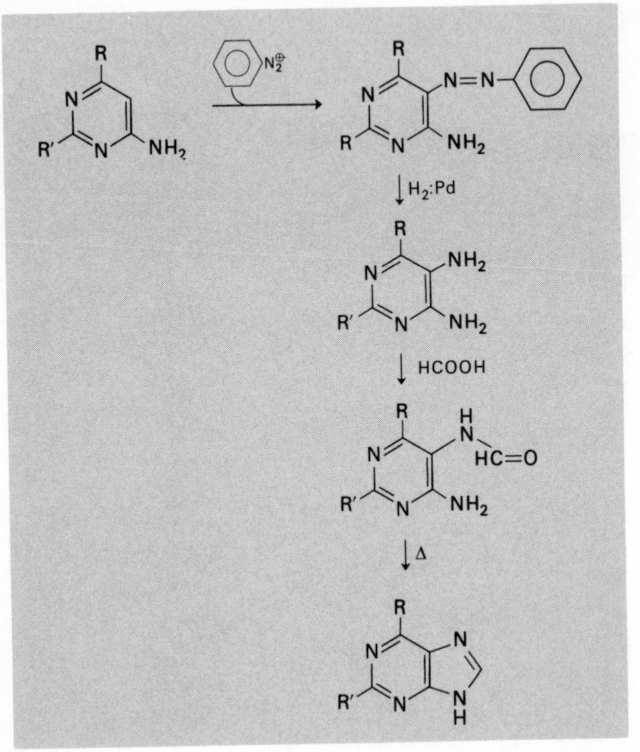

Figure 8.4

The synthesis of the purine ring system. [A. Bendich, *The Nucleic Acids* (E. Chargaff and J. N. Davidson, ed.), Academic Press, Inc., New York 1955, p. 81.]

structure, it is reasonable to infer that the free base must also have this structure. A further comparison is given in Figure 8.6. Only the 2-*O*-methyl compound B must exist in the lactim form and its ultraviolet characteristics are very different from those of C which must exist as the lactam.

The amino-substituted compounds exist primarily in the lactim form, as seen by comparing E and F—the latter cannot exist as a lactam. The tautomeric preferences of the pyrimidine (and purine) bases are vitally important to the correct transmission of genetic information and are discussed in this context in Section 8.3.

The pK_a values of the natural bases and nucleosides are given in Figure 8.7. The differences are very useful in separating the purines and pyrimidines and their derivatives by electrophoresis, as well as ion-exchange chromatography and in identifying them by the changes in ultraviolet spectra associated with changes in ionic forms.

8.2   NUCLEOSIDES AND NUCLEOTIDES

The most abundant nucleosides and nucleotides are *N*-glycosides of the pentose sugars D-ribose and 2-deoxy-D-ribose (more correctly 2-deoxy-D-*erythro*-

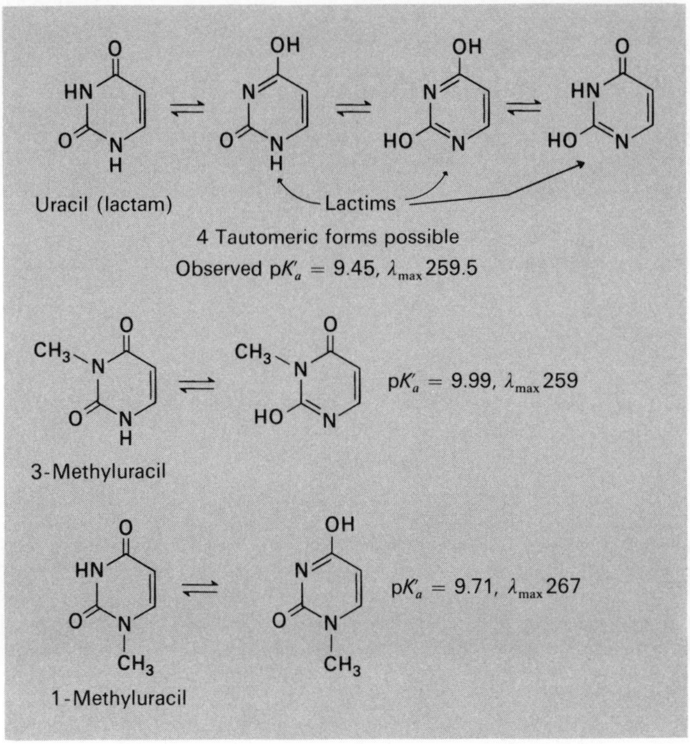

Uracil (lactam)

Lactims

4 Tautomeric forms possible

Observed $pK'_a = 9.45$, λ_{max} 259.5

3-Methyluracil

$pK_a = 9.99$, λ_{max} 259

1-Methyluracil

$pK_a = 9.71$, λ_{max} 267

Figure 8.5

The tautomeric forms of uracil and substituted uracils.

pentose). Typical examples are shown below. The nucleotides are phosphate

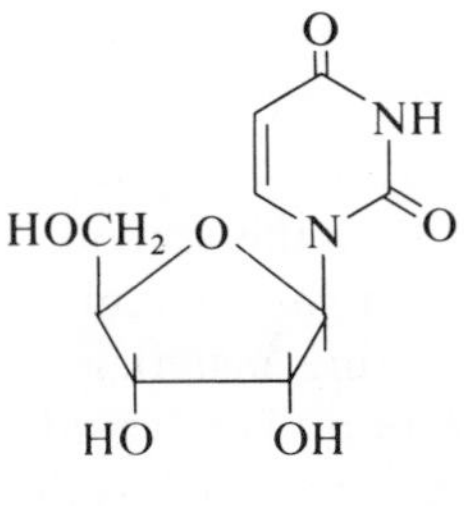

Uridine
1-(β-D-ribofuranosyl)uracil
a nucleoside

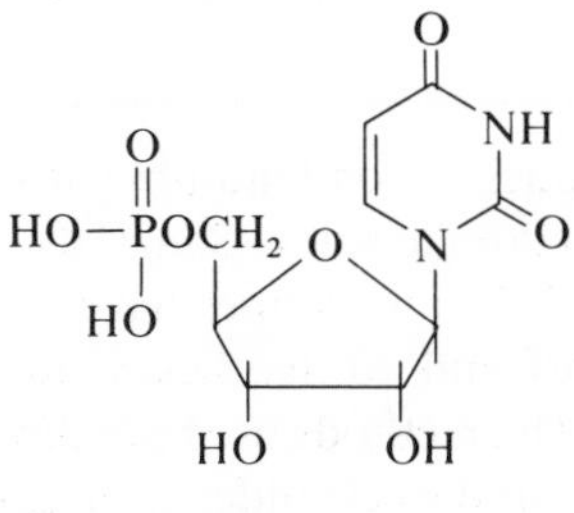

5' Uridylic acid, 5'-UMP
1-(β-D-ribofuranosyl)uracil 5'-phosphate
a nucleotide

ester derivatives of the nucleosides. The nomenclature of the common nucleo-sides and nucleotides is given in Table 8.1.

Although the major proportion of tissue nucleotides are present as compon-ents of the highly polymeric nucleic acids, a small proportion of mono- and dinucleotides can be isolated from all tissues. They are vital for normal cellular

		λ_{max} pH = 7.0	ε_{max}
A.	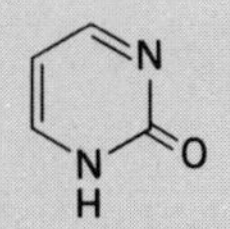	<215 298	>10,000 4,710
B.	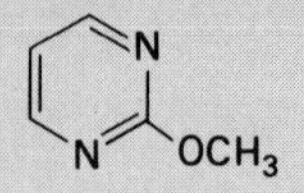	264	4,780
C.	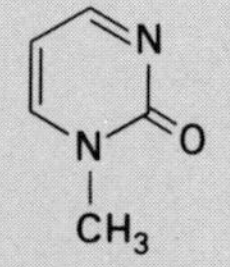	215 302	10,000 5,400
D.	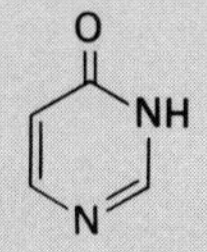	223 260	7,320 3,740
E.	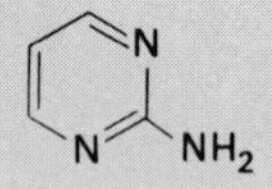	224 292	13,600 3,170
F.	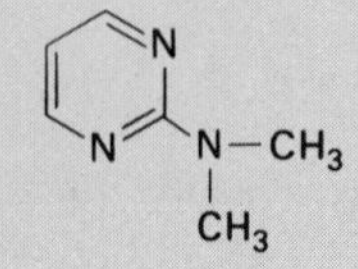	243 318	18,200 2,160

Figure 8.6

The absorption maxima and extinction coefficients of a number of substituted pyrimidines.

function. Nucleotides in combination with other moieties are also present and play important metabolic roles. Some of them are dealt with in this section.

The chemistry of the nucleotides has been extensively investigated mainly to establish structures and to synthesize the naturally occurring materials or analogs of them. The methods used are described in this section. The degradation of nucleosides and nucleotides is discussed first, however, followed by discussion of the properties of the bases and of some of the naturally occurring low-molecular-weight nucleosides to provide a background for the synthetic studies which follow.

A. Hydrolysis

The N-glycoside bond is generally stable toward alkaline hydrolysis and labile toward acid. The ease of acid hydrolysis is very much dependent on the struc-

pK_a values of bases and nucleosides

Pyrimidines

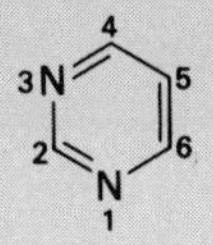

	pK'_a values	Assignment
Uracil (2,4-dihydroxy)	9.45	—NH lactam at 1 *or* 3
Uridine (2,4-dihydroxy, 1-ribofuranosyl)	9.17	—NH lactam at 3
	12.5	2'—OH of ribosyl residue
Cytosine (4-amino-2-hydroxy)	4.45	—NH$_3^{\oplus}$ at 4
	12.2	—NH at 1
Cytidine (4-amino-2-hydroxy, 1-ribofuranosyl)	4.11	—NH$_3^{\oplus}$ at 4
	13.0$^+$	2'—OH of ribosyl residue
Thymine (2,4-dihydroxy-5-methyl)	9.82	—NH lactam at 1 *or* 3

Purines

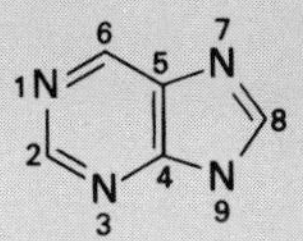

	pK'_a value	Assignment
Adenine (6-amino)	4.15	—NH$_3^{\oplus}$ at 6
	9.8	—NH imidazole at 9
Adenosine (6-amino,9-ribofuranosyl)	3.45	—NH$_3^{\oplus}$ at 6
	12.5	2'—OH of ribosyl residue
Guanine (2-amino,6-hydroxy)	3.3	—NH$_3^{\oplus}$ at 2
	9.6	—NH lactam at 1
	12.3	—NH imidazole at 9
Guanosine (2-amino,6-hydroxy, 9-ribofuranosyl)	1.6	—NH$_3^{\oplus}$ at 2
	9.2	—NH lactam at 1
	12.3	2'—OH of ribosyl residue

Figure 8.7   The pK_a values of the naturally occurring purines and pyrimidines and their assignment. [*Handbook of Biochemistry, Selected Data for Molecular Biology* (H. A. Sober, ed.), The Chemical Rubber Co., Cleveland, Ohio 1968, p. J–51, with permission of the Chemical Rubber Co.].

tures of the carbohydrate and the base components. The naturally occurring *N*-glycosides are furanosides—specifically *β*-D-furanosides. A number of *N*-pyranosides have been synthesized—as expected by analogy with the *O*-glycosides (Section 5.12); they are more stable than the furanosides toward hydrolysis. The effect of structure on the rate of hydrolysis of nucleosides is shown in Figure

TABLE 8.1   NOMENCLATURE OF NUCLEOTIDES, NUCLEOSIDES, AND BASES

| Base | *Ribosides* | | | | *2-Deoxyribosides* | | | |
	Nucleoside	*Abb.*	*Nucleotide[a]*	*Abb.[b]*	*Nucleosides*	*Abb.*	*Nucleotides[a]*	*Abb.[b]*
Uracil	Uridine	Urd	Uridylic acid	UMP Urd-5′-P	deoxyUridine	dUrd	—	—
Thymine	Ribosylthymine	Thd	—	—	Thymidine	dThd	Thymidilic acid	TMP dThd-5′-P
Cytosine	Cytidine	Cyd	Cytidylic acid	CMP Cyd-5′-P	deoxyCytidine	dCyd	deoxyCytidylic acid	dCMP dCyd-5′-P
Adenine	Adenosine	Ado	Adenylic acid	AMP Ado-5′-P	deoxyAdenosine	dAdo	deoxyAdenylic acid	dAMP dAdo-5′-P
Guanine	Guanosine	Guo	Guanylic acid	GMP Guo-5′-P	deoxyGuanosine	dGuo	deoxyGuanylic acid	dGMP dGuo-5′-P

[a] The position of the phosphate group is expressed with the prefix 2′-, 3′-, or 5′- for simple nucleotides and by 2′,3′- or 3′,5′- for cyclic nucleotides.

[b] The first abbreviation listed is usually used to specify the 5′-nucleotide without a prefix. If the 2′- or 3′-nucleotide is specified, the appropriate prefix is used. The second, somewhat longer, abbreviation is useful in that it is more specific. It can be used to specify more complex structures such as uridine diphosphate glucose, Urd-5′-P-P-Glc.

8.8 and the general mechanism of hydrolysis is shown in Figure 8.9. It is observed that the presence of a proton in the base which can be transferred readily to the ring oxygen facilitates hydrolysis [Figure 8.9(a)]. The N-3 atom of the purine ring and an amino substituent at C-2 of the pyrimidine ring can both serve this purpose. The carbonyl oxygen at C-2 in the natural pyrimidines is not easily protonated, however, and it tends to interfere with protonation of the ring oxygen. The position at which the charge in a protonated base is located is probably reflected in the ease of hydrolysis of the nucleoside; if the charge is close to the ring hydroxyl group, it may hinder its protonation. The importance of the N-3 atom of the purine ring is illustrated by the ease with which DNA can be converted to apurinic acid by mild acid hydrolysis [Figure 8.9(b) and Figure 8.35].

The acid lability of the *N*-glycosidic bond of ribosyl derivatives is increased by periodate oxidation, the dialdehyde being significantly more labile.

In a few instances, nucleosides are found to be base-labile, for example, adenosine 3′,5′-cyclic phosphate.

The hydrolysis of the phosphate ester bonds of nucleotides is accomplished as described in Section 6.6. In the case of the 2′- or 3′-ribofuranosyl phosphates, participation by the neighboring hydroxyl group is important.

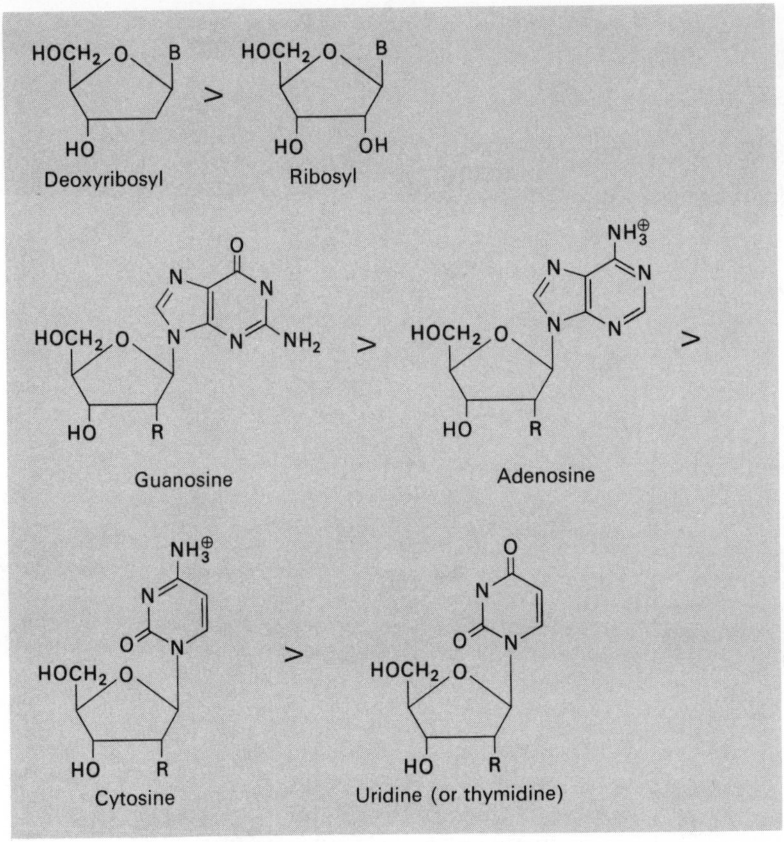

Figure 8.8   The relative susceptibilities of the *N*-glycosidic bond of
nucleosides to acid hydrolysis.

B. The Pyrimidine Ring

The pyrimidine ring system is susceptible to electrophilic substitution at position 5. Some of the derivatives which have been prepared are shown in Figure 8.10. The 5-bromo group can be replaced by alkoxyl or amino functions [Figure 8.10(*a*)]. Reduction of the 5,6 double bond increases the ease of hydrolysis of the glycosidic bond—a fact which proved useful in the demonstration that pyrimidine deoxynucleosides contained 2-deoxy-D-ribose [Figure 8.10(*b*)]. Hydrolysis of the normal nucleoside yields the pyrimidine and levulinic acid; whereas hydrolysis of the dihydronucleoside under much milder conditions allowed 2-deoxy-D-ribose to be obtained.

The 5 position of uracil nucleosides is also susceptible to attack by formaldehyde with the formation of the 5-hydroxymethyl derivative [Figure 8.10(*c*)]. This offers a route to the corresponding thymine derivative since the hydroxy-

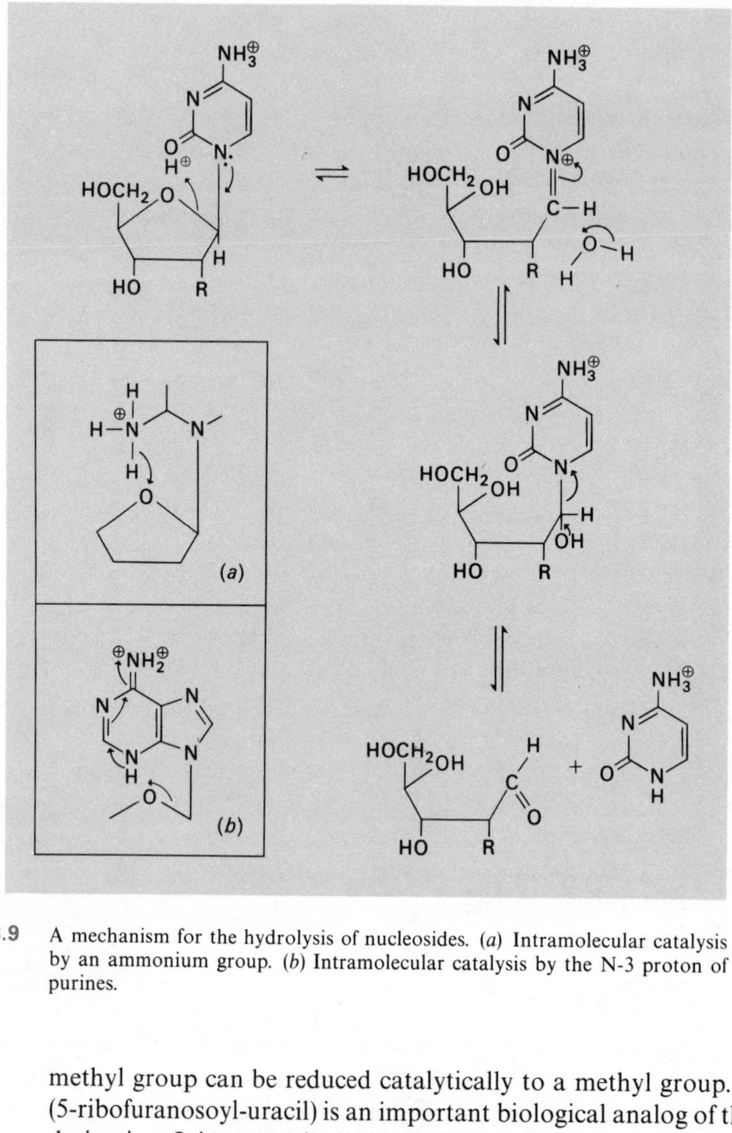

Figure 8.9   A mechanism for the hydrolysis of nucleosides. (a) Intramolecular catalysis by an ammonium group. (b) Intramolecular catalysis by the N-3 proton of purines.

methyl group can be reduced catalytically to a methyl group. Pseudo-uridine (5-ribofuranosoyl-uracil) is an important biological analog of the formaldehyde derivative. It is present in tRNA (Table 8.6) and possesses some unique chemical properties (Figure 8.11). Pseudo-uridine is resistant to acid hydrolysis; neither ribose nor uracil is released even after hydrogenation which in this compound does not result in reduction of the 5,6 double bond but in opening the furanosyl ring. Treatment with acid results in rearrangements in the ribosyl residue but not in cleavage of the "C-glycoside" bond.

    The hydrogen atoms on the ring nitrogens of the pyrimidine nucleosides can be replaced by methyl groups by treatment with diazomethane [Figure 8.10(d)]. This replacement demonstrates the occurrence of the tautomer with the hydrogen atom on the ring nitrogen rather than on the adjacent oxygen. Unless pro-

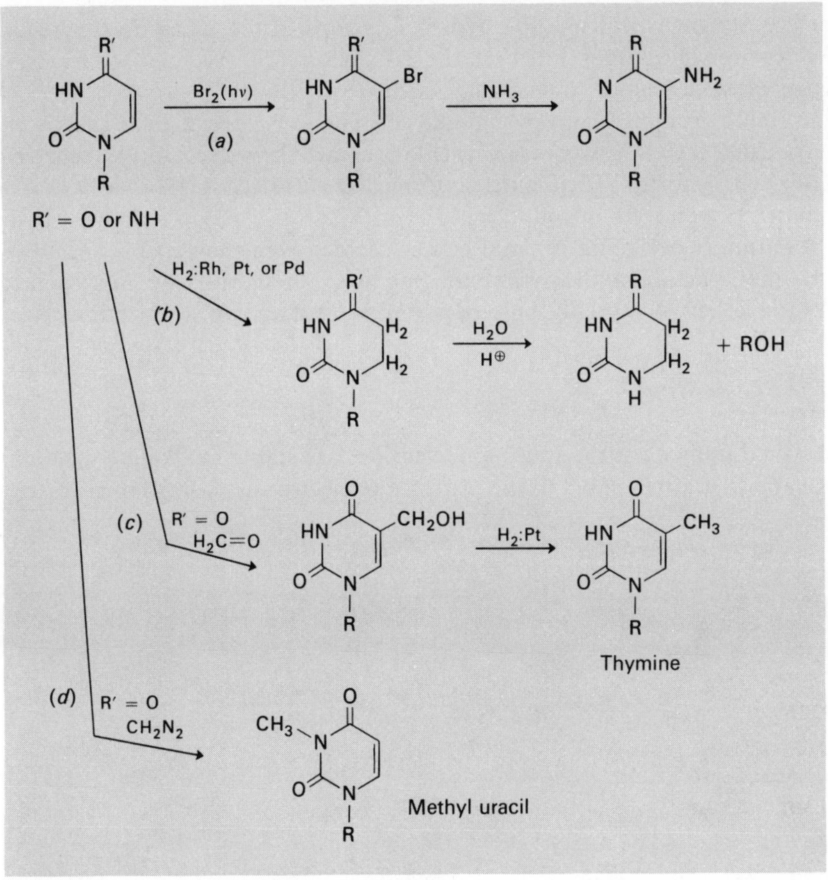

Figure 8.10   The derivitization of the pyrimidine ring in nucleosides.

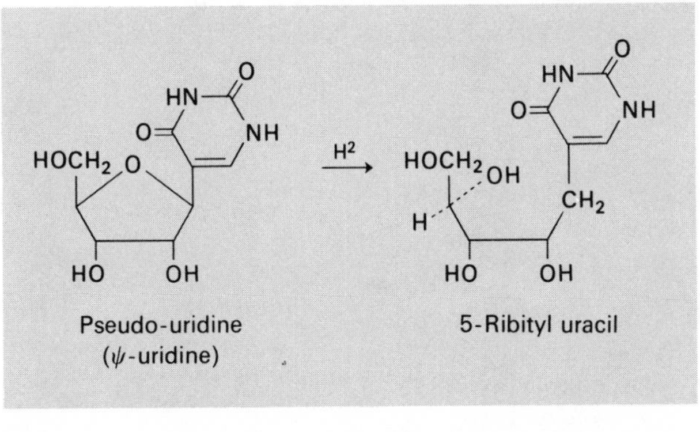

Figure 8.11

Pseudo-uridine and its conversion to 5-ribityl uridine.

longed treatment is utilized, the hydroxyl groups of the sugar do not react. When reaction does occur, however, the 2′-hydroxyl group is found to be more reactive, presumably due to its greater acidity.

The amino groups of both pyrimidine and purine bases are readily acylated with the usual acylating agents (acetyl chloride, carbobenzoxychloride, benzoylchloride, and so forth). The amides so formed are resistant to alkaline hydrolysis but can be removed by aminolysis.

The naturally occurring pyrimidine nucleosides were shown to be β and to involve the 3 position of the pyrimidine ring by synthesis and by comparison of their ultraviolet spectra with those of pyrimidines of known substitution.

C. The Purine Ring

Most naturally occurring purine nucleosides are stable in alkaline medium. This stability is attributable to the presence of electron-withdrawing groups in

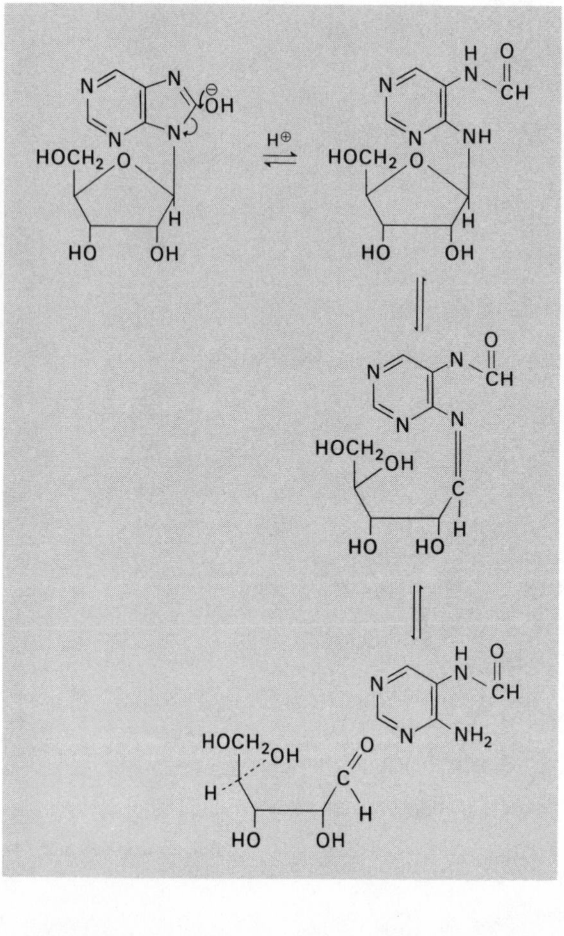

Figure 8.12

A mechanism for the alkaline decomposition of unsubstituted purine nucleosides.

the pyrimidine portion of the ring since unsubstituted purine nucleosides are base-labile (Figure 8.12). The lability depends upon cleavage of the imidazole ring with the formation of a very easily hydrolyzable *N*-glycoside. The products are the sugar and a 5,6-diamino pyrimidine.

The difference in electron distributions between the adenine and guanine rings is illustrated by the fact that adenine nucleosides are alkylated in the pyrimidine ring, whereas the guanosine derivatives are alkylated at N-7 in the imidazole ring.

The glycosidic linkage in the naturally occurring purine nucleosides was shown to be β by unambiguous synthesis of the compounds. That the linkage involved the 9 position of the purine ring was established by synthesis, by comparison of the ultraviolet spectra of the natural compounds with methyl purines, and by demonstrating that other positions could be manipulated without disturbing the glycosidic bond.

D. Adenine Nucleotides

These are worth special attention since they are involved in the energy conserving and utilizing mechanisms of most biological systems. In addition, they occur as components of a number of complex *dinucleotides* which participate in oxidation-reduction processes in the cell. The structures of these compounds are shown in Figure 8.13.

The Nicotinamide Nucleotides. The *pyridine* nucleoside moieties of NAD$^+$ and NADP$^+$ (Figure 8.13) possess some unique properties; the most important in biological systems is their ability to be reversibly reduced. The pyridinium ring is susceptible to attack by nucleophilic agents at positions 2, 4, and 6 (Figure 8.14). Position 2, which is activated by the carboxamide function in the nicotinamide nucleotides, is the most reactive, but in most cases the initial product rearranges to the more stable 4-isomer. The biological oxidation-reduction reactions of the nicotinamide ring of NAD$^+$ and NADP$^+$ are highly specific for both the position on the ring and the face of the ring to which the proton is added (see Section 3.4) and are good examples of the extreme specificity of biological catalysis.

The reduced forms are significantly more acid-labile than the oxidized forms of the nicotinamide coenzymes. The reaction is complex. It involves protonation of the ring oxygen, formation of an acyclic form, addition of H_2O to the 5,6 double bond, and ring closure to a mixture of α and β anomers of the 1,4,5,6-tetrahydro-6-hydroxy derivative prior to degradation (Figure 8.15). The hydration reaction is reversible but the decomposition of the hydrated material is not. Under strongly acidic conditions the decomposition leads to ring opening with formation of an aldehyde. The only well-characterized product of the degradation step is the adenosine diphosphate portion (R in Figure 8.15).

The coenzymes, NAD$^+$ and NADP$^+$, appear to exist in solution with their bases, adenine and nicotinamide, stacked one above the other. This intra-

Adenine-containing nucleotides

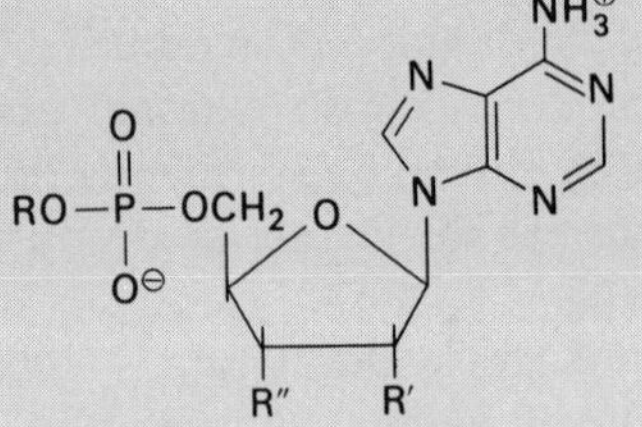

Mononucleotides

5'-Adenylic acid, AMP R = H or $\ominus$; R' = R" = OH
(2' and 3'-adenylic acids also exist)

Adenosine diphosphate, ADP   R = $-\overset{\overset{\displaystyle O}{\|}}{\underset{\underset{\displaystyle O\ominus}{|}}{P}}-O\ominus$; R' = R" = OH

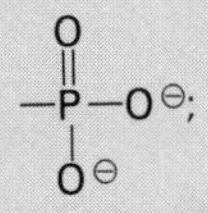

Adenosine triphosphate, ATP   R = $-\overset{\overset{\displaystyle O}{\|}}{\underset{\underset{\displaystyle O\ominus}{|}}{P}}-O-\overset{\overset{\displaystyle O}{\|}}{\underset{\underset{\displaystyle O\ominus}{|}}{P}}-O\ominus$; R' = R" = OH

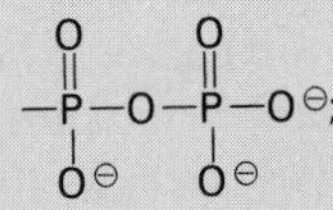

3',5'-Cyclic adenylic acid, AMP! or 3',5'-AMP

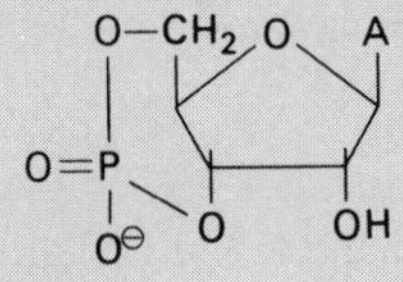

Dinucleotides

Nicotine adenine dinucleotide, NAD$^{\oplus}$
or diphosphopyridine nucleotide, DPN$^{\oplus}$

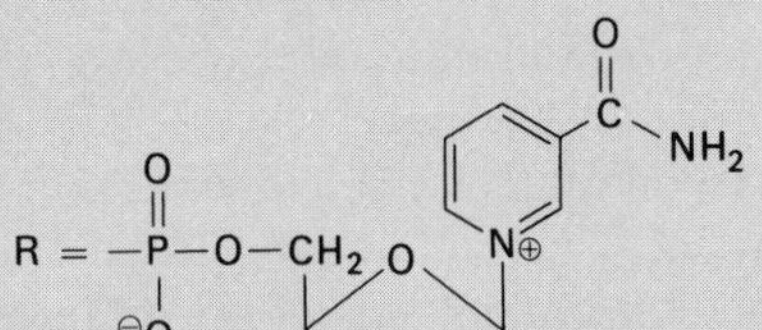

Nicotinamide ribotide

R' = R" = OH

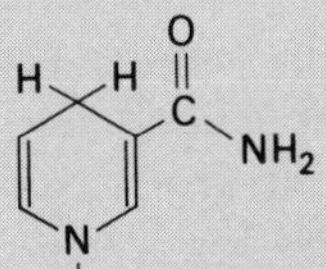

Reduced form

Function: Nicotinamide ring acts as an oxidizing agent; can accept H^{+} + 2e. Reaction is readily reversible.

Nicotine adenine dinucleotide phosphate, NADP⊕
or Triphosphopyridine nucleoside, TPN⊕

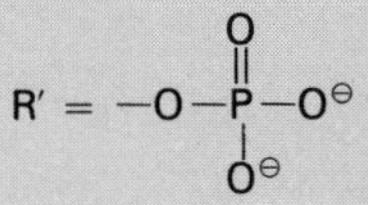

$$R' = -O-\overset{\overset{\displaystyle O}{\|}}{\underset{\underset{\displaystyle O^\ominus}{|}}{P}}-O^\ominus$$

R = nicotinamide ribotide
     (see above)

R″ = OH

Function: Similar to NAD⊕

Flavin adenine dinucleotide

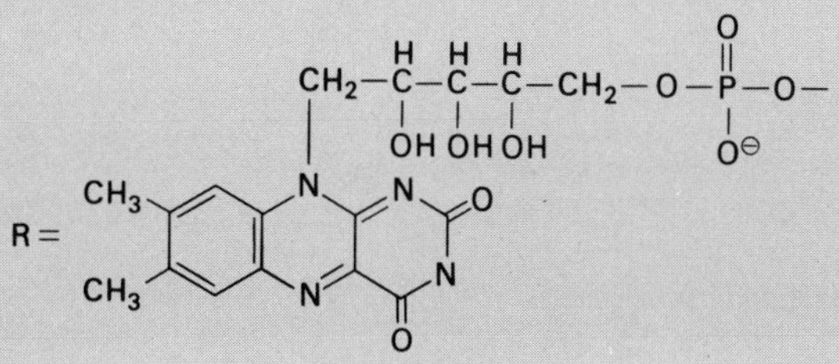

R′ = R″ = OH

Riboflavin phosphate

Function: Similar to NAD⊕; can accept 2H⁺ + 2e

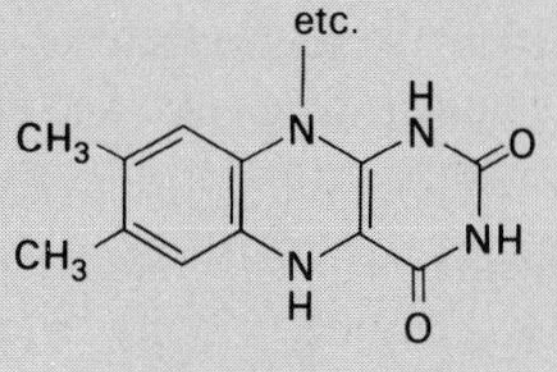

Reduced form

Coenzyme A, CoASH

R′ = −OH

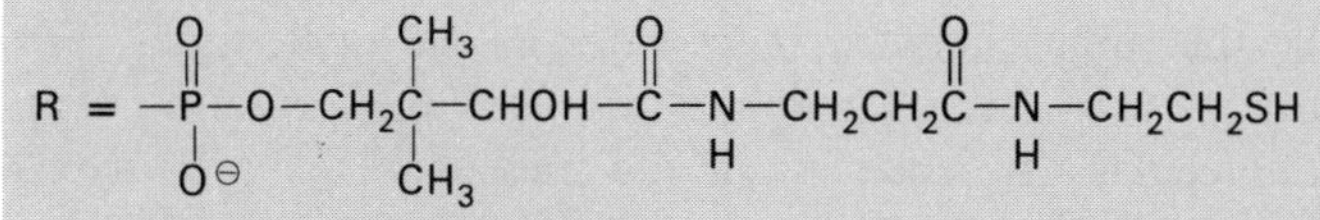

$$R'' = -O-\overset{\overset{\displaystyle O}{\|}}{\underset{\underset{\displaystyle O^\ominus}{|}}{P}}-O^\ominus$$

Pantotheine

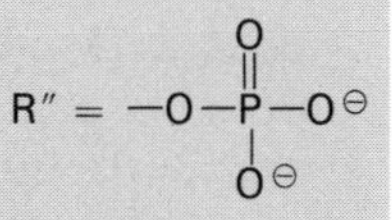

Function: Acceptance and transfer of acyl groups

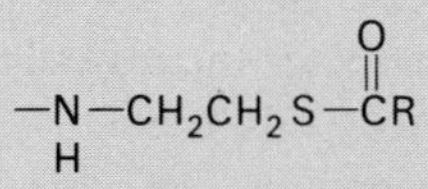

$$-\underset{\underset{\displaystyle H}{|}}{N}-CH_2CH_2S-\overset{\overset{\displaystyle O}{\|}}{C}R$$

Acylated form

Figure 8.13   Adenine-containing nucleotides and dinucleotides.

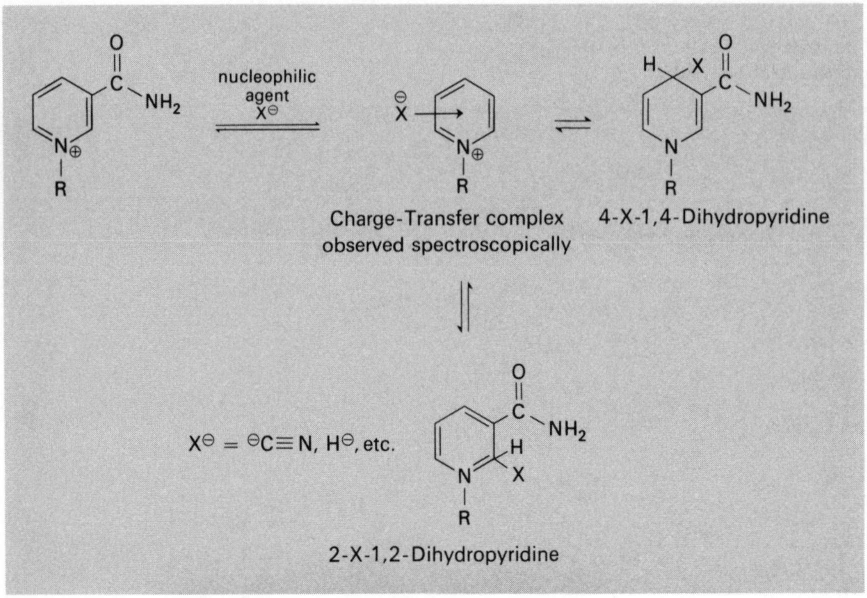

Figure 8.14   Reactions of nicotinamide nucleosides.

molecular interaction causes a decrease in absorbance similar to that observed in the nucleic acids (see Section 8.3). It also causes the compound to be much more compact than would otherwise be the case.

The Flavin Nucleotides. Riboflavin is a vitamin which is converted into a mononucleotide, riboflavin phosphate (flavin mononucleotide), or into a dinucleotide, flavin adenine dinucleotide, in biological systems which utilize these compounds as coenzymes in oxidation-reduction systems. Strictly speaking, the compounds are not nucleotides; the flavin-ribitol bond is not a glycosidic bond and the compounds are *N*-alkyl derivatives of isoalloxazine. The nucleoside terminology is in common use, however, although the compounds do not have the same chemical reactivities as the nucleosides. For example, acid hydrolysis of riboflavin phosphate results in the formation of an anhydride by intramolecular displacement of the phosphate by a hydroxyl group at C-2 of the ribitol (Figure 8.16).

The structure of the isoalloxazine ring of the flavins is shown on page 300. The biological role of the flavins is similar to that of the nicotinamide nucleotides in that it involves the reversible reduction of the isoalloxazine ring system. In many instances this reduction is accomplished by the reduced form of a nicotinamide coenzyme and the oxidation by a quinone or a cytochrome-containing enzyme system (Figure 8.17). The reduction of riboflavin in solution consists of two one-electron steps as shown on page 300, but it is not established that the enzyme-catalyzed reductions proceed in the same way. Most

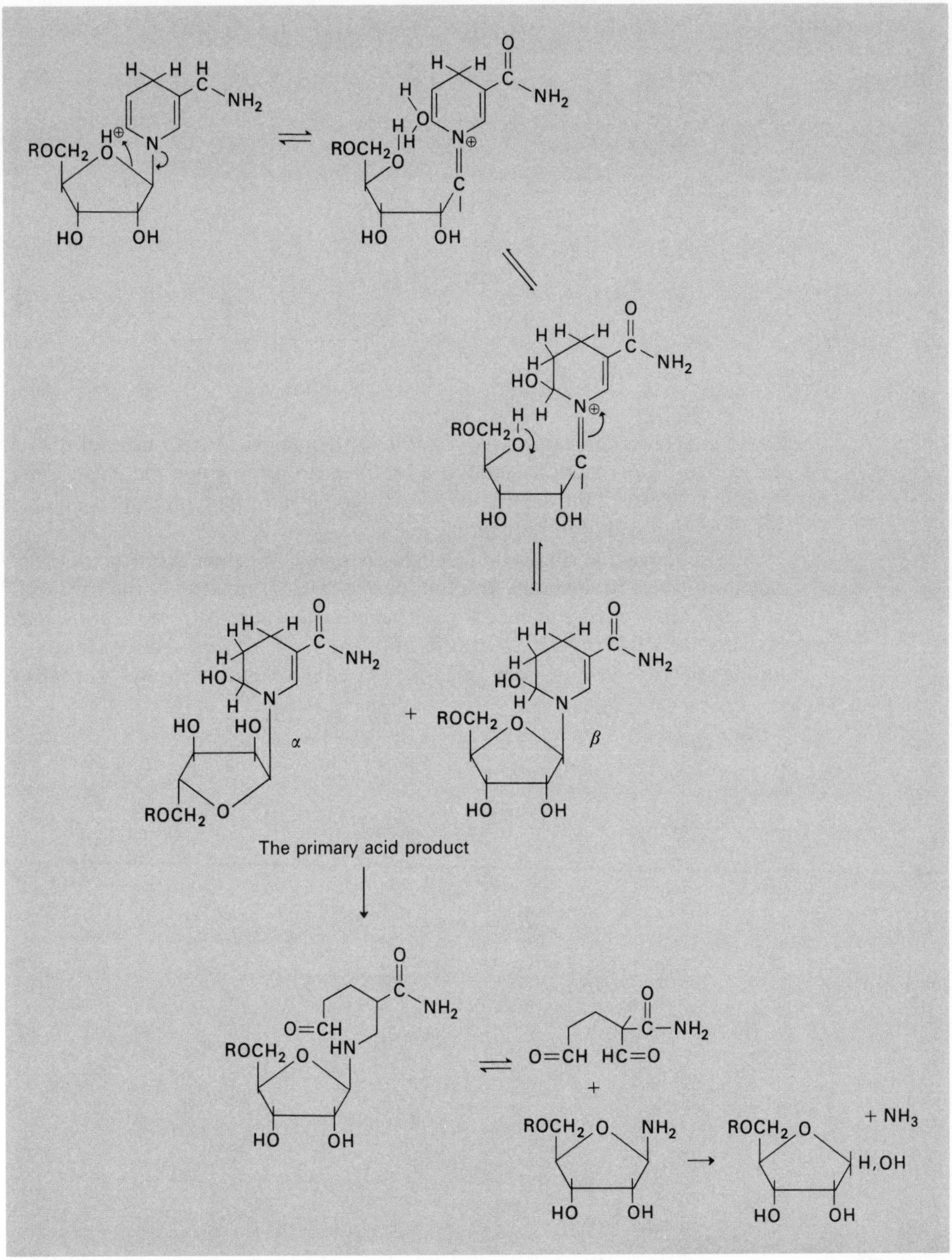

Figure 8.15   Acid-catalyzed hydrolysis of reduced nicotinamide nucleosides.

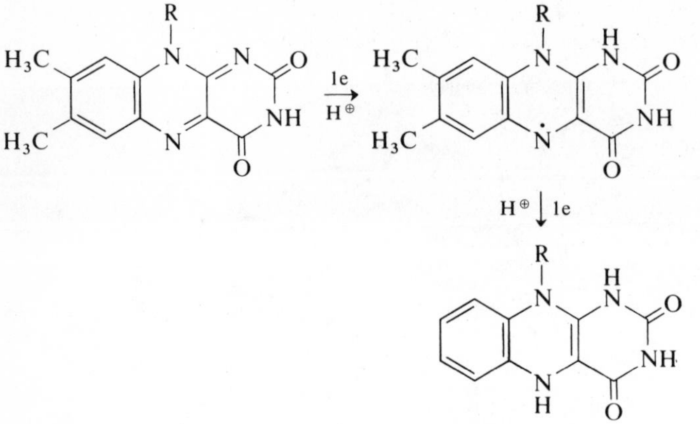

flavin-enzyme complexes are very stable (with dissociation constants of 10^{-8} M or less) and the reactions catalyzed by them do not involve the flavin free in solution. Many of the complexes also contain iron or molybdenum which play a role in the oxidation-reduction cycle.

The isoalloxazine ring system is light-sensitive. Whether this is of biological significance is problematical but the spectroscopic properties of the compound are useful in studying its reactions. The spectrum of most isoalloxazines have maxima at 450, 370, and 260 nm in both the oxidized and reduced forms, although they are less intense in the latter. The oxidized forms fluoresce at 540 nm

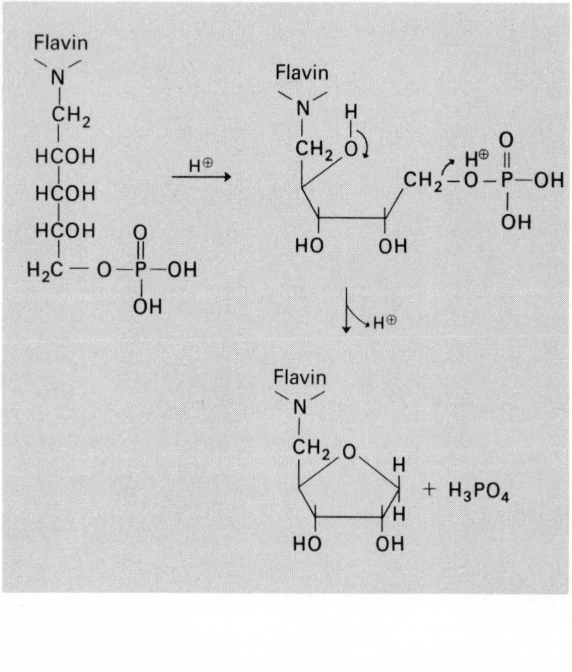

Figure 8.16

The action of acid on flavin "nucleoside."

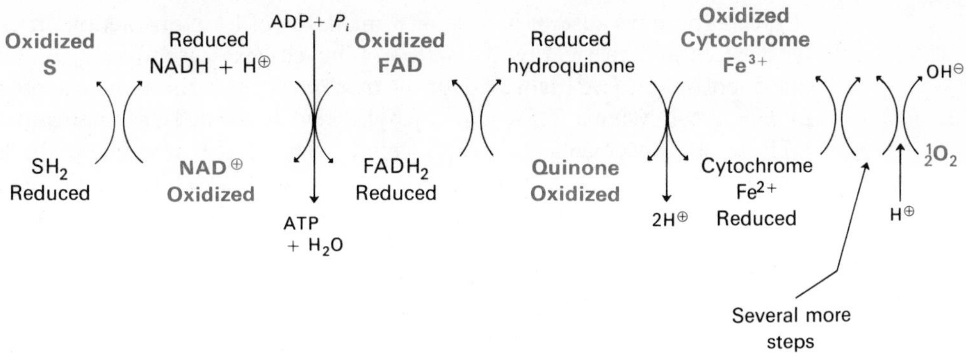

Figure 8.17   A portion of the electron transport system of the mitochondrion. The system allows the controlled release of the energy of oxidation so that some of the substance is trapped for synthetic purposes as ATP. One of the steps at which the phosphorylation of ADP occurs is shown. The mechanism of this process is not known.

$$SH_2 = \text{reduced metabolite, such as succinic acid.}$$
$$S = \text{oxidized metabolite, such as fumaric acid.}$$

when irradiated at 336 nm. The products from the photolysis of riboflavin have not been fully characterized. The reaction involves the side chain as well as the ring and the course of the reaction is strongly influenced by the conditions used. Some of the products are shown in Figure 8.18.

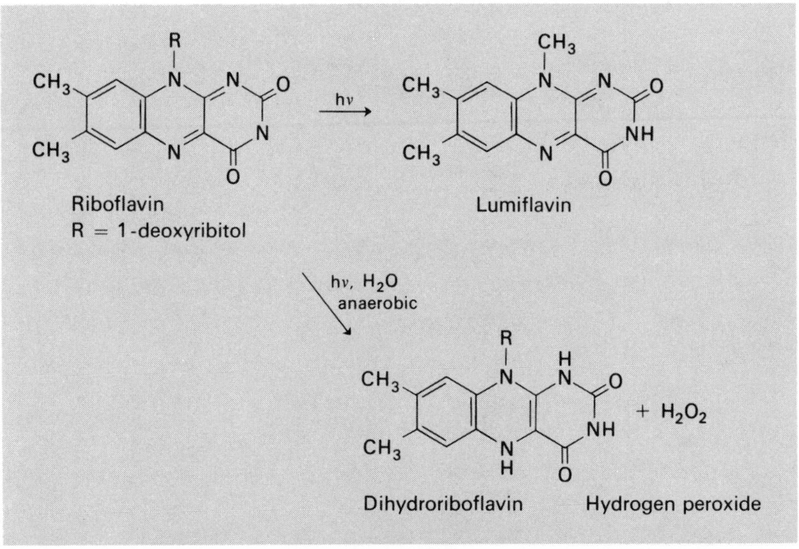

Riboflavin
R = 1-deoxyribitol

Lumiflavin

Dihydroriboflavin   Hydrogen peroxide

Figure 8.18   The action of light on riboflavin.

3′,5′-Cyclic Adenylic Acid. This compound (Figures 8.13 and 8.19) is worthy of special note because of its role as a mediator of hormone action. It can be synthesized in high yield by the action of dicyclohexylcarbodiimide on adenosine 5′-phosphate (see Figure 4.6 for the mechanism of carbodiimide activation). In biological systems 3′,5′-cyclic adenylic acid is formed enzymatically from ATP by displacement of pyrophosphate. The cyclic phosphate diester is slightly more labile than an acyclic diester and alkaline hydrolysis occurs more readily at the 5′ position to give the 3′-monoester. In contrast, the enzyme which catalyzes the hydrolysis is entirely specific for the 3′ bond and produces only the 5′-ester. There is a possible utility in this specificity since the 5′-nucleotide can be reconverted to ATP in most tissues.

Coenzyme A. The structure of coenzyme A is shown in Figure 8.13 and its chemical synthesis in Figures 6.7 and 6.8. It, like the nicotinamide nucleotides, contains a portion derived from a vitamin, in this case, pantothenic acid.

The function of coenzyme A is very different from that of the nicotinamide coenzymes. It is an alkylating and acylating agent. It is capable of activating acetyl groups and transferring them to suitable acceptors such as oxaloacetate or acetate with the formation of citrate and acetoacetate, respectively (Figure 8.20). Although the citrate-forming reaction is formally analogous to a Claisen condensation which proceeds via a carbanion, there is no evidence for a carbanion intermediate in the enzymatic process. The α hydrogen of the acetate is lost only in the presence of the substrate oxaloacetate and the enzyme, and it occurs only to the extent that condensation occurs. In simpler model compounds, however, it has been shown clearly that the α hydrogens of thiol esters are considerably more acidic than those of the corresponding oxygen esters and

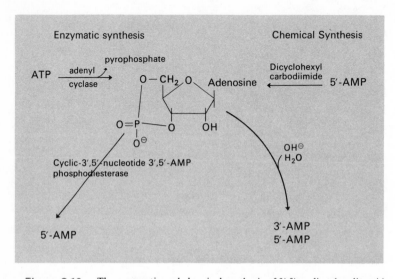

Figure 8.19   The enzymatic and chemical synthesis of 3′,5′-cyclic adenylic acid.

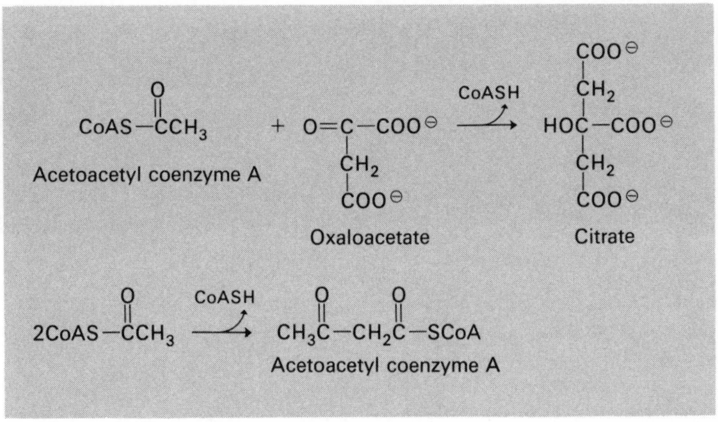

$$\underset{\substack{\text{Acetoacetyl coenzyme A}}}{\text{CoAS}-\overset{\displaystyle O}{\overset{\|}{C}}\text{CH}_3} \quad + \quad \underset{\substack{\text{Oxaloacetate}}}{\overset{\displaystyle \text{COO}^{\ominus}}{\underset{\text{COO}^{\ominus}}{\overset{\text{O}=\text{C}-\text{COO}^{\ominus}}{\underset{\text{CH}_2}{}}}}} \quad \xrightarrow{\text{CoASH}} \quad \underset{\substack{\text{Citrate}}}{\overset{\displaystyle \text{COO}^{\ominus}}{}}$$

$$\underset{\substack{\text{Acetoacetyl coenzyme A}}}{2\text{CoAS}-\overset{\displaystyle O}{\overset{\|}{C}}\text{CH}_3} \quad \xrightarrow{\text{CoASH}} \quad \text{CH}_3\overset{\displaystyle O}{\overset{\|}{C}}-\text{CH}_2\overset{\displaystyle O}{\overset{\|}{C}}-\text{SCoA}$$

Figure 8.20 Reactions involving coenzyme A. The structure of the coenzyme is given in Figure 8.13.

that thiol esters are much more reactive toward electrophilic agents. This reactivity appears to be due to lack of resonance stabilization in the thiol esters (Figure 8.21). Thus, the oxygen ester (*a*) is considerably more stabilized by resonance than the thiol ester (*b*) relative to acetic acid (the product of hydrolysis). In addition, the sulfur atom is less electronegative than oxygen and the carbonyl group is more readily polarizable in the thiol ester than in the oxygen ester and the carbon of the carbonyl is more susceptible to nucleophilic attack because of the presence of the species (*c*). This form also facilitates loss of $H^{\oplus}$ from the α carbon with the formation of the carbanion (*d*) which is stabilized by contributions from the structures shown.

There are five types of enzyme-catalyzed reaction in which thiol esters participate:

(1) Condensations at the α carbon of the ester (see Figure 8.20).

(2) Attack by a nucleophile at the carbonyl group, for example,

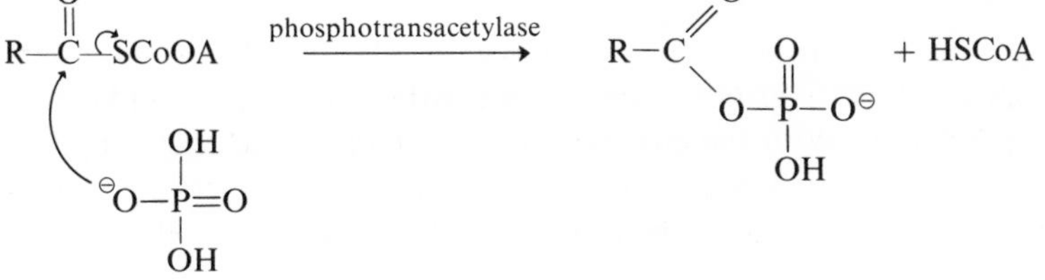

(3) 1,4 addition to α-β unsaturated thiol esters

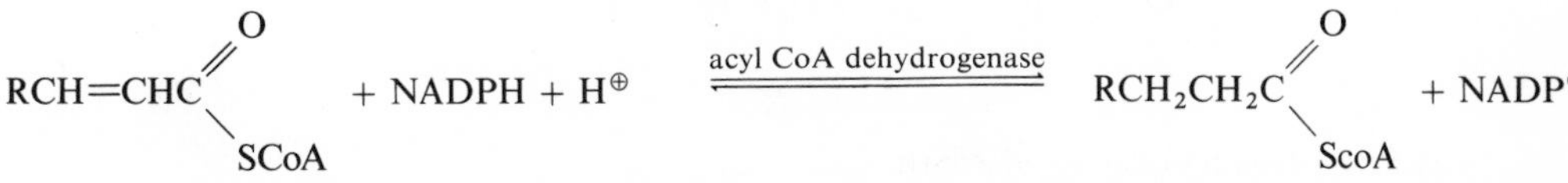

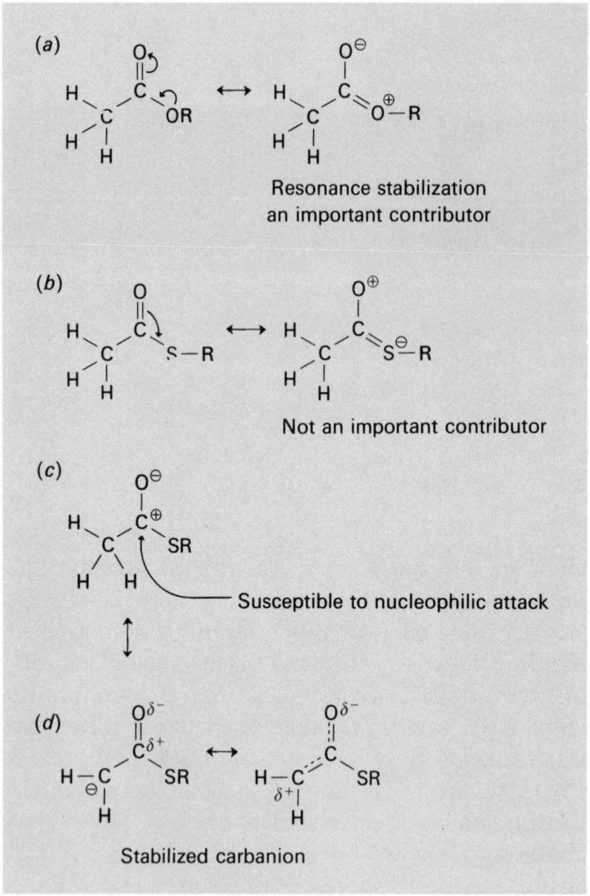

Figure 8.21

The resonance forms of thiolacetate esters and their role in reactions of the compound.

(4) Oxidation reduction. These reactions usually involve thiol esters formed from a sulfhydryl group on an enzyme surface or of lipoic acid rather than of coenzyme A. With the enzyme glyceraldehyde phosphate dehydrogenase, the reaction involves a cysteine on the enzyme, is reversible, and is coupled by the same enzyme to the formation of an acyl phosphate:

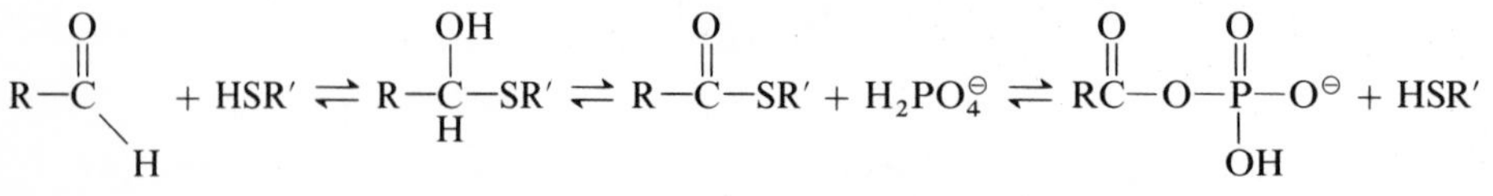

enzyme thiohemiacetal thiolester

(5) Thiol ester exchange:

$$\underset{\text{acetoacetyl CoA}}{CH_3\overset{O}{\overset{\|}{C}}CH_2\overset{O}{\overset{\|}{C}}-SCoA} + \underset{\text{succinate}}{{}^{\ominus}OOCCH_2CH_2COO^{\ominus}} \xrightleftharpoons{\text{thiophorase}} \underset{\text{acetoacetate}}{CH_3\overset{O}{\overset{\|}{C}}CH_2\overset{O}{\overset{\|}{C}}-O^{\ominus}}$$

$$+ \underset{\text{succinyl CoA}}{{}^{\ominus}OOCCH_2CH_2\overset{O}{\overset{\|}{C}}-SCoA}$$

E. Other Nucleotides

There is no value in attempting to formulate a comprehensive list of nucleo-tides found in nature; however, there is another group which deserves mention. These are the substituted nucleotide diphosphates that are used in the transfer of many organic residues in the synthesis of complex cell structures. Some representative examples are given in Figure 8.22.

F. The Synthesis of Nucleosides

This can be accomplished in several ways. The preformed carbohydrate and base can be condensed using suitably activated forms; the nucleoside can be built up from a simple N-glycoside precursor; or a preformed nucleoside can be modified to give the desired structure.

In the first example (Figure 8.23), a glycosyl halide is condensed with an alkoxy derivative of a base. The alkoxy derivatives fixes the base in the lactim form and limits the reaction to nitrogens since the other potential site of reac-tion, the hydroxyl group, is blocked. The intermediate formation of a quatern-ary amine is proposed. The condensation of glycosyl halides with purines or pyrimidines which have reactive amino groups or which can exist in the lactam form can be accomplished best by acylation of the amino functions and by using the mercuri- or chloromercuri- derivatives of the bases (Figure 8.24). The stereochemistry of the glycosidic bond is determined by the character of the substituents at C-2' of the sugar. When a 2'-O-acyl group is present, it can participate in the reaction so that a C1-C2 *trans* arrangement is produced (Figures 8.23 and 24). A *cis* arrangement can be achieved if a nonparticipating substituent is present at C-2.

The successful synthesis of 2-deoxy ribonucleosides was accomplished using 2-deoxy-3,5-di-O-p-nitrobenzoyl-D-ribofuranosyl chloride and the chloromer-curi- derivative of a base. Both α- and β-N-glycosides are formed and must be separated by chromatography.

The condensation of a glycosylamine with β-ethoxy-N-ethoxycarbonyl-acrylamide is an example of the synthesis of pyrimidine nucleosides by the

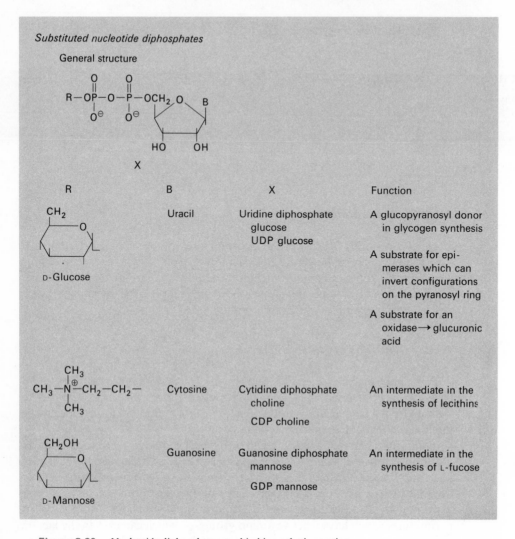

Figure 8.22   Nucleotide diphosphates used in biosynthetic reactions.

formation of the base on a preformed carbohydrate moiety (Figure 8.25). Both α and β isomers are produced. The synthesis of purine glycosides has been accomplished by building up the pyrimidine ring portion of the system using a preformed substituted glycosylimidazole. There is a strong analogy between this type of synthesis and the biosynthesis of purine nucleotides.

A very large number of syntheses based on modifications of preformed nucleosides have been reported. Inversions, oxidations, reductions, and substitutions have been performed. A single example is shown which was chosen to show that the base can participate in alterations of the sugar (Figure 8.26). Modifications of the base in a nucleoside can be accomplished provided the

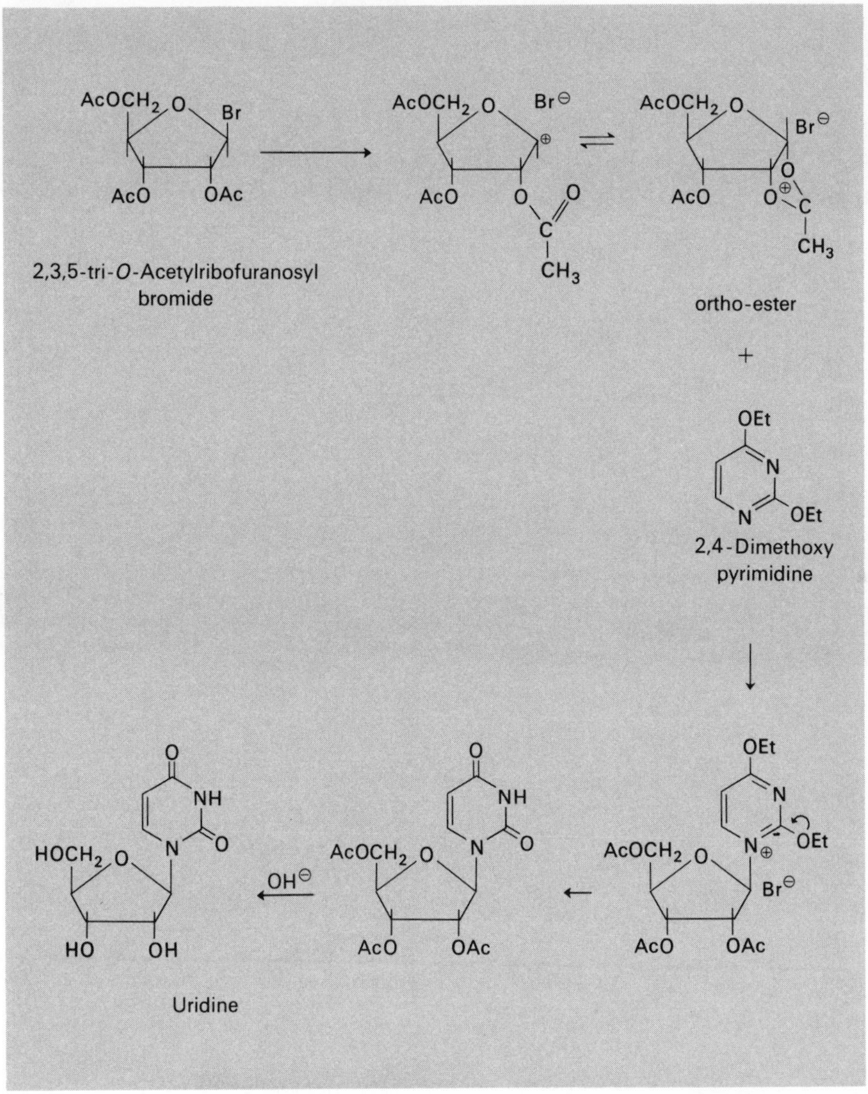

Figure 8.23   The synthesis of uridine. [G. A. Howard, B. Lythgoe, and A. R. Todd, *J. Chem. Soc.*, 1,052 (1947).]

procedures are mild enough to retain the glycosidic bond. The conversion of uridine to thymidine is shown in Figure 8.10.

Among the most versatile derivatives of the purine and pyrimidine bases are the chloro compounds. Some of the interconversions that can be accomplished starting with the 6-chloropurine nucleoside are shown in Figure 8.27.

None of the synthetic processes described above is universally applicable. The detailed structure of both the base and the sugar play roles in determining

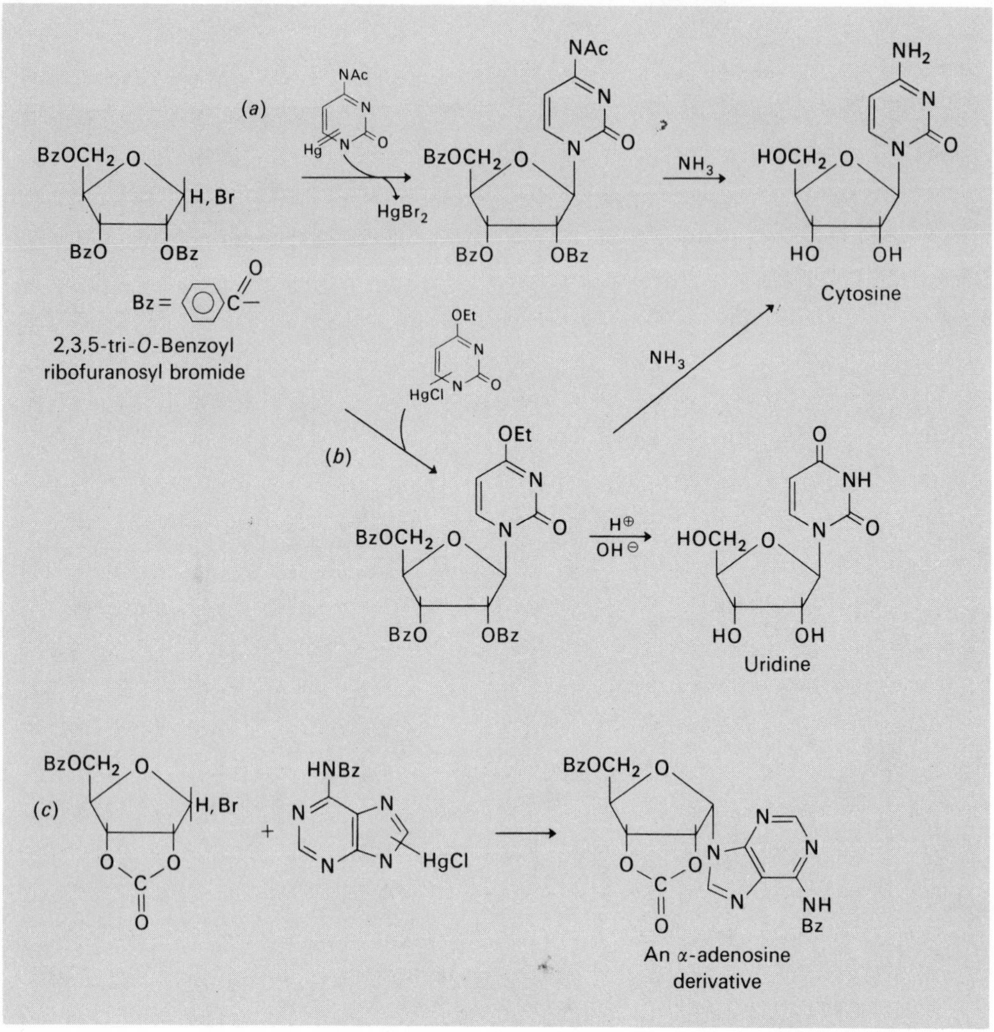

Figure 8.24   The use of mercuri- and chloromercuri-derivatives of bases in the synthesis of nucleosides. (*a*) and (*b*) J. J. Fox, N. C. Yung, I. Wempen, and I. L. Doerr, *J. Am. Chem. Soc.*, **79**, 5,060 (1957). (*c*) P. Blumbergs and C. L. Stevens, Abstr. 139th Am. Chem. Soc. Meeting, 30N (1961).

reactivity. However, the examples do illustrate the methods which have been successful.

The synthesis of nucleotides utilizes the phosphorylation of suitably blocked nucleosides. The choice of phosphorylating agent depends on the goal of the synthesis. Some of the procedures have been described in Chapter 6.

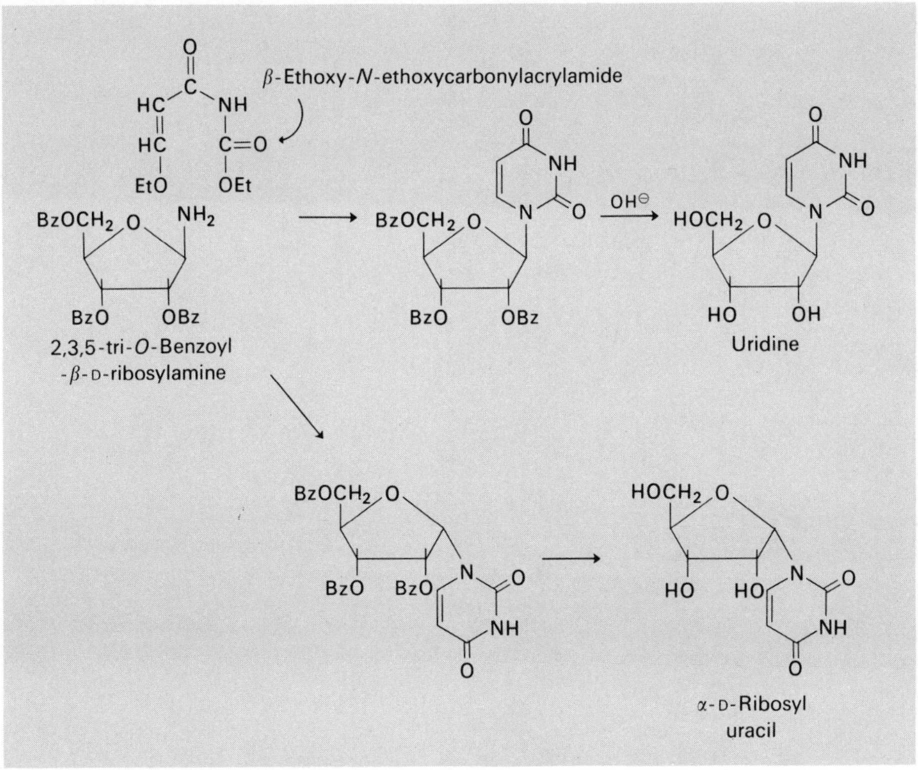

Figure 8.25   The synthesis of a nucleoside by formation of the base on a preformed carbo-hydrate moiety. [G. Shaw, R. N. Warrener, M. H. Maguire, and R. K. Ralph, *J. Chem. Soc.*, 2,294 (1958).]

8.3   THE POLYNUCLEOTIDES AND NUCLEIC ACIDS

As with the polypeptides and proteins, there is no clear distinction between polynucleotides and nucleic acids. In general, the former term is used to refer to synthetic materials and the latter to the naturally occurring high-molecular-weight compounds. Since synthetic methods are now being applied to the synthesis of large compounds, this distinction is disappearing. The term *oligonucleotide* is used for compounds having a few nucleotide residues.

In nature two types of polynucleotide occur; one contains D-ribose and is termed *ribose nucleic acid* or RNA and the other contains 2-deoxy-D-ribose and is termed *deoxyribose nucleic acid* or DNA. As seen later, these two types play different roles and have slightly different base content. Nucleic acids containing both ribose and deoxyribose are not found. Both DNA and RNA are formed from nucleotides joined in phosphodiester linkage between the 3'- and 5'-

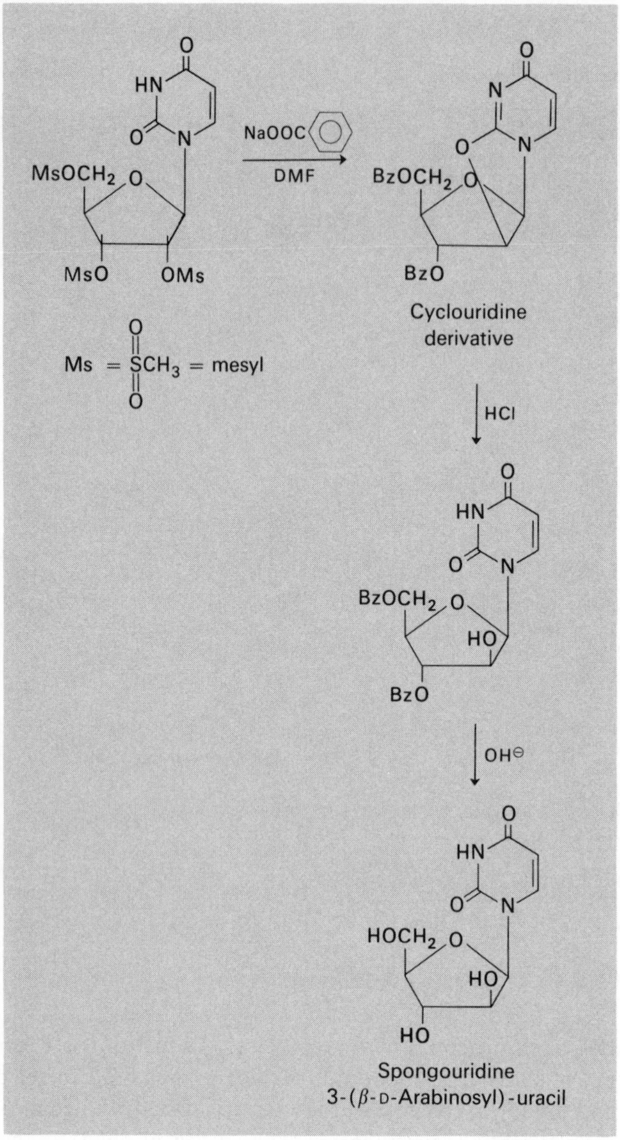

Figure 8.26

The synthesis of spongouridine. [N. C. Yung, J. H. Buchenal, R. Fecher, R. Duschinsky, and J. J. Fox, *J. Am. Chem. Soc.*, **83**, 4,060 (1961).]

hydroxyl groups of the carbohydrate. Examples of the two are shown in Figure 8.28 along with a convenient shorthand representation of the structures. In either class a specific structure can be specified by a series of letters which stand for the bases—the remainder of the structure is a repeating unit of ribosyl (or deoxyribosyl) phosphate moieties. By convention, sequences are always written with the 5′ end to the left. Note the similarity to the convention for specifying amino acid sequences in proteins.

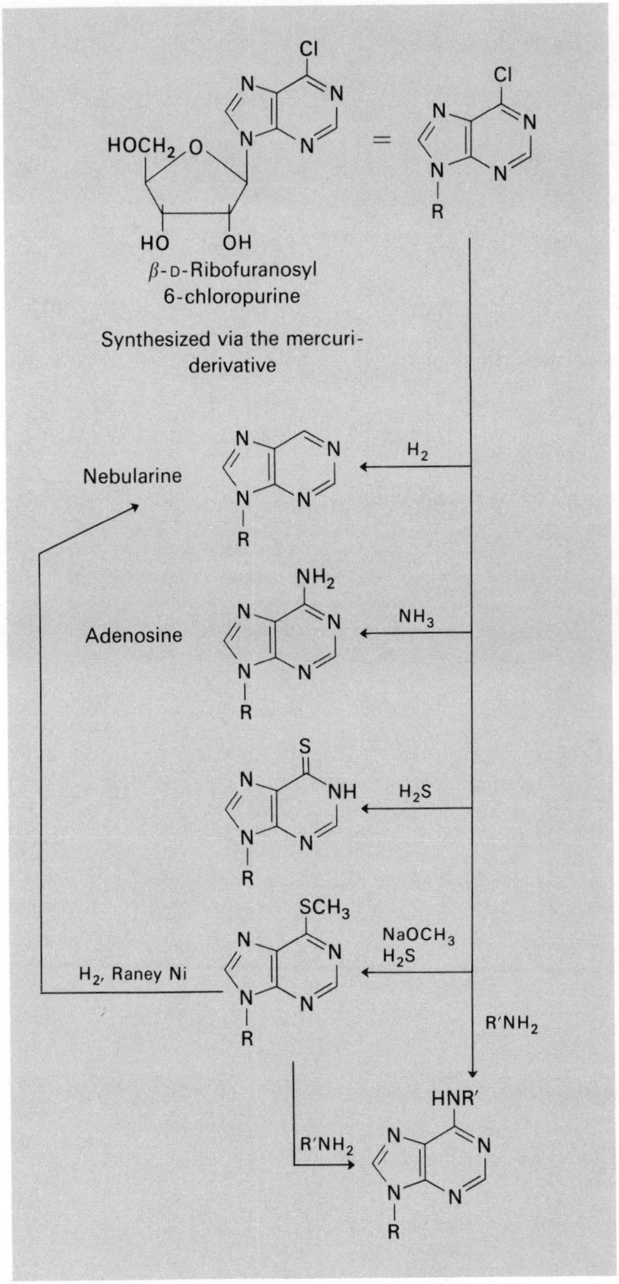

Figure 8.27

Reactions of 6-chloropurine nucleosides.

In this section, the synthesis of oligonucleotides and polynucleotides is described and the conformations that they can adopt are considered. The structures of the common nucleic acids are presented along with a brief description of their biological role.

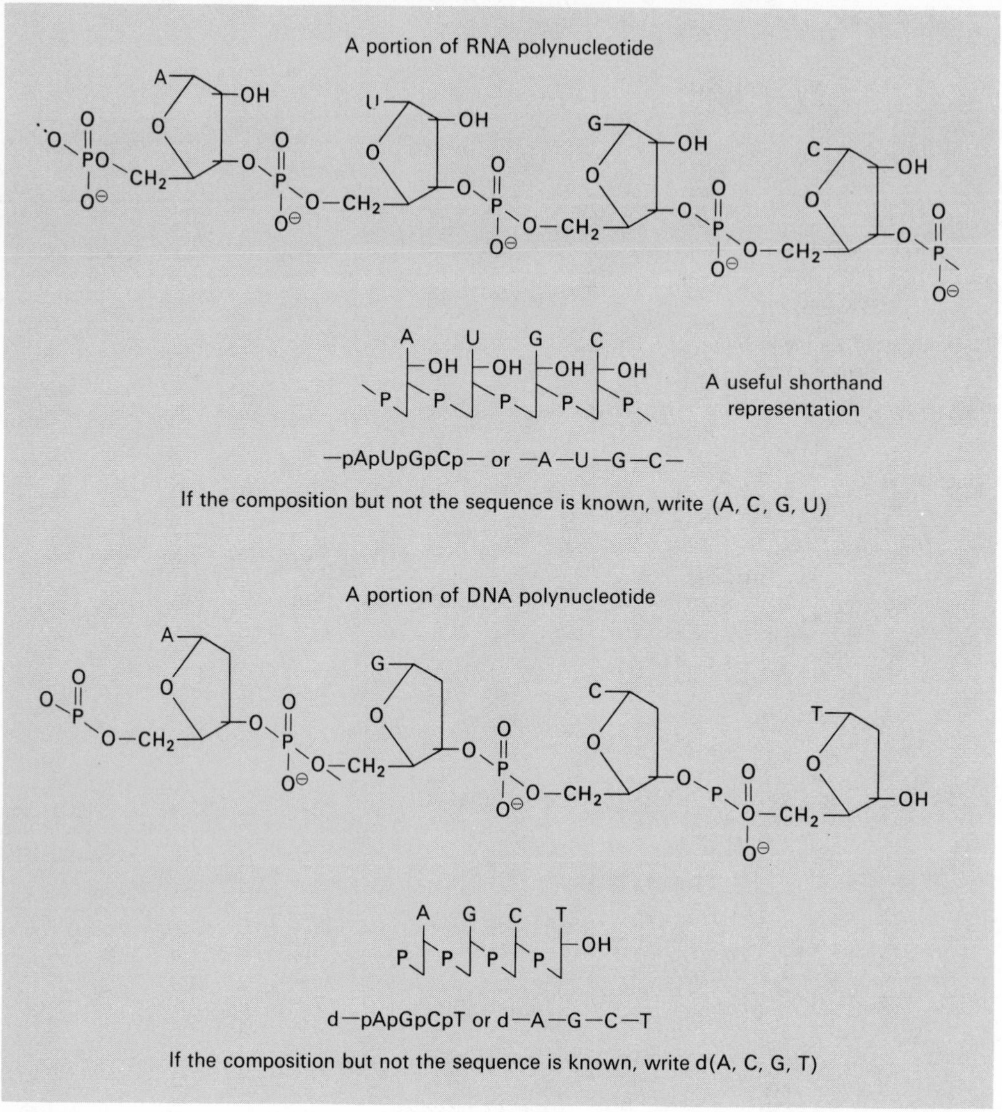

A portion of RNA polynucleotide

A useful shorthand representation

—pApUpGpCp— or —A—U—G—C—

If the composition but not the sequence is known, write (A, C, G, U)

A portion of DNA polynucleotide

d—pApGpCpT or d—A—G—C—T

If the composition but not the sequence is known, write d(A, C, G, T)

Figure 8.28   The structures and shorthand notations for ribose nucleic acids (RNA) and deoxyribose nucleic acids (DNA).

A. The Synthesis of Polynucleotides

The process for the synthesis of polynucleotides is similar to that for forming ordinary phosphate esters. The reaction can be performed by activating the phosphate group of a nucleotide and then offering it an opportunity to react

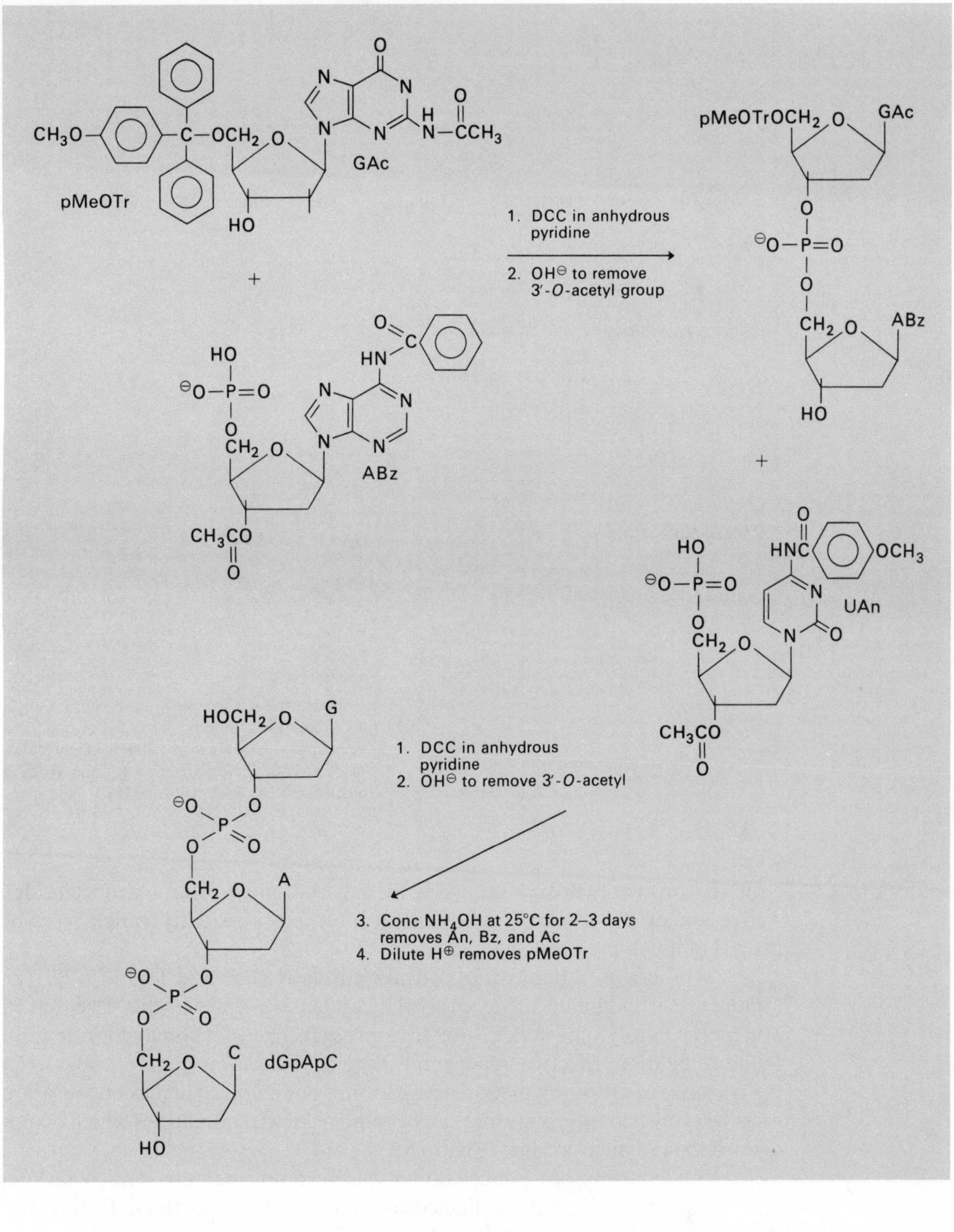

Figure 8.29 The synthesis of a deoxyribosyl trinucleotide. [S. A. Narang, T. M. Jacobs, and H. G. Khorana, *J. Am. Chem. Soc.*, **89**, 2,158 (1967), and other papers in the series.]

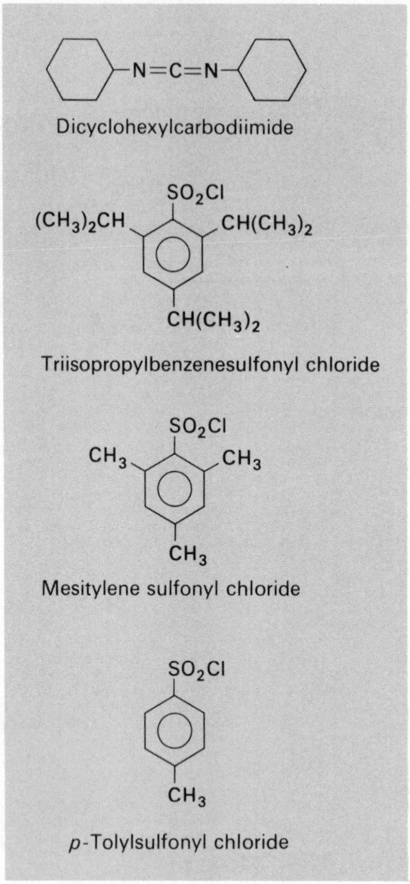

Dicyclohexylcarbodiimide

Triisopropylbenzenesulfonyl chloride

Mesitylene sulfonyl chloride

p-Tolylsulfonyl chloride

Figure 8.30

Reagents used to form phosphate ester bonds in polynucleotide synthesis. (H. G. Khorana, *Some Recent Developments in the Chemistry of Phosphate Esters of Biological Interest*, John Wiley & Sons, Inc., New York, 1961, p. 121.)

with the proper hydroxyl group in an adjacent nucleotide (or nucleoside). Activation of the hydroxyl group is an alternate possibility which has not proved of much practical value.

The synthesis of a deoxyribosyl trinucleotide is shown in Figure 8.29. Dicyclohexylcarbodiimide was used as the condensing agent in this case, but in other cases p-tolyl-, misitylene-, or triisopropylbenzene sulfonyl chloride were found to be more effective in accomplishing condensation (Figure 8.30). As in the synthesis of other polymers, reactive groups not involved in the condensation must be protected by groups that can be removed under conditions which do not disturb the polymer product (Figure 8.31).

The condensation of ribonucleotides was accomplished in a similar fashion utilizing suitably blocked ribonucleosides and 3′-ribonucleotides. In this case the 5′-hydroxyl group is condensed with the 3′-phosphoryl group; that is, the chain is built up in the opposite direction from the deoxynucleotide chain. These and similar synthetic procedures have been used to synthesize all the 64

Position blocked	Derivative	Abbreviation	Removal
3'-OH	$-\overset{\overset{O}{\|}}{C}CH_3$ Acetate	Ac	OH^-
2'-OH	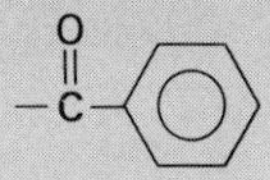 $-\overset{\overset{O}{\|}}{C}$–(phenyl) Benzoate	Bz	OH^-
5'-OH	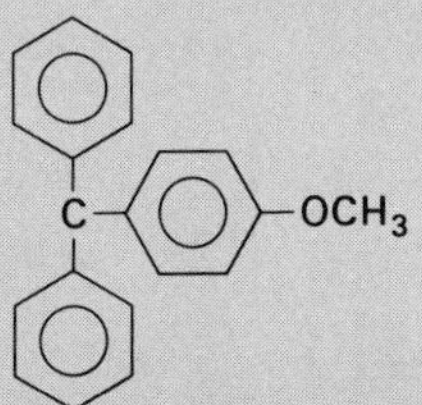 C(phenyl)(phenyl)(phenyl–OCH_3) p-Methoxytrityl (monomethoxytrityl)	MMTr	$H^{\oplus}$
$-NH_2$ of guanine	$-\overset{\overset{O}{\|}}{C}CH(CH_3)_2$ or $-\overset{\overset{O}{\|}}{C}CH_3$ Isobutyryl Acetyl	iBu, Ac	Conc. NH_4OH
of cytosine	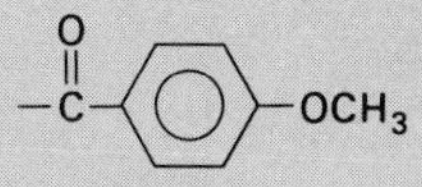 $-\overset{\overset{O}{\|}}{C}$–(phenyl)–$OCH_3$ Anisoyl	An	Conc. NH_4OH
of adenine	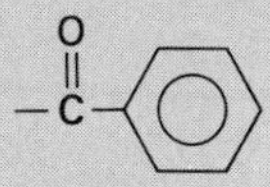 $-\overset{\overset{O}{\|}}{C}$–(phenyl) Benzoyl	Bz	Conc. NH_4OH

Figure 8.31   Blocking groups used to protect reactive sites during polynucleotide synthesis.

possible triribonucleotides which can be formed from the four bases: uracil, cytosine, adenine, and guanine.

In addition, the techniques have been extended to create polynucleotides of known squence. Two methods are outlined in Figure 8.32. The first (a) shows the synthesis of a polymer of high molecular weight of known repeating sequence. The second (b) shows the synthesis of a hexadecanucleotide of known sequence. As seen later in this chapter, these polynucleotides have been useful in confirming the mechanism of information transfer in biological systems. The extension of

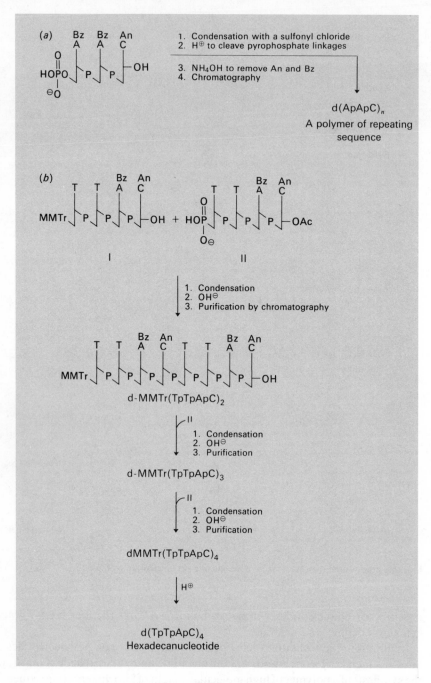

Figure 8.32   (a) The synthesis of polynucleotide of known repeating sequence but unknown degree of polymerization. [S. A. Narang, T. M. Jacobs, and H. G. Khorana, *J. Am. Chem. Soc.*, **89**, 2,158 (1967).] (b) The synthesis of a hexadeca nucleotide. [E. Ontsuka and H. G. Khorana, *J. Am. Chem. Soc.*, **89**, 2,195 (1967).]

the synthetic approach to even larger polynucleotides of known sequence has utilized specific enzymes as synthetic tools and an approach to the synthesis of a relatively large specific polymer is shown in Figure 8.45. Before this synthetic approach can be described, it is necessary to consider the structures of the naturally occurring polymers which are the substrates for the enzymes used in the synthesis.

As with the proteins the description of the nucleic acids as linear polymers of specific sequence conveys only a hint of the complexity of the natural structures. There are two major sites for interaction in the nucleic acids: the phosphate diester backbone which is negatively charged in aqueous solution and the purine and pyrimidine bases which have aromatic character and functional groups capable of hydrogen bonding.

The following sections give brief descriptions of the structures of the two types of nucleic acids along with some consideration of their chemical and physical properties important to their biological function.

B. Deoxyribose Nucleic Acids

Structure. The nuclei of nucleated cells contain relatively large amounts of DNA, which accounts for the name of the material. DNA is present in smaller amounts in other structures and in nonnucleated cells capable of reproduction (many microorganisms); the DNA is distributed throughout the cell. In nuclei it is present in the chromatin material of the chromosomes as a nucleoprotein. The proteins of nucleoprotein have a very high content of arginine and lysine and are combined through ionic linkages involving these positively charged residues and the negative charges of the phosphate diester backbone of DNA. The protein can be removed by treatment with sodium chloride.

The isolated DNA has an enormous molecular weight (up to 10^9 daltons) and contains phosphate, 2-deoxy-D-ribose, and thymine; cytosine, adenine, and guanine. The bases are present in pairs such that the amount of adenine equals the amount of thymine, and the amount of guanine equals the amount of cytosine; that is, $A:T = G:C = 1:1$, but $A \neq G$. Thus the bases of DNA are one-half purines and one-half pyrimidines. The DNA molecules are highly regular rod-like structures formed by the combination of two polynucleotide strands joined together by hydrogen bonds between the bases. The two strands have opposite polarity; the 5' end of one is paired with the 3' end of the other. Also, they are exactly complementary in the sense that an adenine moiety in one strand is always paired with a thymine moiety in the other. Similarly, guanine always complements cytosine. The geometry of the hydrogen-bonded nucleotide pairs is such that the two strands are formed into a helix which has a right-handed screw axis. The structure is referred to as a *double helix*. The base pairs involved are shown in Figure 8.33. Note that the deoxyribose residues are not exactly opposite one another across the helix axis. This placement of the carbohydrate

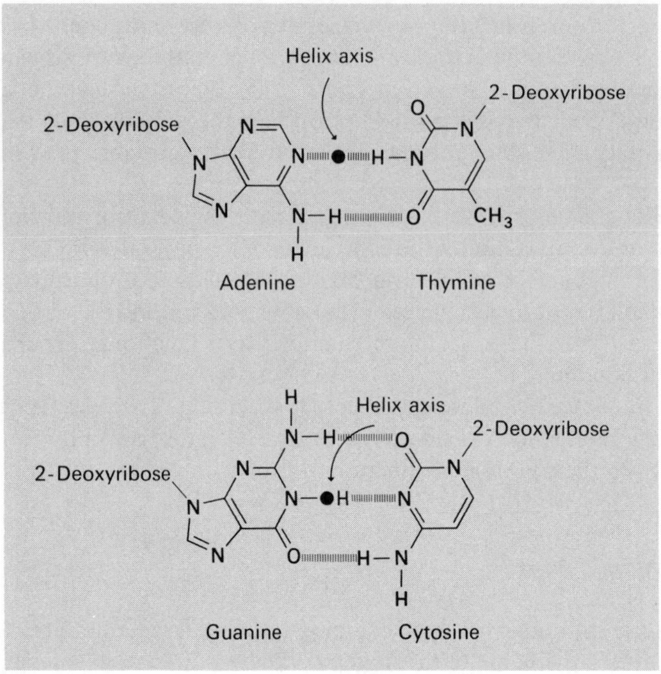

Figure 8.33
The base pairs present in DNA.

phosphate backbones of the two chains produces a large and a small groove on the outer surface of the helix. The DNA molecule resembles the globular proteins in having its polar groups to the outside and its nonpolar groups to the inside. In Figure 8.34(*a*) a small portion of the double-stranded structure is shown in a highly diagramatic way. In (*b*) the structure is diagrammed to emphasize the relationship between the two phosphate diester backbones. In this representation the base pairs are not shown.

The structure of DNA was elucidated by X-ray diffractometry with the knowledge of the overall composition of the material and many of its functions. The patterns obtained could not be interpreted in terms of the placement of individual atoms since the interior of the DNA molecule is highly hetereogeneous. The patterns were interpreted in terms of the regularity of the overall structure, however, and the structure described above was deduced largely by building molecular models to determine how the parts could be fitted together to give the regularity observed.

The structure of DNA therefore was not proved in the usual sense. However, all the properties of the molecule—physical, chemical, and biochemical—serve to verify the proposed structure.

Naturally occurring DNA molecules should possess unique structures since each cell contains one or more molecules of DNA which carry its genetic information, and these molecules must possess unique base sequences. The determination of DNA sequences presents very much the same kind of problems as the

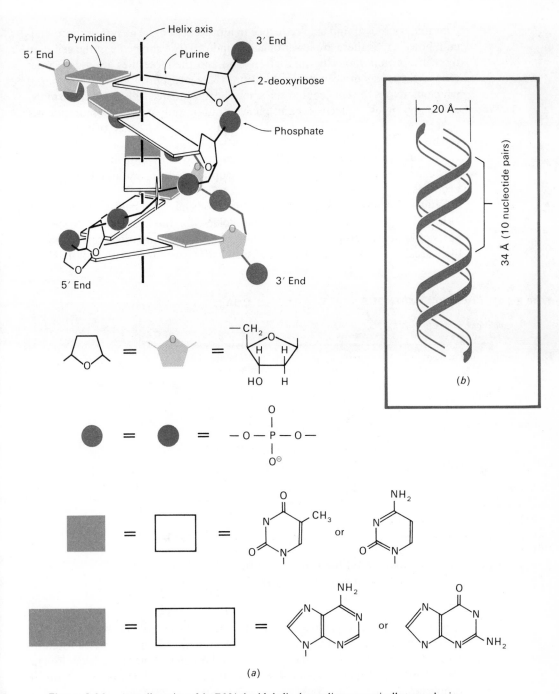

Figure 8.34   A small portion of the DNA double helix shown diagrammatically to emphasize the relative orientation of the component bases, deoxyribose, and phosphate units. (b) The DNA double helix represented as a pair of ribbons to demonstrate the relative location of the two ribosyl phosphate backbones.

determination of amino acid sequence in a protein does. The task is more difficult, however, as there are fewer of each kind of DNA molecule in the cell than there are protein molecules of a given kind and the molecules are much larger than the protein molecules. In addition, there are only four different bases, which increases the chance that the same short sequence will occur in the molecule many times. The viruses and phages contain only one DNA molecule per particle and can be used to obtain pure DNA, but the other two difficulties remain.

The base composition of DNA is determined by acid hydrolysis followed by chromatographic separation and spectrophotometric examination of the products.

Sequence determinations by chemical methods have not been very successful, although a few procedures have been evolved for the destruction of specific parts of DNA structure and the isolation of the remaining fragments. Some of these are described in Table 8.2. The use of hydrolytic enzymes to explore DNA

TABLE 8.2   CHEMICAL METHODS FOR DEGRADING DNA

Conditions	Result
Diphenylamine—formic acid[a]	Hydrolysis of purine N-glycosidic bonds, then β elimination to give pyridine oligonucleotides.
Hydrazine—alkali[b]	Removal of pyrimidines, then β elimination to give purine oligonucleotides.
Permanganate—pH 9[c]	Oxidation of all bases except adenine, then β elimination to give adenine oligonucleotides.
OsO$_4$ (osmium tetroxide)[d] Alkali	Oxidizes pyrimidines (thymine, 40 times faster than cytosine).
Diphenylamine—formic acid	Cytosine oligonucleotides can be isolated.

[a] K. Burton and G. B. Petersen, *Biochem. J.*, **75**, 17 (1960).

[b] V. Habermann, S. Habermannová, and M. Cerhová, *Biochim. Biophys. Acta*, **76**, 310 (1963).

[c] A. S. Jones and R. T. Walker, *Nature*, **202**, 1,108 (1964).

[d] K. Burton and W. T. Riley, *Biochem. J.*, **98**, 70 (1966).

structure is also complicated by the fact that those isolated so far are not completely specific for the base of the nucleotide acted on. A few are listed in Table 8.3. One of the ways in which these and other enzymes have been used in DNA structure elucidation is described below in the section on chemical properties.

Physical Properties. The proposed structure predicts that DNA is a rigid rod-like molecule which will form highly viscous solutions, that the length of a DNA

TABLE 8.3 ENZYMES USED TO CLEAVE DNA AND RNA[a]

I. DNA

Enzyme	Specificity	Product
A. Endonucleases		
Deoxyribonuclease I DNase I (pancreas)	Cleaves 3'-phosphate ester bond between Pu and Py —Pu—p—Py— ↑	Nucleotides with terminal 5'-phosphate
DNase II	Cleaves 5'-phosphate ester bond. Between Pu and C preferred —Pu—p—C— ↑	Nucleotides with terminal 3'-phosphate
Phosphodiesterase	Cleaves 3'-phosphate ester bond —X—p—X— ↑	Nucleotides with terminal 5'-phosphate
B. Exonucleases		
Phosphodiesterase I (snake venom)	Requires 5'-phosphate group on polynucleotide Cleaves 3'-phosphate ester bond —p—X↑p—X	5'-Mononucleotides
Phosphodiesterase II (beef spleen)	Requires 5'-hydroxyl group on polynucleotide Cleaves 5'-phosphate ester bond X—p↑X—	3'-Mononucleotides

II. RNA

Enzyme	Specificity	Product
A. Endonucleases		
Ribonuclease RNase (pancrease)	5'-Phosphodiester linkage adjacent to a pyrimidine nucleotide —Py—p—Pu— ↑	Oligonucleotides or nucleotides with 3'-phosphate
5'RNase (rat liver)	3'-Phosphodiester linkage —X↑p—X—	Nucleotides with terminal 5'-phosphate
RNase T1 (takadiestase)	5'-Phosphodiester linkage adjacent to guanine nucleotide —G—p—X— ↑	Nucleotides with terminal 3'-guanylate
RNase T2	5'-Phosphodiester linkages adjacent to adenine nucleotide —A—p—X ↑	Nucleotides with terminal 3'-adenylate
B. Exonucleases		
The enzymes listed for DNA can also cleave RNA.		

[a] *Handbook of Biochemistry, Selected Data for Molecular Biology* (H. A. Sober, ed.) The Chemical Rubber Co., Cleveland, Ohio, 1968, pp. H–20.

molecule should be 3.4 Å times the number of base pairs it contains, that it should be highly negatively charged, and that it should be dissociated into randomly coiled strands by agents which disrupt hydrogen bonds, such as urea, guanidinium chloride, and heat. All of these predictions agree with experimental observations.

The spectral properties of the native molecule also agree with the proposed structure. The molar extinction coefficient of the native molecule at the wavelength of maximum absorption of the component bases is significantly lower than that of unfolded material. This effect is attributed to the stacking of bases in the native molecule, which results in the absorption of light by one base decreasing the probability that neighboring bases will absorb light. In the more random denatured state, this restriction is removed and the denaturation or hydrolysis of native DNA can be followed by observing the increase in absorbance at 260 nm (hyperchromic effect). This increase parallels changes in the other physical properties such as viscosity, which is decreased by hydrolysis and unfolding. Native DNA also shows a strong anomolous optical rotatory dispersion (Cotton effect) in the region of the absorption maxima of the bases. When the helical structure is disrupted, there are very significant changes in the dispersion curves. The optical rotation at the sodium D line is also altered by denaturation. It decreases from $+100$ to 150 in the helical structure to a value close to zero in the denatured state.

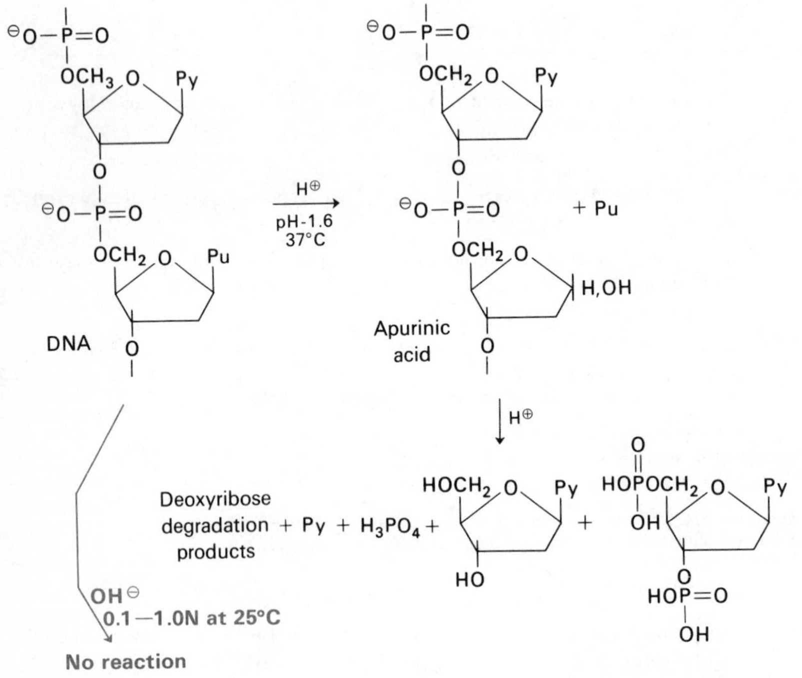

Figure 8.35 The acid-catalyzed hydrolysis of DNA. [C. Tamm and E. Chargaff, *J. Biol. Chem.*, **203**, 689 (1953).]

Chemical Properties. DNA, having 3′,5′-phosphodiester linkages and *N*-glycosidic bonds, should be hydrolyzed by acid to give phosphoric acid, the component bases, and deoxyribose (or a degradation product of deoxyribose) (Figure 8.35). However, the *N*-glycoside bonds involving purines are more labile than those involving pyrimidines and mild acid hydrolysis (pH 1.6 at 37°C) produces a depurinated material termed "apurinic acid." The lability of the purine *N*-glycosidic bond is due to intramolecular catalysis by the 3-NH of the purine moiety (Figure 8.9). In addition, the phosphate monoesters which are formed by cleaving the diester bonds are relatively stable and under certain conditions of acid hydrolysis deoxyribonucleosides and their 3′,5′-diphosphates can be isolated. DNA is stable to alkali (see Figure 6.12) since it has no groups adjacent to the phosphate diester group to assist in its hydrolysis.

The highly specific base pairing proposed for DNA requires that certain chemical properties of the material will vary with the base composition. For example, the stability of the double helix should increase as the proportion of guanine plus cytosine increases since the pairing of these bases involves three hydrogen bonds, whereas the adenosine-thymine pair has only two hydrogen bonds (Figure 8.33). The stability of DNA is expressed in terms of a "melting temperature," this is the temperature at which the molecule in solution is 50

TABLE 8.4 THE DEPENDENCE OF POLYNUCLEOTIDE MELTING TEMPERATURES ON BASE CONTENT[a]

Helical structures	*Melting temperature, Tm, °C*
polyrA : polyrU in 0.1 *M* NaCl	56
polyrA : polyrT in 0.1 *M* NaCl	75
polyrG : polyrC in 0.001 *M* NaCl	98
polydA : polydT in 0.1 *M* NaCl	68
polydG : polydC in 0.05 *M* NaCl	91

NATURAL DNA OF VIRAL ORIGIN

Source	*Tm, °C*	*Moles Percent, (Guanosine and Cytosine)*
Human Adenovirus 1	92.8	58.5
Human Adenovirus 2	92.5	55
Human Adenovirus 18	88.8	48
Human Yamavirus	82.5	30
Fowl pox	83.5	35

[a] *Handbook of Biochemistry, Selected Data for Molecular Biology* (H. A. Sober, ed.) The Chemical Rubber Co., Cleveland, Ohio, 1968, pp. H–16.

percent unfolded as judged by measurements of absorbance at 260 nm. That this expectation is realized is shown by the data in Table 8.4.

Most of the other tests of the structure proposed for DNA require some knowledge of its function. A brief description is attempted here; a much more complete description is presented in the volume by Wold in this series. A section of DNA structure is shown diagrammatically in Figure 8.36. As stated

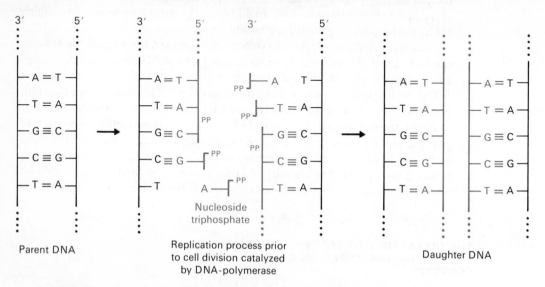

Parent DNA

Replication process prior
to cell division catalyzed
by DNA-polymerase

Daughter DNA

Figure 8.36   The replication of DNA and a diagrammatic representation.

above, it contains the information that is transmitted from generation to generation and that determines the biochemical capabilities of the cell. It is reproduced prior to cell division in a process which can be written formally as involving separation of the strands and the assembly on each of them of the nucleoside triphosphates which will reproduce the complementary sequence. The nucleotides are then joined together to generate two DNA molecules identical with the parent molecule—each containing one of the parent strands. The way in which DNA transmits its information to the cell involves the formation of RNA complementary to one of the DNA strands in a somewhat analogous fashion to that described above. In this case, UTP and not TTP complements A in the parent strand. The RNA (messenger RNA) is transported to the protein-synthesizing apparatus in the cytoplasm (Figure 8.37).

The conversion of the message in the RNA molecule to a biochemical capability requires that the message be translated into the amino acid sequence for a specific protein since the proteins determine the capabilities of the cell. The alphabet in which the messages are written has four letters [A, U, G, and C]. The words, which each specify an amino acid, have three letters; for

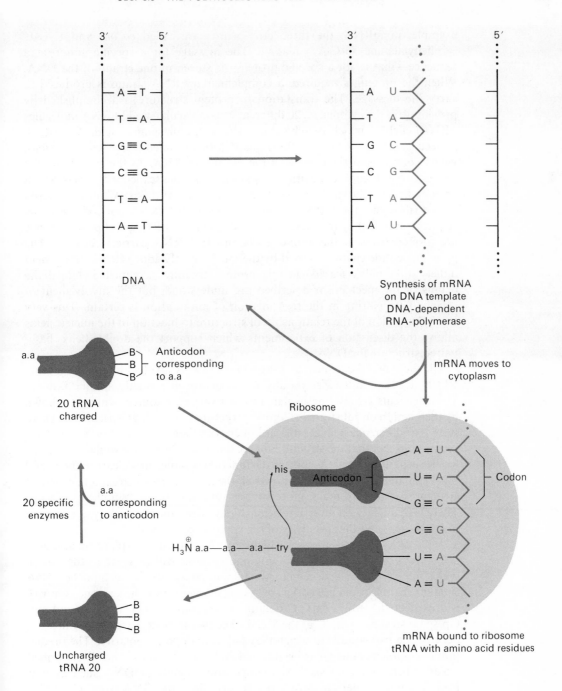

Figure 8.37 The transmission of genetic information in the cell. The formation of mRNA and its utilization in protein synthesis.

example, in mRNA, the three-letter words are called *codons* and U-U-U = phenylalanine, G-C-A = alanine. The message for a specific amino acid sequence—that is, for a specific protein—is stored in one strand of the DNA. When the protein is required, a complementary RNA strand is produced to carry the message. The translation to protein structure is accomplished by providing, in the cytoplasm, 20 different low-molecular-weight RNA molecules (tRNA) each of which is "charged" with a specific amino acid. The tRNA molecules have sequences of three bases in them, called *nodocs* or *anticodons*, which complement the codons of the messenger RNA. In the cytoplasm, the messenger is complexed with cytoplasmic particles called *ribosomes* which provide sites for the enzymes necessary for protein synthesis. While complexed to the ribosome, the mRNA codons recognize the tRNA anticodons in the sequence specified by the sequence of codons in the mRNA, and peptide bonds are formed between the amino acids and the tRNA carriers released. The growing peptide chain is carried by the most recently added tRNA-amino acid moiety and is only released when the protein structure is completed. Not all the details of the mechanism described are understood, but the involvement of specific base pairing in the transmission of information is certain. This very brief description of the relationship of structure to function in the nucleic acids allows the discussion of experiments which support the double-helix, base-paired structure for DNA.

The most useful chemical investigations of DNA involve the use of reagents which cleave or otherwise modify the structure in a highly specific fashion. These reagents are enzymes isolated from a variety of sources which are shown by their action on substrates of known structure to be of high specificity. There is no real difference between the use of enzymes and the use of more common reagents, except that the structures of the latter may be known while those of the former usually are not. As long as the function is sufficiently clearly defined and purity (of function) established, however, the use of enzymes to explore chemical structures is usually much more precise than the use of "simpler" reagents. There is frequently an inverse relationship between the complexity of structure in the reagent and the specificity of the reaction performed.

A list of hydrolytic enzymes which have been used to explore nucleic acid structure is given in Table 8.3. Only one example will be given of the way in which these reagents have been used. As pointed out above, the model of DNA predicts that if a sequence of ApTpC exists in one strand, the sequence GpApT will exist in the other. ApTpC denotes a structure with A at the 5′ end and GpApT a structure with G at the 5′ end. The two sequences are complementary because the two strands in which they fall have opposite polarity. The prediction of complementarity has been tested and the process used has been termed "nearest-neighbor analysis." A homogeneous sample of DNA such as that from a simple bacterium or a virus is used since the DNA from a complex organism contains so many different molecular species, with every possible sequence of bases and with bases in almost equal proportion, that the results of the kind of experiment to be described are likely to be uninterpretable.

The DNA to be examined is used as a template for the formation of new DNA having the same structure utilizing the enzyme DNA polymerase along with dATP, dGTP, dCTP, and dTTP (Figure 8.38). One of these triphosphates has radioactive phosphorous in the 5'-phosphate group. The enzyme catalyzes the displacement of pyrophosphate from triphosphate by the 3'-hydroxyl of the entering nucleoside triphosphate, forming a 3'-5' linkage. The new DNA is then hydrolyzed with two enzymes—calf spleen diesterase and micrococcal DNAase which result in conversion of the DNA to 3'-mononucleotides in high yield. By this cycle of events, the radioactivity which was introduced in the 5' position of one nucleotide is transferred to the 3' position of the nucleotide which was its neighbor in the DNA strand. The 3'-nucleotides can be separated by electrophoresis and the ^{32}P content of each estimated. The physical properties of the newly synthesized DNA must also be examined to demonstrate its similarity to the template material.

A completely random sequence of nucleotides in DNA would be expected to result in each base being present in the same proportion. In addition, each of the 3'-nucleotides should have exactly the same ^{32}P content after the sequence of reactions shown in Figure 8.38 had been carried out. The proportions of 3'-nucleotides would be different if the proportion of a given nucleotide in a DNA sample differed from 25 percent. For example, if a given DNA sample contained

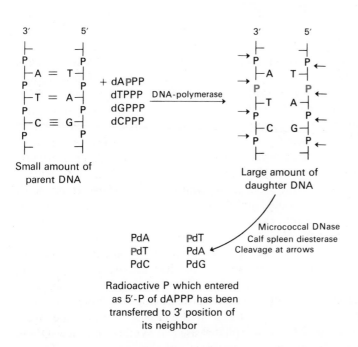

Figure 8.38   The nearest-neighbor analysis.

33.5 percent each of Cp and Gp and 16.5 percent each of Ap and Tp, a completely random arrangement of these bases would predict that the 16 pairs of neighbors which could exist in the molecule would occur with the frequencies shown in Table 8.5. The frequency of base pairing which would be generated by random

TABLE 8.5   THE NEAREST-NEIGHBOR ANALYSIS OF DNA FROM
MICROBACTERIUM PHLEI: **PREDICTED OF NEAREST NEIGHBORS BASED ON A RANDOM ARRANGEMENT AND EXPERIMENTALLY DETERMINED FREQUENCY**[a]

Base composition: dG = dC = 33.5%; dA = dT = 16.5%

Pair	Frequency		Pair	Frequency		Pair	Frequency		Pair	Frequency	
	Predicted[b]	Found		Predicted[b]	Found		Predicted[b]	Found		Predicted[b]	Found
GpC	11.2	12.2	GpA	5.5	6.3	CpT	5.5	4.5	ApT	2.72	3.1
Cpg	11.2	13.9	ApC	5.5	6.4	TpC	5.5	6.1	TpA	2.72	1.2
GpG	11.2	9.0	GpA	5.5	6.5	GpT	5.5	6.0	ApA	2.72	2.4
CpC	11.2	9.0	ApG	5.5	4.5	TpG	5.5	6.3	TpT	2.72	2.6

[a] J. Josse, A. D. Kaiser, and A. Kornberg, *J. Biol. Chem.*, **236**, 864 (1961), with permission of the *Journal of Biological Chemistry*.

[b] The predicted frequency is equal to the product of the percentage of each base in the DNA.

sequences is compared with that found in *Microbacterium phlei* DNA, which has the same base composition, by nearest-neighbor analysis. These results demonstrate clearly that the arrangement of bases in DNA is not random. When compared to the results on DNA from other sources, they show that DNA varies in this regard depending on its source—a result which could be predicted simply from knowledge that the base compositions of DNA vary with its source. Most important, they clearly demonstrate that the individual strands have opposite polarity. If the strands have the *same* polarity, then the sequence ApC in one strand will give the sequence TpG in the other. If the strands have *opposite* polarity, then ApC should give GpT, and these pairs should occur with the same frequency—they do (Table 8.6). When the template DNA and the product DNA are compared for nearest neighbors, it is clear that the former does indeed control the structure of the latter; although the in vitro duplication is not perfect. The starting material and the product usually have somewhat different physical properties and appear different in the electron microscope.

Additional evidence for the validity of the DNA structure and mode of replication has been obtained by modifying the structure of bases either in the intact molecule or in the nucleotide triphosphates used in synthesizing new DNA. Treatment of DNA with nitrous acid deaminates the amino-substituted bases and results in an altered ability of the deaminated bases to form H bonds

TABLE 8.6   DINUCLEOTIDE PAIRS WHICH SHOULD OCCUR IN EQUAL AMOUNT IN DNA AND RESULTS OF NEAREST-NEIGHBOR ANALYSIS

A *Sequence in One Strand[b]*	B *Sequence in Complementary Strand[b]*	*Ratio A/B[a] Observed*
GpG	CpC	9.0/9.0
CpA	TpG	6.3/6.3
ApC	GpT	6.4/6.0
GpA	TpC	6.5/6.1
ApG	CpT	4.5/4.5
ApA	TpT	2.4/2.6

The sequences produce the sequences:

ApT	ApT
TpA	TpA
GpC	GpC
CpG	CpG

[a] Data from Table 8.5.

[b] All sequences are written $5' \rightarrow 3'$—i.e., strands have opposite polarity.

(Figure 8.39). The altered DNA produces a different DNA product when used as a template.

Hydrogen bonding between base pairs requires a specific tautomeric form of each base. Although the pairing found in DNA involves the most stable tautomers of the bases, other tautomers are possible and their occurrence may occasionally lead to a mistake in replication of DNA or in the transcription of RNA (mutations). Some of the possible alternate tautomers are shown in Figure 8.40. It should be pointed out that the observed frequency of mutation is lower than would be predicted from the relative stabilities of the tautomeric forms.

DNA is also sensitive to treatment with ultraviolet light; this can cause adjacent thymidines to condense to give a dimer which may bind the two strands together. A similar result can be obtained using difunctional reagents such as nitrogen mustards $(ClCH_2CH_2)_2NR$ which can alkylate suitably situated bases.

Dimer produced by ultraviolet irradiation of DNA

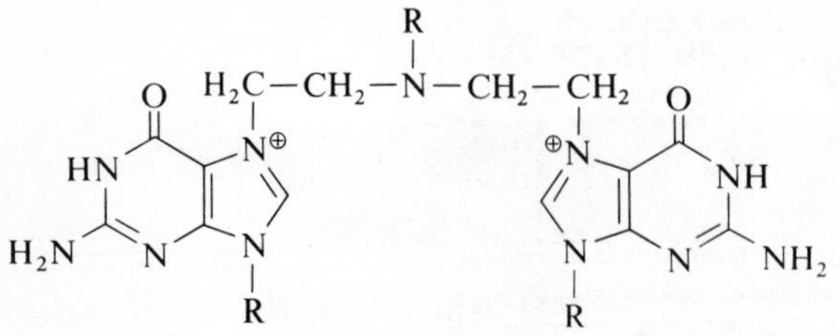

R
|
H₂C—CH₂—N—CH₂—CH₂

Dimer produced by treatment with nitrogen mustard

The best proof that the genetic code consists of a series of three-letter words (that is, three adjacent bases in the DNA code for one amino acid in a protein)

The deamination of amino purines and its effect on base-pairing.

NH₂

HNO₂

Adenine
pairs with thymine
or uracil

Hypoxanthine
pairs with cytosine

HNO₂

Guanine
pairs with cytosine

Xanthine
pairs with cytosine

Pairs with thymine
or uracil

NH₂

HNO₂

Cytosine
pairs with guanine

Uracil
pairs with adenine

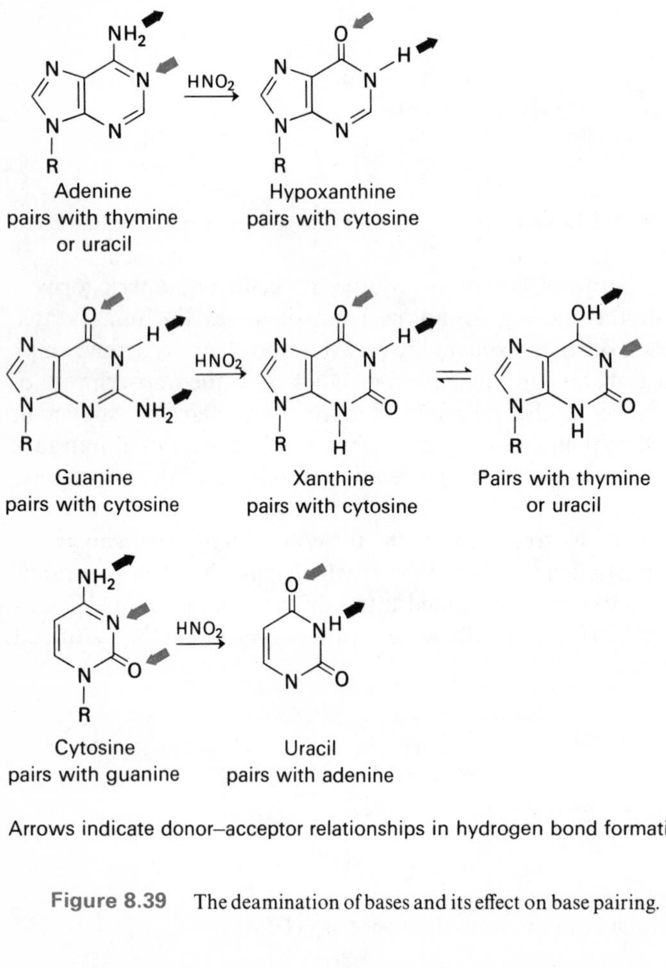

Arrows indicate donor–acceptor relationships in hydrogen bond formation.

Figure 8.39   The deamination of bases and its effect on base pairing.

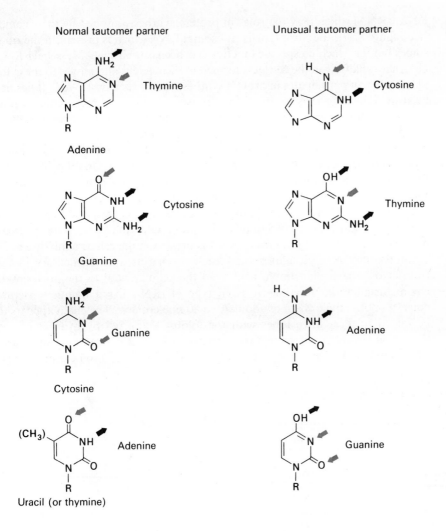

Normal tautomer partner

Adenine — Thymine

Guanine — Cytosine

Cytosine — Guanine

Uracil (or thymine) — Adenine

Unusual tautomer partner

Cytosine

Thymine

Adenine

Guanine

Figure 8.40   Tautomeric forms of bases and their potential for hydrogen-bond formation.

was provided by H. G. Khorana and his co-workers by the synthesis of poly-nucleotides of known base sequence of both deoxyribose and ribose (see Section 8.2A and Figure 8.32) and the demonstration that they could serve as templates for the enzyme systems.

C. Ribose Nucleic Acids

There are at least three types of RNA present in all cells which have nuclei and/or can synthesize protein. They are messenger RNA (mRNA) and transfer

RNA (tRNA) which play the roles in protein synthesis implied by their names (see Figure 8.37); the third type is ribosomal RNA which is present in the ribosomes but for which no specific function has been established. *Messenger RNA* has a short half-life in the cell; it is formed in the nucleus and moves to the cytoplasm where it participates in protein synthesis and is then destroyed. It has not been well characterized because of its transient nature and the small amounts present at any time. It has a molecular weight and purine-pyrimidine base composition which depend on the protein for which it codes. An "average" mRNA might contain 600 to 1,000 nucleotides. mRNA contains ribose, phosphate, and the bases uracil, cytosine, adenine, and guanosine in ratios which are fixed for a specific molecule but which can (in theory) have any value. Unlike DNA, there is not a 1:1 relationship between the amounts of purines and pyrimidines. The polymer is probably threadlike and corresponds to a fairly random coil which can have a small amount of interaction between the bases by base pairing or stacking. *Transfer RNA* is present in the cell as a mixture of 20 or more different molecular species each serving as an intermediary in the transfer of a particular amino acid from the cellular pool to the polypeptide chain during protein synthesis. Each type of tRNA has a unique structure consisting of a single chain of about 70 to 80 nucleotides. The chain appears to be folded to form three lobes with the folding stabilized by hydrogen-bond formation between complementary base pairs. There are several unusual bases present in tRNA (Figure 8.41 and Table 8.7). The amino acid to be transferred is

TABLE 8.7   UNUSUAL BASES PRESENT IN tRNA

N(6)-Dimethyladenosine	= DMA
1-Methyladenosine	= MA
N(6)-Isopentyladenosine	= NPA
N(2)-Dimethylguanosine	= DMG
N(2)-Methylguanosine	= NMG
1-Methylguanosine	= 1MG
2'-*O*-Methylguanosine	= OMG
N(7)-Methylguanosine	= 7MG
Inosine (hypoxanthine or 6-hydroxy purine)	= I
1-Methylinosine	= MI
5-Methylcytidine	= 5MC
N(6)-Acetylcytidine	= NAC
2'-*O*-Methylcytidine	= OMC
4,5-Dihydrouridine	= DHU
2'-*O*-Methyluridine	= OMU
Pseudo-uridine	= PSU
Ribothymidine	= RT

attached to ester linkage to the 3' end of the tRNA (Figure 8.41 insert). *Ribosomal RNA* is a constitutive part of the ribosomes (the site of protein synthesis), has a long lifetime, but has the common bases. It is mainly single stranded but has a small amount of base pairing. The structure of one has been determined.

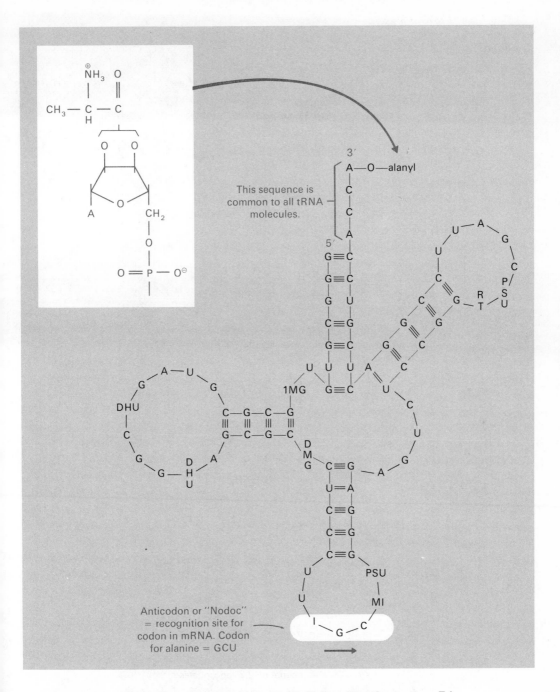

Figure 8.41 The structure of alanyl tRNA. [R. W. Holley, *Scientific American*, Feb., 1966; and R. W. Holley, J. Agpar, G. A. Everett, J. T. Madison, M. Marquissee, S. H. Merrill, J. R. Penswick, and A. Zamir, *Science*, **147**, 1,462 (1965). Copyright © 1966 by Scientific American, Inc. All rights reserved.]

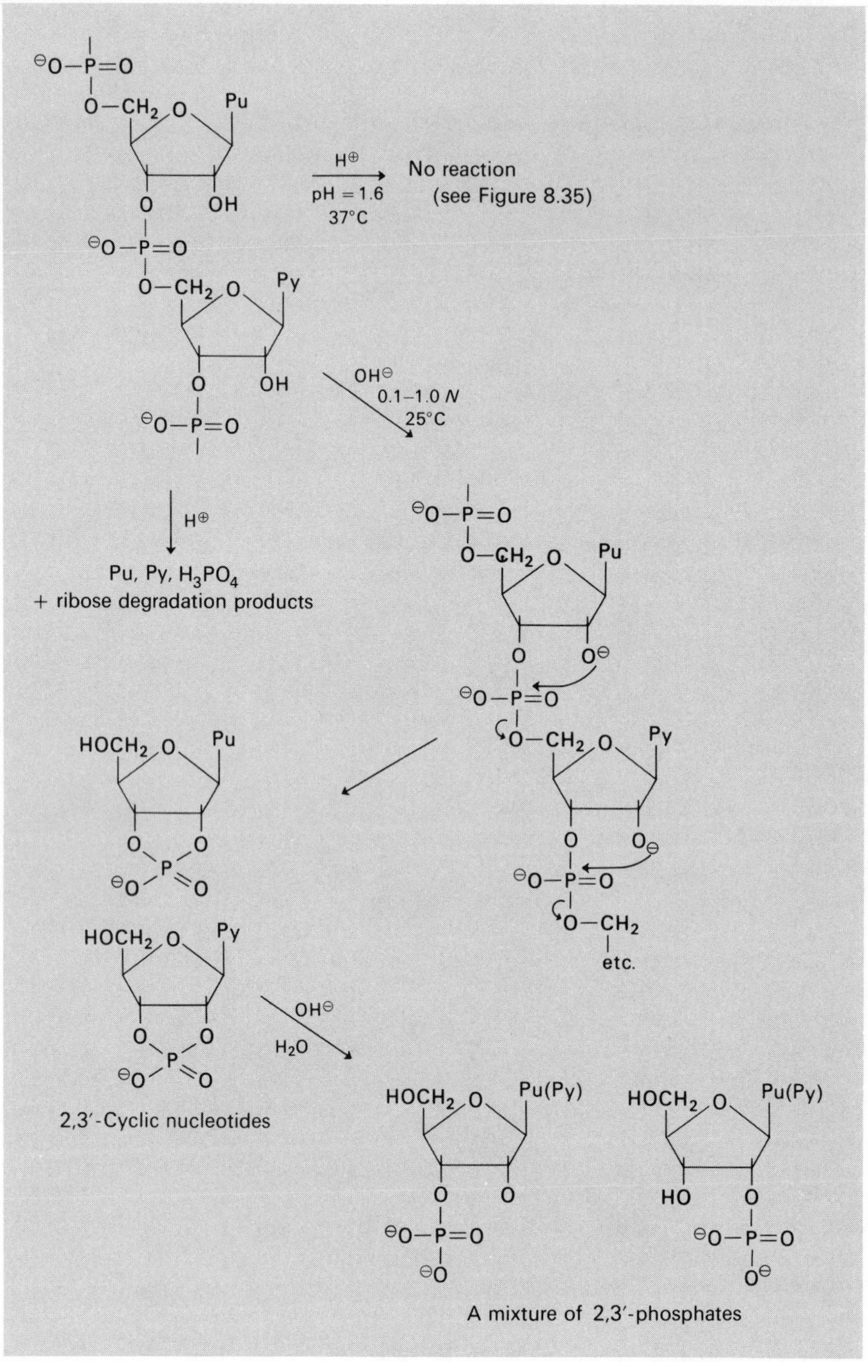

Figure 8.42   The hydrolysis of RNA. [D. M. Brown and A. R. Todd, *J. Chem. Soc.*, **52** (1952).]

It is possible that this RNA provides binding sites in the ribosome for mRNA and tRNA although this has not been established.

The chemical properties of RNA are quite different from those of DNA. RNA is unstable in alkaline medium due to intramolecular catalysis of phosphodiester hydrolysis by the 2′-hydroxyl group of the ribose (Figure 8.42). In addition, RNA is stable under the conditions which give apurinic acid from DNA due to the decreased lability of the N-glycosidic bond. The 2-hydroxyl probably produces this stability by withdrawing electrons from the ring oxygen.

There are a relatively large number of unusual bases in tRNA. Most of them are methylated derivatives of the normal bases, but three of them are not. Ribothymidine is quite normal except that the base thymine is usually present in DNA. It is not present in mRNA where uracil replaces it. Inosine is the hydroxy analog of adenosine—that is, its deamination product. Inosine is a nucleoside; the base in it is hypoxanthine. Its structure and base-pairing characteristics are shown in Figure 8.39. Note that it is present in the anti-codon region of alanine tRNA (Figure 8.41). It has been proposed that it functions to allow degeneracy in the triplet base coding described above. The spatial requirements for base pairing are much less stringent in the functioning of RNA than they are in the duplication of DNA and one can hypothesize that the base hypoxanthine can base pair with uracil, cytosine, guanine, or adenine in this relaxed situation (Figure 8.43). Since inosine is in the anticodon region of alanine-tRNA, this adapter molecule will recognize codons containing any of these bases. Thus, alanine should be specified by four codons—GCU, GCA, GCC, and GCG. This has been proved to be the case, although whether the explanation given above for the role of inosine is correct remains to be seen. Pseudo-uridine has already been discussed (Figure 8.11). Its role in tRNA is not known.

The determination of the sequence of alanine-tRNA represented a major breakthrough in nucleic acid chemistry. The problems of determining nucleic acid structures have been mentioned above; however, in tRNA the molecules contain a relatively small number of nucleotides and a relatively large number of unusual bases. Both are factors which simplify the task of sequence determination. The approach used is basically the same as that used in protein sequencing that is, to devise methods for end-group analysis and apply them to fragments obtained by a variety of hydrolytic procedures. Then by looking for common features in different oligonucleotides and overlapping established sequences, a unique sequence can be deduced. The techniques used are illustrated in Figure 8.44.

D. The Chemical Synthesis of Large Polynucleotides

The attempt to synthesize a polynucleotide of known sequence sufficiently long to code for a functional protein has not been accomplished. However, encouraging results have been obtained in attempts to synthesize a double-

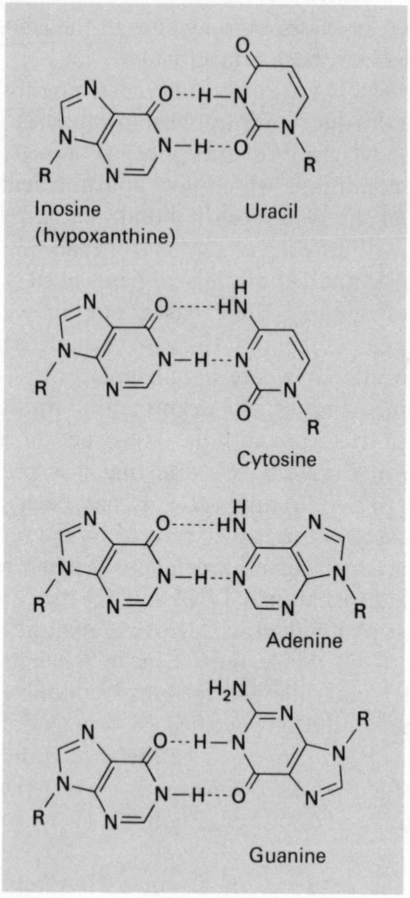

Figure 8.43

The base pairing of inosine with the four bases normally found in RNA.

stranded DNA with a sequence corresponding to the base sequence of alanine-tRNA (tRNA^{ala}) Figure 8.45. The chemical synthesis of several single-stranded oligo-deoxynucleotides which have sequences complementary to the tRNA sequence has been accomplished. Oligo-deoxynucleotides with the same sequence as the tRNA have also been prepared. These fragments, when mixed, form small double-stranded regions through base pairing. The breaks between adjacent nucleotides are then repaired by an enzyme called DNA-ligase. Although this work is not completed at the time of writing, it is sufficiently advanced to demonstrate its practicality and opens up many possibilities for the exploration of the relationship of structure to biological function in the nucleic acids.

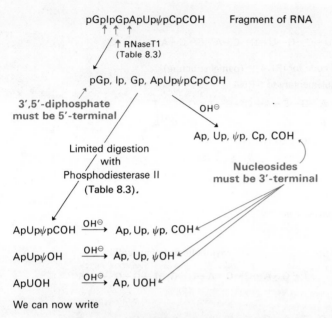

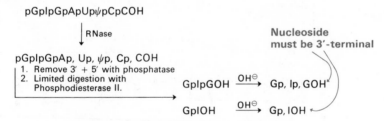

We can now write the unique sequence.

Figure 8.44   The sequencing of a tRNA fragment. (a) The action of RNase T1 which cleaves adjacent to guanylate or inosinate nucleotides as shown gives a diphosphate which must be on the 5' terminus. The remaining oligonucleotide can be cleaved with alkali to give the 3'-terminal residue as a nucleoside. Limited digestion of the fragment with Phosphodiesterase II gives smaller fragments which are separated and treated with alkali to identify the 3'-terminal nucleosides. (b) The original fragment treated with RNase is cleaved as shown and the residual fragment treated with phosphatase to allow the application of a limited digestion by Phosphodiesterase II to give small fragments that can be treated with alkali to identify their 3'-terminal nucleosides. Isolation of the two fragments shown allows a complete sequence to be written. [R. W. Holley, *Scientific American*, Feb., 1966; K. Burton, *Essays in Biochemistry*, **1**, 57 (1965).]

(*a*) Partial t-RNA[ala] sequence

—G—A—U—U—C—C—G—G—A—C—U—C—G—U—C—C—A—C—C—A   3′

(*b*) Double stranded DNA presumed to code for tRNA[ala] (partial structure)

DNA strand complementary to t-RNA

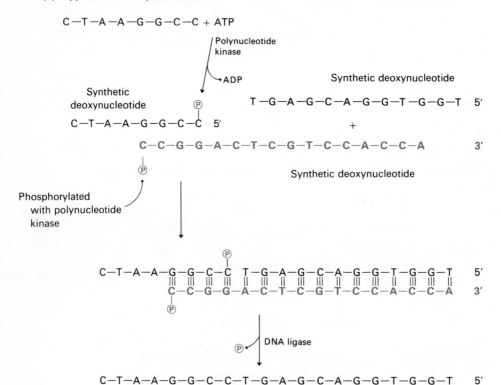

—C—T—A—A—G—G—C—C—T—G—A—G—C—A—G—G—T—G—G—T   5′

—G—A—T—T—C—C—G—G—A—C—T—C—G—T—C—C—A—C—C—A   3′

DNA strand equivalent to t-RNA

(*c*) Approach to the synthesis of DNA

C—T—A—A—G—G—C—C + ATP

Polynucleotide
kinase

ADP

Synthetic
deoxynucleotide

Ⓟ

Synthetic deoxynucleotide

T—G—A—G—C—A—G—G—T—G—G—T   5′

C—T—A—A—G—G—C—C   5′

+

C—C—G—G—A—C—T—C—G—T—C—C—A—C—C—A   3′

Ⓟ

Synthetic deoxynucleotide

Phosphorylated
with polynucleotide
kinase

C—T—A—A—G—G—C—C   Ⓟ T—G—A—G—C—A—G—G—T—G—G—T   5′

C—C—G—G—A—C—T—C—G—T—C—C—A—C—C—A   3′

Ⓟ

DNA ligase

Ⓟ

C—T—A—A—G—G—C—C—C—T—G—A—G—C—A—G—G—T—G—G—T   5′

C—C—G—G—A—C—T—C—G—T—C—C—A—C—C—A   3′

Ⓟ

—G—T—T—C—G—A—T—T   Double stranded structure
                          DNA equivalent to tRNA

Synthetic nucleotide with
overlapping sequence

REFERENCES

Davidson, J. N., and W. E. Cohn, *Progress in Nucleic Acid Research and Molecular Biology*, Vol. I, Academic Press, Inc., New York, 1963. An occasional publication presenting essays in circumscribed areas of nucleic acid biochemistry. Volume 9 of the series was published in 1969 and contains papers dealing with the physicochemical aspects of nucleic acids as well as a paper dealing with the regulation of RNA polymerase action.

Michelson, A. M., *The Chemistry of Nucleosides and Nucleotides*, Academic Press, Inc., New York, 1963. A well-referenced coverge of the chemistry and biochemistry of the nucleosides, nucleotides, and nucleic acids. Good coverage is given to the synthesis of nucleosides and nucleotides.

The Nucleic Acids (E. Chargaff and J. N. Davidson, ed), Academic Press, Inc., New York, 1955. Although a relatively old compilation, it contains very useful reviews of the chemistry of the purine and pyrimidine rings and of the evidence for the basic chemical structure and composition of the nucleic acids.

Proceedings of the National Academy of Sciences. Almost 50 percent of each issue of this publication has been concerned with developments in fields related to biochemistry, especially the biochemistry of nucleic acids, during the last several years.

Figure 8.45

The approach to the synthesis of a portion of the DNA corresponding to t-alanyl RNA. (*a*) The sequence of a portion of tRNA[ala] which can be used to predict the desired sequence in the DNA molecule. (*b*) The double-stranded DNA structure. (*c*) The synthesis begins with the preparation of a number of polynucleotides by the methods shown in Figure 8.32(*b*). Two of the polynucleotides are converted to the 5'-phosphorylated derivatives as shown. (*d*) When mixed together, the polynucleotides complex as shown by base pairing between complementary segments. The longest polynucleotide acts as a template to bring the two shorter ones into the correct relationship so that the enzyme DNA ligase can link them to give a partial double-stranded structure with a few bases in a single strand available for base pairing with further small polynucleotides. (*e*) By alternate additions of polynucleotides to the two strands, a number of large fragments can be built up and finally joined by the action of DNA ligase. H. G. Khorana and co-workers. See K. L. Agrawal *et al.*, *Nature*, **227**, 27 (1970).

NINE

PORPHYRINS, VITAMINS, AND ACETOGENINS

In the preceding chapters, the compounds which can be placed easily into categories are discussed. There are many other compounds of natural origin which are not so easily categorized but which are of importance. In this chapter, a series of brief statements, some structural formulas for the compounds, and references to more detailed treatments are given for some of the more frequently encountered types of compounds. No attempt has been made to be inclusive in this treatment, and it should be borne in mind that the importance of a compound to a biological system cannot be related to the amount of that compound present. This point is clearly illustrated by the vitamins, some of which are discussed briefly in this chapter.

9.1 PORPHYRINS

The organic skeleton of porphyrin is shown with the numbering system in Figure 9.1. It is a fully conjugated system in which each carbon atom is trigonal so that the whole structure is planar. Two of the nitrogens carry hydrogens and two do not. The structure, which is formed biologically from succinate and

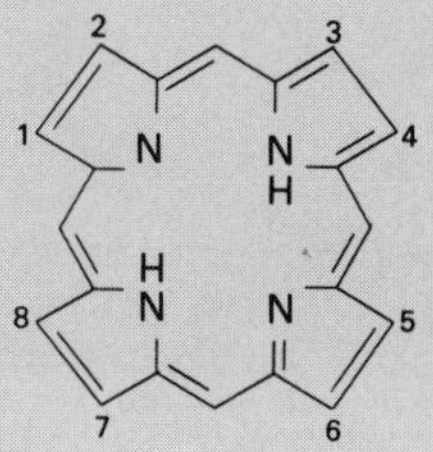

If two nitrogens across the ring from each other are written with H, then any fully conjugated set of double bonds is a "correct" set.

Classes of Porphyrins	Side Chains
Etioporphyrins	4 Methyl, 4 ethyl
Mesoporphyrins	4 methyl, 2 ethyl, 2 propionic acid
Protoporphyrins	4 methyl, 2 vinyl, 2 propionic acid
Coproporphyrins	4 methyl, 4 propionic acid
Uroporphyrins	4 acetic acid, 4 propionic acid

Figure 9.1   The porphyrin skeleton and numbering system.

glycine, is a tetrapyrrole. The biosynthesis involves pyrrole units called *porphobilinogen* which have side chains of acetic and propionic acid residues at positions which will occupy positions 1 through 8 of the finished porphyrin. With this substitution, it is termed a *uroporphyrin*. There are four possible arrangements of the side chains; two of them are naturally occurring—uroporphyrin I with Ac, Pr, Ac, Pr, Ac, Pr, Ac, Pr and urophorphyrin III with Ac, Pr; Ac, Pr; Pr, Ac; Ac, Pr (Ac = carboxymethyl and Pr = 2-carboxyethyl). Since the porphobilinogen molecule supplies not only the substituents but also the methine bridge carbon, the formation of uroporphyrin III as the major product is difficult to explain. The formation of cyclic tetrapyrrole from porphobilinogens would appear to be achieved most easily when each monomer supplies one of the methine bridges and this *requires* that the uroporphyrin I arrangement of side chains result. The formation of uroporphyrin III is rationalized as proceeding through a normal acyclic tetrapyrrole which in the process of ring closure reverses one of the pyrrole rings (Figure 9.2). Following ring closure, a metal atom is inserted (see below) and the methine bridges are dehydrogenated.

There are a variety of porphyrins in nature which serve many functions, although all are chelating agents for metals. The complexes so formed are reactive cofactors for proteins, many of which are enzymes. In Table 9.1 are listed some of the porphyrin proteins, their porphyrin and metal components, and the valence change, if any, associated with their action. The metals in protoporphyrins are capable of forming two more coordinate-covalent linkages than are utilized in complex formation with the porphyrin ring. The bonds so formed

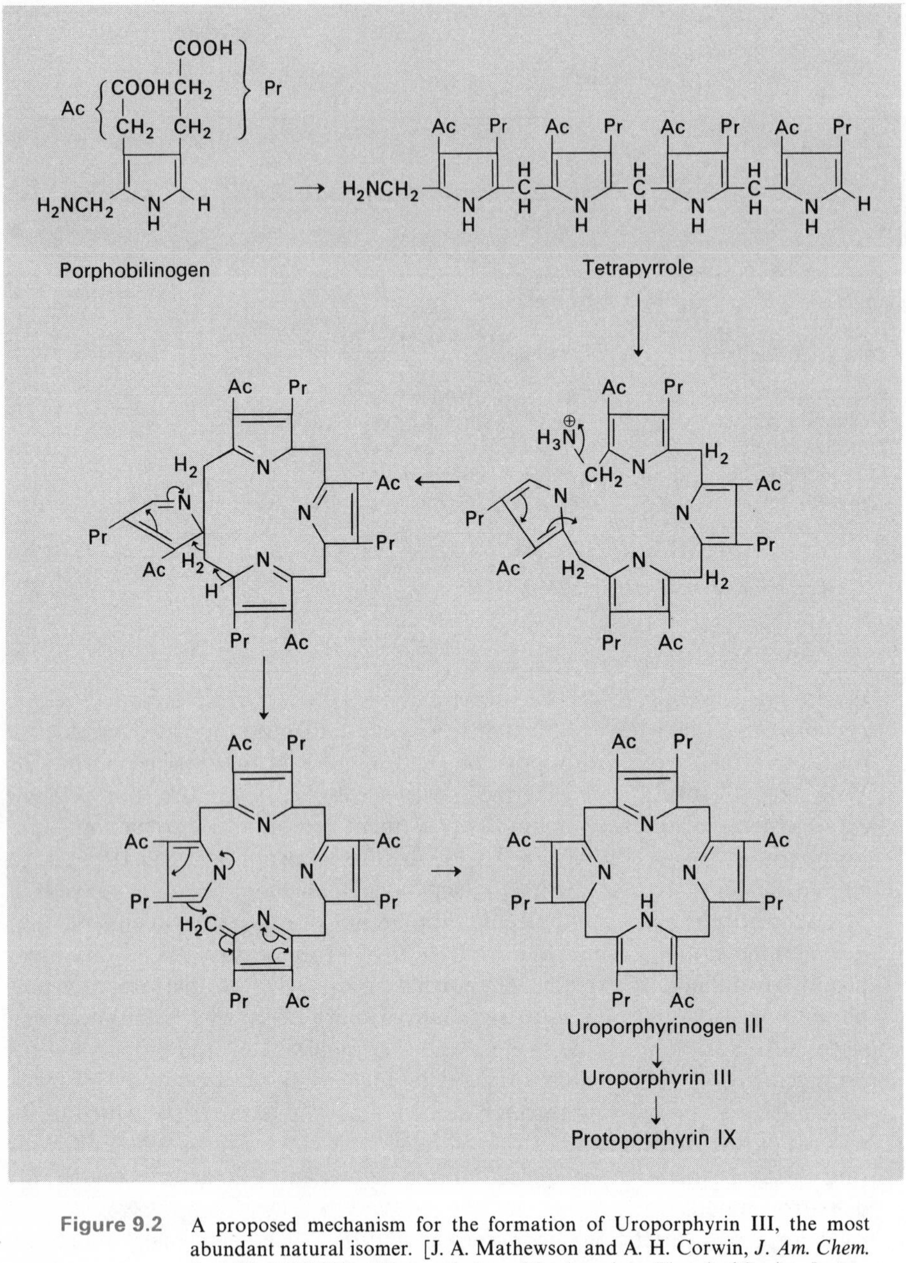

Figure 9.2   A proposed mechanism for the formation of Uroporphyrin III, the most abundant natural isomer. [J. A. Mathewson and A. H. Corwin, *J. Am. Chem. Soc.*, **83**, 135 (1961), with permission of the American Chemical Society.]

are important in the function of the complex. In the case of heme, which contains $Fe^{2\oplus}$, histidine is involved at the fifth site and binds the porphyrin to globin to form hemoglobin; the sixth site is occupied by water. In the oxygenated state the water is displaced by O_2 which can bond more tightly through π bonds. In

TABLE 9.1   SOME PORPHYRIN PROTEINS

Porphyrin		*Metal Ion*	*Protein*	*Function*	*Valence Change*
Heme	CH_3 $HC=CH_2$	$Fe^{2\oplus}$	Hemoglobin	O_2 transport	None
			Myoglobin	O_2 storage	None
		$Fe^{3\oplus}$	Catalase	Reduction of H_2O_2	None
		$Fe^{2\oplus} \rightleftharpoons Fe^{3+}$	Class B cytochromes	Electron transport	Yes

(*Note:* There are 15 isomers of heme possible because of the 4 methyl, 2 vinyl, and 2 propionic acid side chains. The natural isomer is arbitrarily designated IX.)

Heme C
1,3,5,8-Tetramethyl-2,4-di(1-*S*-cysteinyl)ethyl-
 6,7-dipropionic acid porphyrin

$Fe^{2\oplus} \rightleftharpoons Fe^{3\oplus}$ Class C   Electron   Yes
                                                cytochromes   transport

[*Note:* Two cysteinyl residues of the protein have added to the vinyl groups of heme forming a covalent thioether
 (C—S—C) bond to the protein.]

Heme A
 1,3-Dimethyl-8-formyl-6,7-dipropionic acid
 4-vinyl-2-(15 carbon-1-hydroxy branched
 side chain) 5—no substituent

$Fe^{2\oplus} \rightleftharpoons Fe^{3\oplus}$ Class A   Electron   Yes
                                                cytochromes   transport

Heme a_2
 1,3,5,8-Tetramethyl-6,7-dipropionic acid-
 2-vinyl-4-(1-hydroxyethyl)-7,8-dihydroporphyrin

$Fe^{2\oplus} \rightleftharpoons Fe^{3\oplus}$ Class a_2 Electron   Yes
                                                cytochromes   transport

| Chlorophyll a | CH_3 CH_2CH_3 | $Mg^{2\oplus}$ | Chlorophyll | Energy transduction | None |

Chlorophyll b—3-formyl—the rest is the same.

the case of the cytochromes, both of the available coordination sites are bonded to histidines.

There are two oxidation states for the iron; in the ferrous ($Fe^{2\oplus}$) state, the protoporphyrin complexes are termed *protoheme* or *heme*. In these, the $Fe^{2\oplus}$ replaces the two hydrogens of the protoporphyrin ring and the structure is electrically neutral. In the $Fe^{3\oplus}$ state, the complexes are termed *ferriprotoporphyrins*, or *hemins*; they are positively charged and are always associated with an anion (*hemin chloride* or hematin and hemin hydroxide). In both oxidation states, the hexacoordinate complex is readily formed with additional ligands such as H_2O, nitrogen bases, cyanide, and so forth. Cyanide has a particularly strong affinity for *ferri*protoporphyrins, and cyanide poisoning is due to its interaction with the *ferric* form of cytochromes. The antidote is amyl nitrite which converts some of the *heme* (*ferro*) of hemoglobin to *hemin* (*ferric*) which has a similar high affinity for cyanide. Since the loss of a small proportion of the oxygen transport system is not as critical as the inhibition of the cytochrome system of oxidative phosphorylation, cyanide poisoning can be treated—but you have to be fast!

There is abundant evidence that the iron of the hemes and probably the magnesium of the chlorophylls are the "active centers" of the molecules, and the identification of these metal complexes as important in biological function makes the determination of their fine structure and the way in which it is altered during reaction a matter of concern to biochemists. The following brief discussion of coordination compounds (that is, metal complexes) is not rigorous and it is intended only to introduce the concepts which are fundamental to the field. There are two ways of approaching the analysis of the interaction between a metal and its ligands. They are treated here separately under the headings "Crystal Field" and "Ligand Field Approaches." Both give useful insight into the interactions between ligand and metal but stress different facets of that interaction.

The transition elements which are relatively abundant in biological systems and which are engaged in complex formation are iron (Fe) and cobalt (Co). Calcium, nickel, copper, and zinc also occur, but whether they engage in the type of complexes described here is not known.

Crystal Field Theory. The transition elements can form a wide variety of complexes utilizing the *d* orbitals of the third shell of the metal. In the isolated atom, there are five *d* orbitals available, arranged in space as shown in Figure 9.3; all have the same energy (that is, they are *degenerate*). When the atom is surrounded by ligands (such as protoporphyrin IX), however, the orbitals available to its *d* electrons are no longer equivalent. The orbitals will have different energies depending on the location and characteristics of the ligand atoms which are engaged in the complex. Consider the simple case of an atom with four ligands located on the *x* and *y* axes as shown in Figure 9.4. This is a *square-planar* arrangement of the ligands. Depending on their identity, the ligands have either a net negative charge or a pair of electrons which create an

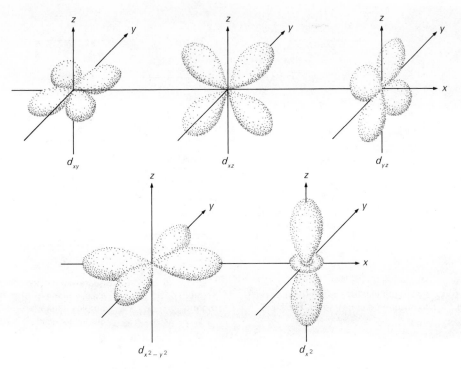

Figure 9.3   The five degenerate d orbitals of the transition metals.

electrostatic field that repels electrons in the closest orbitals of the transition metal. In this case, the $d_{x^2-y^2}$ and d_{xy} orbitals which lie in the xy plane are most strongly influenced. If the energy levels of the five d orbitals are now compared, it is found that they have been split into four levels as shown in Figure 9.4(e). The d_{xz} and d_{yz} orbitals, which are perturbed only slightly, have equal energy (are degenerate) and the other three orbitals have higher energy levels. The magnitude of the splitting (i.e., the energy difference between the orbitals) is dependent on the strength of the electrostatic field of the ligands. However, the ratio between the splittings depends only on the relative positions of the ligands, that is, the symmetry of the field they produce, and the particular metal involved. The simplest systems which provide this type of geometrical arrangement between transition elements and negatively charged elements are crystals and the separation of the d-orbital energy levels described above is referred to as *crystal field splitting*. The approximate distribution of energy levels for the d orbitals in complexes of different geometry are shown in Figure 9.4(b), (c), (d) and (e). The differences between energy levels can be evaluated from spectroscopic data ($E = h\nu$ where ν is the frequency of the absorption band corresponding to the transition). The symbol Δ is used to refer to energy differences.

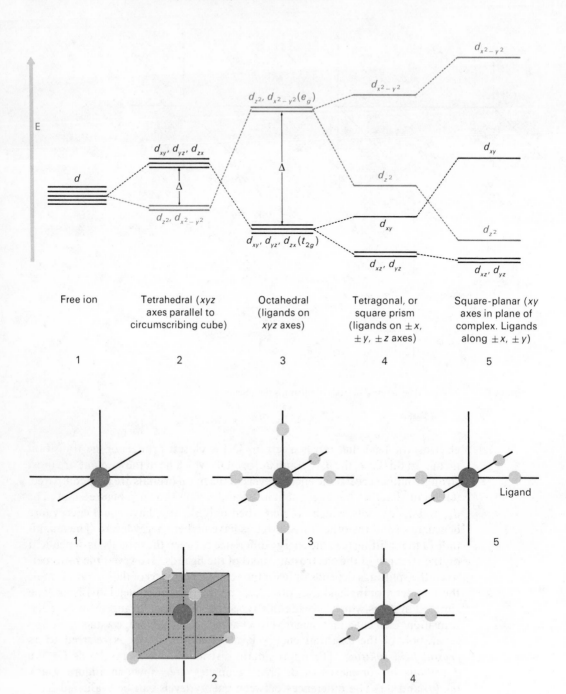

Figure 9.4   The splitting of degenerate d orbitals in complexes of differing geometry. The energy levels for the five orbitals are represented by lines in the upper part of the figure and the diagrammatic representations below show the relative position of the ligands which produce the splittings.

The common ligands are ranked below in decreasing order of their ability to cause splitting.

$$CO > CN^- > NH_3 > CNS^- > H_2O > F^- > RCO_2^- > OH^- > Cl^- > Br^- > I^-$$

(strong-field ligands)   (weak-field ligands)

When a specific case of complex formation is considered, the electrons can be thought of as being added one at a time to the system. According to Hund's rule, electrons with the same spin are put into different members of degenerate orbitals until each member of the set has one electron. When the next available orbital has a higher energy than the members of the degenerate set, however, the next electron added can go either into that orbital and have a parallel spin or into one of the occupied orbitals of the degenerate set with a spin opposite to that of the electron already occupying that orbital. Whether the electron goes into the higher energy orbital or the occupied orbital depends on the energy difference between the orbitals. If the energy difference is small, all the d orbitals will be occupied by one electron before pairing occurs. If the energy

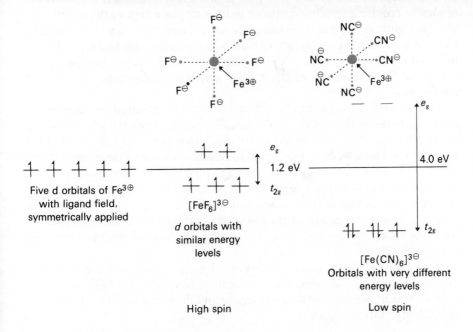

Figure 9.5   The placement of electrons into vacant d orbitals of Fe^{3+} in $[FeF_6]^{3-}$ and $[Fe(CN)_6]^{3-}$. With $[FeF_6]^{3-}$ the energy difference between the low and high energy orbitals is small and all orbitals are occupied by one electron. In $[Fe(CN)_6]^{3-}$ the pairing of electrons is forced by the large energy difference between the orbitals.

difference is large, pairing will occur as soon as the low energy orbitals have one electron each. This difference is illustrated by the two octahedral complexes of iron shown in Figure 9.5. In both complexes, the $d_{x^2-y^2}$ and d_{z^2} orbitals have a ligand in line with them (compare Figure 9.4(c)) and have higher energy levels. They are termed the e_g orbitals in this complex. The other three orbitals are lower and degenerate (have equal energy levels) and are termed the t_{2g} orbitals. In the fluoride complex the difference is not large enough to force the fourth and fifth electrons into pairing with the first and second electrons. Thus, there is an electron in each d orbital and the distribution of electrons around the iron is almost spherical. In the cyanide complex, the energy difference is large enough to force pairing and the d electrons are distributed in the d_{xy}, d_{yz}, and d_{xz} orbitals.

The number of unpaired electrons in a complex can be estimated from the magnetic susceptibility of the material so that the distribution of electrons in a complex can be determined. If they are distributed so as to be maximally unpaired, the condition is termed *high spin*; if they are maximally paired, the state is termed *low spin*. Only those forms of the transition elements with from four to seven electrons can exist in high-spin or low-spin states in octahedral complexes. This follows from the fact that there are three essentially degenerate orbitals with low energy and two with high energy. Thus, four electrons are needed to require a selection between pairing or placement in the e_g orbitals; also the eighth, ninth, and tenth electrons *must* be paired or enter the e_g orbitals. In complexes with metals having either four or six electrons, pairing results in the complex having no magnetic susceptibility—that is, becoming *diamagnetic*. Complexes with unpaired electrons and magnetic susceptibility are termed *paramagnetic*.

The crystal field treatment of coordination focuses attention on electrostatic repulsion between the metal and the ligands and describes the ways in which the metal d orbitals are deformed by the close approach of the electrons of the ligands. The process can be pictured if the electron clouds of the metal orbitals are represented by a soft balloon with the metal nucleus at its center. The ligands push in toward the center of the balloon along the x, y, and z axes, deforming the surface of the balloon and causing it to bulge out in the spaces between. The bulges represent the low energy orbitals in the complex and the locations between the ligands and the center of the balloon represent the high energy orbitals.

The crystal field treatment considers only repulsive electrostatic interactions between electrons of the ligand and the metal. In fact, there is usually some more or less covalent bonding between the ligand and the metal which contributes to the stability of the complex.

The Ligand Field Theory. This takes into account the bonding between metal and ligand and its contribution to the stability and electron distribution in the complex. The formation of bonds between the ligand and the metal involves the 3d, 4s, and 4p orbitals of the latter which are shown in Figure 9.6.

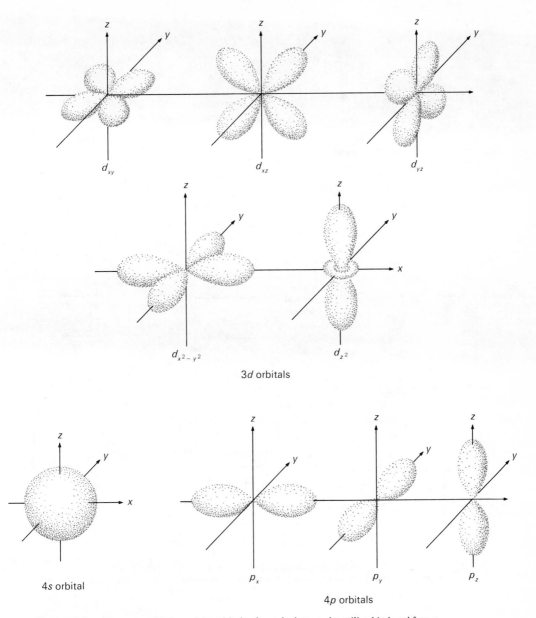

Figure 9.6   The vacant $3d$, $4s$, and $4p$ orbitals of metals that can be utilized in bond formation with ligands.

For clarity, the $3d$ and $4p$ orbitals are shown on separate sets of axes; however, they are in fact simultaneously present in the metal. In octahedral complexes these orbitals are correctly positioned to interact with ligand σ *orbitals* as shown in Figure 9.7. The result is a set of 12 hybrid orbitals, 6 with lower

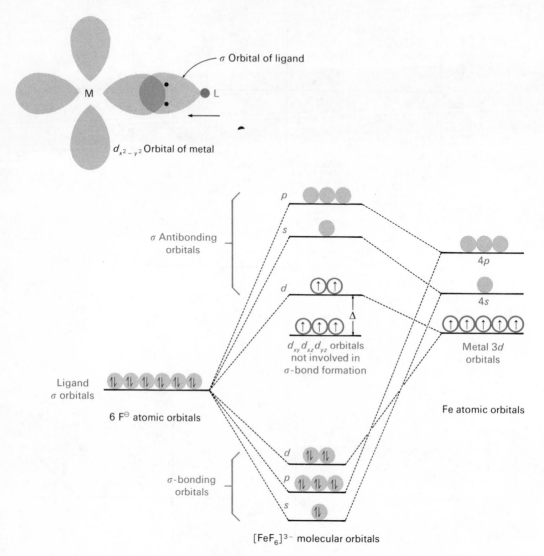

Figure 9.7   The interaction of metal orbitals with ligand orbitals to form a set of hybrid molecular orbitals available for σ-bond (covalent) formation.

(bonding) and 6 with higher (antibonding) energies than the original set. Note that the d_{xy}, d_{xz}, and d_{yz} orbitals of the metal are not properly oriented to participate in the hybridization. The result of the interaction is the release of energy by σ-bond formation between the metal and the ligands and the splitting of the d orbitals of the metal into two sets—that set having the higher energy being an antibonding set. A diagram of the orbitals of the ligands, the metal, and the complex is shown in the lower part of Figure 9.7. In the figure, the electron distribution that occurs in $[FeF_6]^{3-}$ is used as an example. Note that the electrons in the bonding orbitals are all supplied by the ligand. The lowest energy antibonding orbital is sufficiently close in energy to the d_{xy}, d_{xz}, and d_{yz} orbitals (Δ is small) so that each contains one electron and a high-spin complex is formed. With other ligands (for example, NH_3), Δ is large and the electrons pair up in the lower energy d_{xy}, d_{xz}, and d_{yz} orbitals. The energy difference between bonding and antibonding molecular orbitals is dependent on the degree of overlap between the contributing atomic orbitals. The greater the overlap, the greater the energy difference and the more likely that low-spin complexes will be observed.

Some ligands, and $CN^\ominus$ is an example, are capable of forming π bonds as well as σ bonds with the metal as shown in Figure 9.8. In this case, the metal supplies the electrons for the π bond formed by hybridizing the d_{xy}, d_{xz}, or d_{yz} orbitals of the metal with antibonding (π^*) orbitals of the $CN^\ominus$. The result is that the energy difference between the $t_{2g}(d_{xy}, d_{xz}, d_{yz})$ orbitals and the e_g ($d_{x^2-y^2}, d_{z^2}$) orbitals is increased since the electrons of the t_{2g} orbitals are delocalized into the $CN^\ominus$ antibonding orbitals and significant stabilization results. Therefore, all six electrons in the $[Fe(CN)_6]^{4\ominus}$ complex occupy the t_{2g} orbitals and a low-spin condition exists.

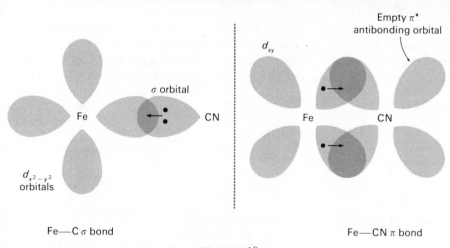

Figure 9.8 σ and π bonding in $[Fe(CN)_6]^{4\ominus}$.

The Naturally Occurring Iron-Protoporphyrin Complexes. These have four ligands supplied by the nitrogens of the protoporphyrin forming a square-planar complex which becomes octahedral when two other ligands are added. Addition of one of these ligands greatly facilitates the addition of the second. These hexacoordinate complexes are termed *hemochromes*. In Table 9.2, a number of iron-protoporphyrin protein complexes are listed with the fifth and sixth ligands and the number of unpaired electrons. In all the examples listed, the imidazole

TABLE 9.2   OCTAHEDRAL IRON-PROTOPORPHYRIN COMPLEXES

Compound	Fe	Ligands		Unpaired electrons	Spin state
		5th	6th		
Hemoglobin	$2^{\oplus}$	His	H_2O	4	High
Myoglobin	$2^{\oplus}$	His	H_2O	4	High
Oxhyemoglobin	$2^{\oplus}$	His	O_2	0	Low
Cyanhemoglobin	$2^{\oplus}$	His	$CN^{\ominus}$	0	Low
Methemoglobin	$3^{\oplus}$	His	$F^{\ominus}$	5	High
Cyanmethemoglobin	$3^{\oplus}$	His	$CN^{\ominus}$	1	Low

nitrogen of a histidine of the protein supplies the fifth coordination group and a variety of other ligands are involved at the sixth position. The characteristics of the complex depend a great deal on the identity of the sixth ligand. For example, hemoglobin coordinated with H_2O has four unpaired electrons. The water molecule cannot participate in π bonding. On the other hand, cyanohemoglobin has no unpaired electrons due to the strong interaction with the cyanide. In crystal field terms, cyanide is a strong-field ligand; in ligand field terms, it is capable of π bonding to give significant delocalization of the electrons favoring their location in the t_{2g} orbitals.

The complexes listed in Table 9.2 all have the same protoporphyrin supplying four ligands and, with the exception of myoglobin, have the same protein supplying the fifth ligand. Similar types of complexes occur, however, in which the protophorphyrin and the protein differ from those described here. In

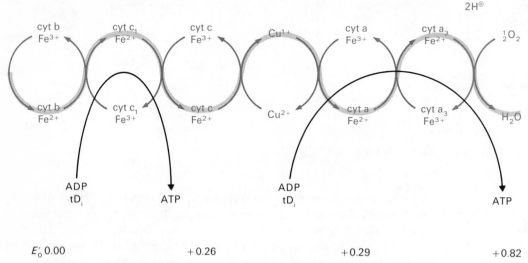

Figure 9.9   A portion of the cytochrome electron transport system of mitochondria. The colored lines show the direction of flow of electrons. E_0' equals the potential relative to the hydrogen electrode at pH 7.0.

addition, the pH, ionic strength, dielectric constant, and so forth of the environment will influence the characteristic of the complex.

In biological systems, a structurally very similar set of iron-porphyrin compounds exists, the members of which have significantly different chemical properties. This set is found in the *cytochrome system*, which is an electron-transporting system present in most cells. The components of the system undergo reversible oxidation and reduction reactions, some of which are coupled to the synthesis of ATP from ADP and inorganic phosphate. Although the mechanism of this process is not known, the point to be made here is that the cytochromes have the same or very similar iron-porphyrin complexes but the iron atoms in them have very different oxidation-reduction potentials. A part of the system is shown very diagrammatically in Figure 9.9. Two of the differences in E_0' are sufficient to provide the energy for ATP synthesis and these differences in E_0' illustrate the effect of environment on the complexes and the efficiency available to the cell because of it. Instead of having to produce distinctly different coordination complexes with different ligands or metals, the cell needs only to synthesize a set of different proteins which combine with the same complex to produce the spread in oxidation-reduction potentials.

There is a useful correlation between the visible absorption spectra of heme compounds and the high-spin/low-spin state of the system. Ferroprotoporphyrins with low spin have three absorption bands as shown in Figure 9.10; those with high spin have only two bands. Ferriprotoporphyrins have less characteristic spectra. In addition to the Soret band in the 420 nm region, there are usually three weaker bands. There is a band at approximately 555 nm

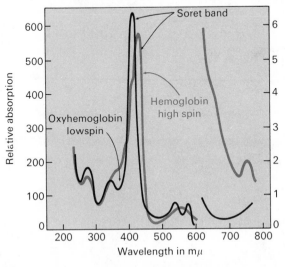

Figure 9.10   The absorption spectra of human hemoglobin and oxyhemoglobin. The scale on the right is multiplied 100 times in order to show the absorption curves above 600 nm. [A. E. Sidwell, Jr., R. H. Munch, E. S. G. Barron, and T. R. Hogness, *J. Biol. Chem.*, **123**, 335, 1938.]

which is significant in low-spin complexes and bands in the region of 500 and 600 nm which increase in intensity as the electronegativity of the ligands decreases. There is another band at approximately 490 nm which is due to a transition of the electrons of $Fe^{3\oplus}$.

In conclusion, it can be stated that whether complexes are analyzed in terms of the crystal field or the ligand field approach, insight can be obtained into the processes by which changes in the ligand produce structural changes, as in the conversion of hemoglobin to oxyhemoglobin (Chapter 4) or changes in the properties of the metal as in the cytochrome system. In both cases, the metal atom serves as the focus for the change.

9.2   COENZYME B$_{12}$, A CORRINOID

Structurally, coenzyme B_{12} (adenosyl cobalamin) bears a strong resemblance to the protoporphyrins described above (Figure 9.11). The tetrapyrrole is

Coenzyme B$_{12}$

formed into a *corrin* ring system which differs from the protoporphyrins in that one of the methine groups is missing and the numbers and kinds of substituents are different. Amide, rather than carboxyl groups, and cobalt III ($Co^{3\oplus}$), rather than iron, are present.

The mechanism of action of the coenzyme has not been established, but it may involve the cobalt acting as a very potent electron sink in the transfer of a hydrogen or a methyl group. The structure of vitamin B_{12} (cyanocobalamin) in which the 5'-adenosyl moiety of Figure 9.11 is replaced by cyanide (an artifact of the isolation procedure) was established by X-ray crystallography.

9.3   THIAMINE DIPHOSPHATE

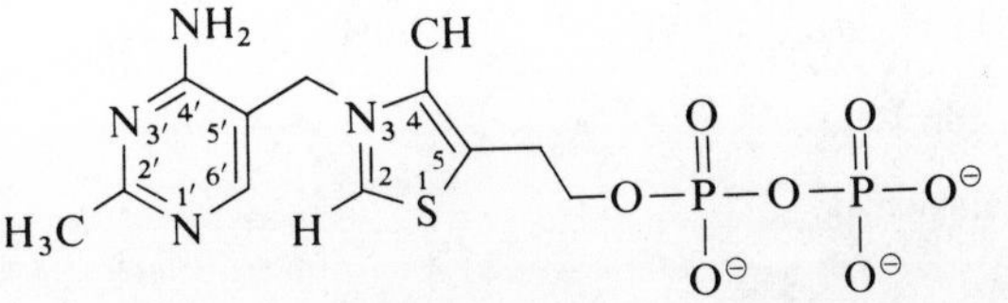

Thiazoles are catalysts (rather poor ones) of the acyloin condensation shown below and of decarboxylation reactions. Their effectiveness depends on the

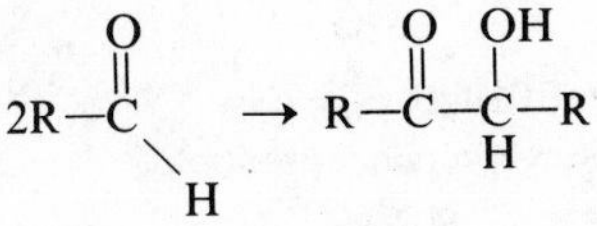

The acyloin condensation

presence of a labile hydrogen at C-2. The accepted mechanism for pyruvate decarboxylation is shown in Figure 9.11. The nucleophilic character of the thiazolium ion is similar to that of a cyanide ion, which is also an effective catalyst of the acyloin condensation.

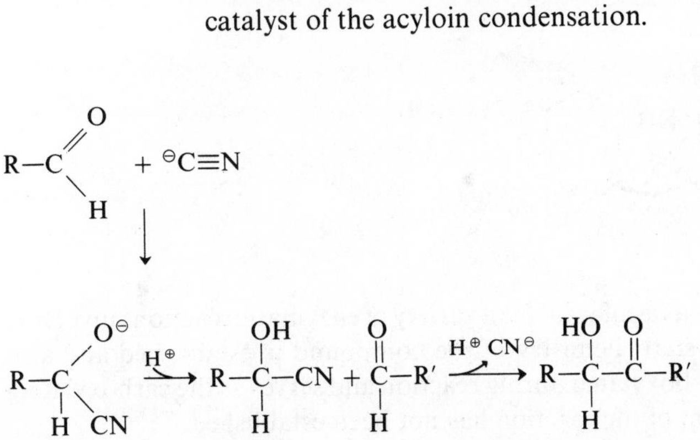

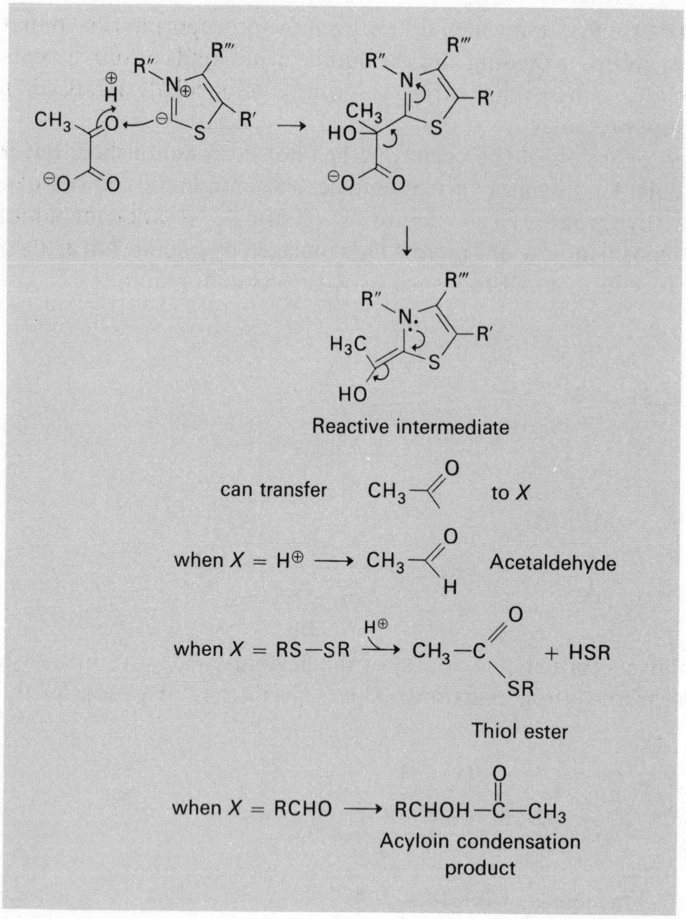

Reactive intermediate

can transfer $CH_3-\overset{O}{\underset{}{C}}$ to X

when $X = H^{\oplus} \longrightarrow CH_3-\overset{O}{\underset{H}{C}}$ Acetaldehyde

when $X = RS-SR \xrightarrow{H^{\oplus}} CH_3-C\overset{O}{\underset{SR}{}} + HSR$

Thiol ester

when $X = RCHO \longrightarrow RCHOH-\overset{O}{\underset{}{C}}-CH_3$

Acyloin condensation
product

Figure 9.11

The decarboxylation of pyruvate catalyzed by a thiazole.

9.4 BIOTIN

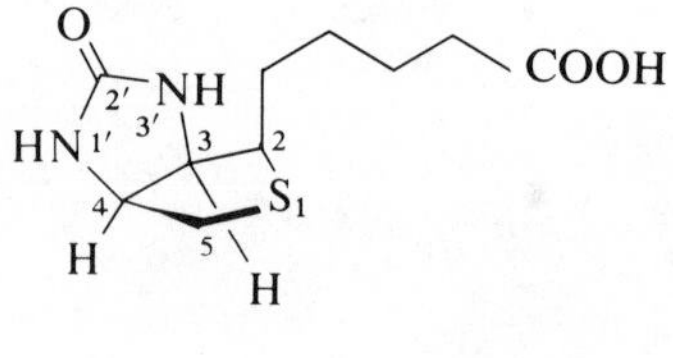

This compound serves as a cofactor for a variety of enzymatic reactions involving carbon dioxide. The sterochemistry of the compound is established and also the fact that N-1′ is carboxylated during reaction and serves as the carboxylating agent. The mechanism of the reaction has not been established.

9.5 FOLIC ACID

The biologically active forms of the compound are shown below. They are involved as an intermediate in the transfer of one-carbon fragments at the oxidation level of formate, formaldehyde, or methyl.

a

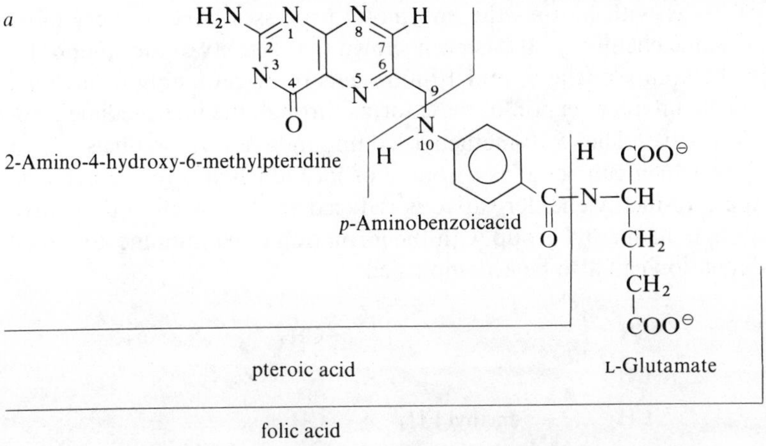

2-Amino-4-hydroxy-6-methylpteridine

p-Aminobenzoicacid

pteroic acid   L-Glutamate

folic acid

b

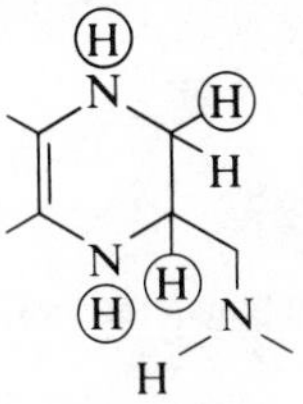

Partial structure

tetrahydrofolic acd = FH_4

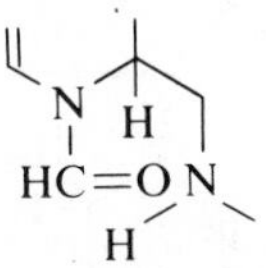

5-formyltetrahydrofolic acid = f^5FH_4

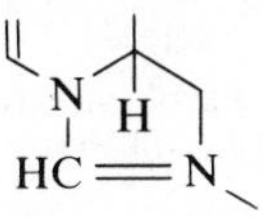

5,10-methenyltetrahydrofolic acid = $f^{5,10}FH_4$

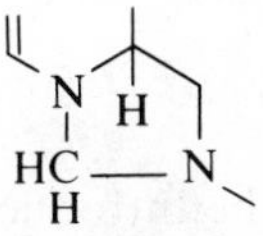

5,10-methylenetetrahydrofolic acid   = $hf^{5,10}FH$

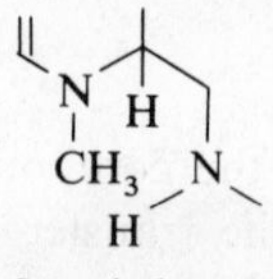

5-methyltetrahydrofolic acid = 5-methyl-FH$_4$

In this case—as with biotin—the enzymatic processes have no very close analogy in organic chemistry. It has been shown that activity of the compound depends on the ability of the 5- and 10-nitrogens to be reversibly formylated and on the facile interconversion of these forms through the intermediate 5,10-methenyl derivative. The 5,10-methenyl compound can be reduced to a 5,10-methylene which can serve as a source of formaldehyde. In methylation reactions, the 5,10-methylene derivative is reduced to the 5-methyl derivative which can donate its methyl group with the formation of methionine as shown below. Methylation can also be accomplished

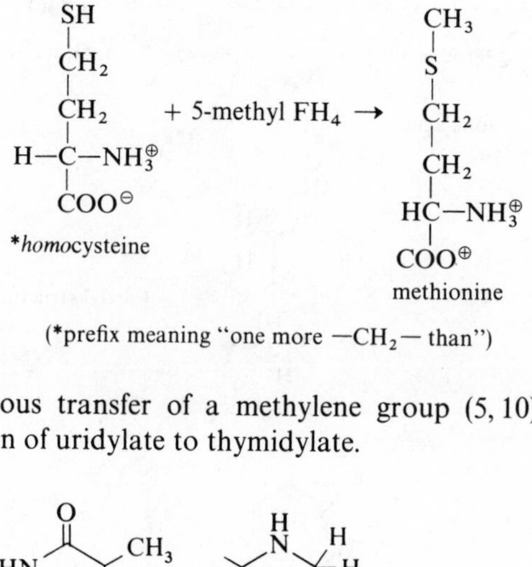

$$
\begin{array}{l}
\text{SH} \\
|\\
\text{CH}_2 \\
|\\
\text{CH}_2 \qquad + \text{ 5-methyl FH}_4 \rightarrow \\
|\\
\text{H—C—NH}_3^{\oplus} \\
|\\
\text{COO}^{\ominus}
\end{array}
\qquad
\begin{array}{l}
\text{CH}_3 \\
|\\
\text{S} \\
|\\
\text{CH}_2 \\
|\\
\text{CH}_2 \\
|\\
\text{HC—NH}_3^{\oplus} \\
|\\
\text{COO}^{\oplus}
\end{array}
$$

*homocysteine   methionine

(*prefix meaning "one more —CH$_2$— than")

by a simultaneous transfer of a methylene group (5, 10) and hydrogen as in the conversion of uridylate to thymidylate.

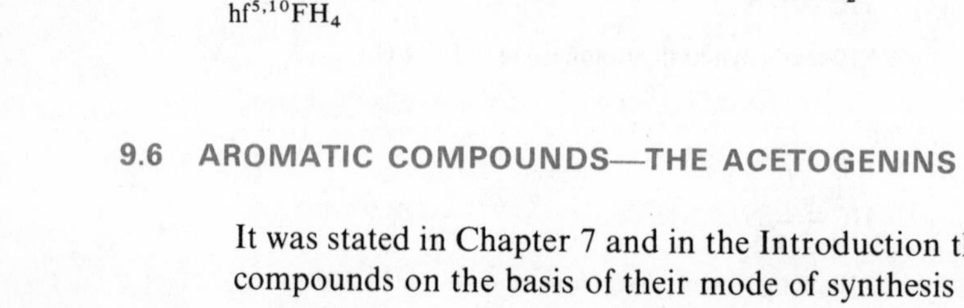

PO$_4$-ribosyl   H$_2$C —— N—R   PO$_4$-ribosyl
5'-UMP   Partial   5'-TMP   Partial structure
                     structure of   of FH$_2$
                     hf^{5,10}FH$_4$

9.6   AROMATIC COMPOUNDS—THE ACETOGENINS

It was stated in Chapter 7 and in the Introduction that it is possible to classify compounds on the basis of their mode of synthesis in biological systems. The fundamental building blocks available are carbon dioxide, amino acids, carbohydrates, and acetate. The last mentioned is the biological precursor of the major portions of all lipids. In addition, it serves as the carbon source for

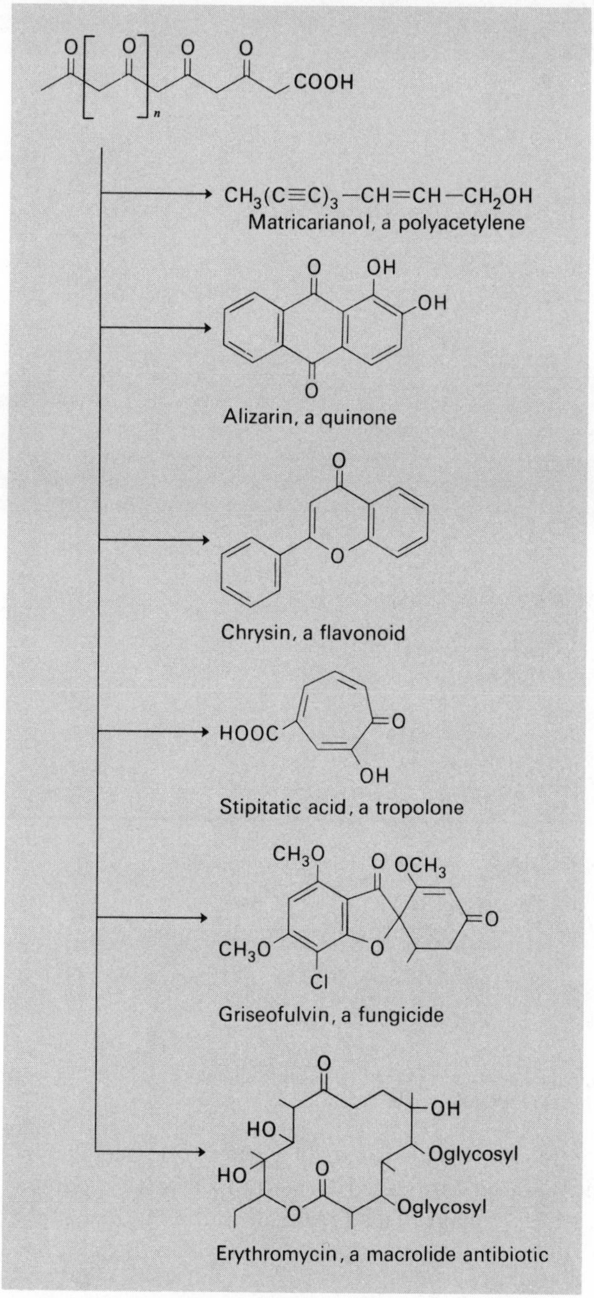
COOH
$CH_3(C{\equiv}C)_3{-}CH{=}CH{-}CH_2OH$
Matricarianol, a polyacetylene
Alizarin, a quinone
Chrysin, a flavonoid
HOOC
OH
Stipitatic acid, a tropolone
CH_3O
OCH_3
Cl
Griseofulvin, a fungicide
HO
OH
Oglycosyl
Erythromycin, a macrolide antibiotic

Figure 9.12

A polyacetyl chain and the structures of some compounds that can be considered to derive from it.

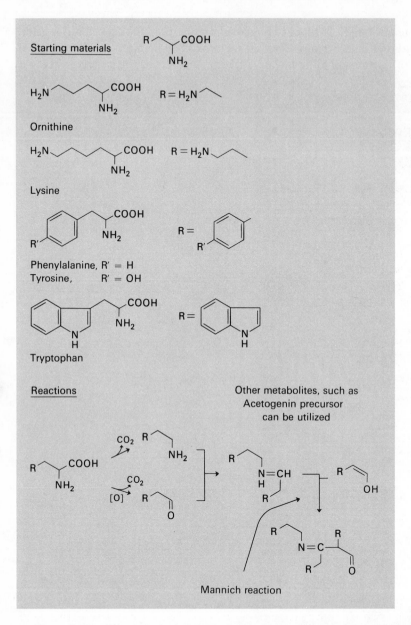

Figure 9.13 The synthesis of alkaloids showing the known starting materials and some of the products and the reactions which are presumed to be involved.

many—but not all—aromatic compounds found in nature. The acetogenins are those compounds which can be considered formally to involve the intermediate formation of a polyacetyl chain. This precursor and a number of the

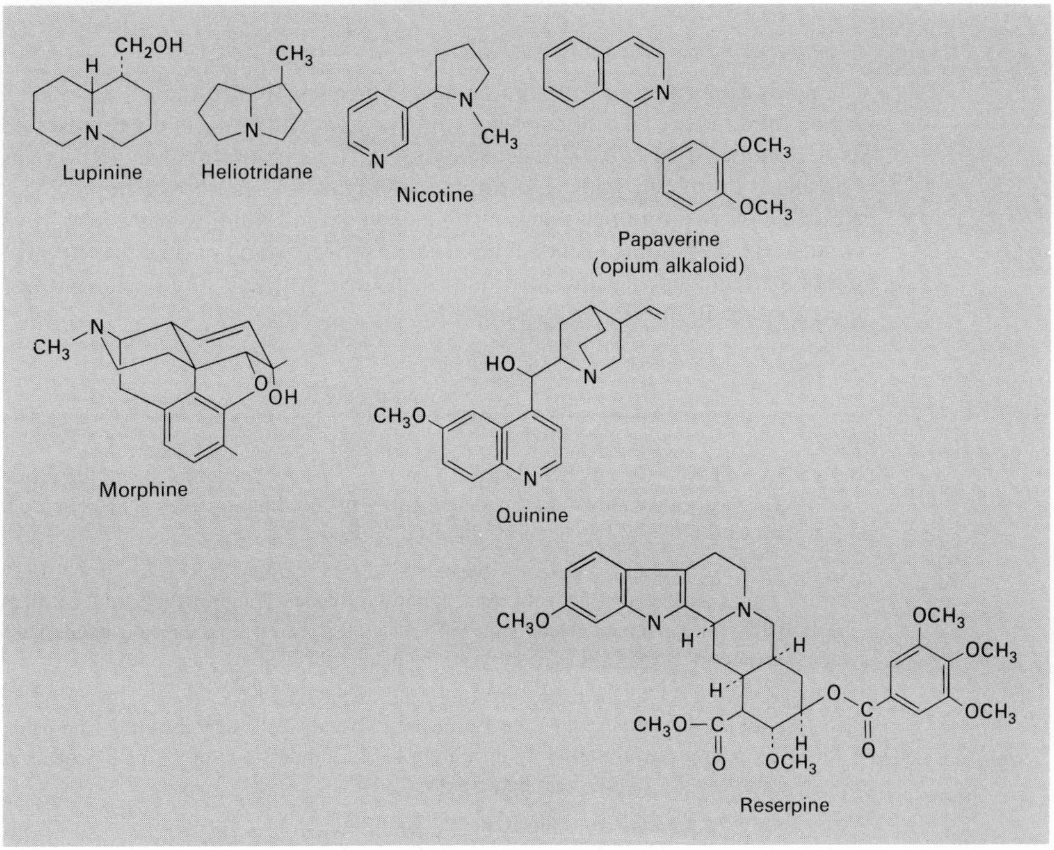

Figure 9.13

natural products which are considered to arise from it are shown in Figure 9.12. Each methylene group in the polyacetyl chain is activated by the adjacent carbonyl groups and can be readily alkylated or hydroxylated ($OH^{\oplus}$). In addition, interchain coupling, reduction, decarboxylation, and particularly cyclization reactions can occur. The natural products in this class have been extensively studied and the organic chemistry of the aromatic systems is well understood. Strangely, these compounds have been studied because they tend to accumulate but their functions, if any, are poorly understood. In many cases they do not seem to be concerned with the day-to-day economy of the systems producing them, although this conclusion eventually may be proved to be incorrect. A number of them have become important because they possess antibiotic activity or are used in commerce.

The aromatic amino acids are not formed from acetate but from carbohydrate precursors.

9.7 ALKALOIDS

Alkaloids are produced from amino acids by conversion to the corresponding amines and aldehydes and condensation via what can be formally represented as a Mannich reaction. The usual biological precursors are phenylalamine, tyrosine, tryptophan, lysine, or ornithine. In Figure 9.13 are shown the precursor amino acids, the Mannich reaction, and a few of the alkaloids which have been characterized. To date, 2,500 alkaloids are known. Many of them have potent pharmacologic activity and for this reason their synthesis and chemical manipulation has been extensively pursued.

REFERENCES

Basolo, F., and R. Johnson, *Coordination Chemistry*, W. A. Benjamin, Inc., New York, 1964. An excellent treatment of the fundamentals of coordination chemistry. There are few, if any, examples of biochemical interest discussed.

Biogenesis of Natural Compounds (P. Bernfield, ed.), Pergamon Press, New York, 1963. Many topics, such as the biosynthesis of tannins, terpenes, lignins, rubber, and carotenoids, that are not given much coverage in the usual text are covered in some detail. Each chapter is contributed by an authority in the field and is heavily referenced.

Blakley, R. L., "The Biochemistry of Folic Acid and Related Pteridenes," *Frontiers in Biology*, **13** (A. Neuberger and E. L. Tatum, eds.), North-Holland Publishing Company, 1969. An authoritative review of our present level of understanding of the biochemical roles and chemistry of folic acid derivatives.

Hemes and Hemoproteins (B. Chance, R. W. Estabrook, and T. Yonetani, ed.), Academic Press, New York, 1966. A report of a colloquium of the Johnson Research Foundation containing many original articles dealing with the chemistry, physics, and biochemistry of the hemes and hemoproteins. Each article deals with a topic of current interest, data and interpretation are presented, and references to earlier work given. The comments of a number of discussants is also presented.

Hogenkamp, H. P. C., "Enzymatic Reactions Involving Corrinoids," *Ann. Rev. of Biochem.*, **37**, 225 (1968). This article focuses on the enzymatic rather than chemical reactions of the corrinoids (B_{12} and its analogs). Chemistry and biochemistry are inseparable in this field, however, and much of the chemistry of these compounds is considered.

Krampitz, L. O., "Catalytic Function of Thiamin Diphosphate," *Ann. Rev. Biochem.*, **38**, 213 (1969). An excellent review of the chemical evidence relating to the mechanism of thiamin-catalyzed reactions and a survey of the work on several of the enzyme systems in which thiamin pyrophosphate plays a part.

Lynen, F., "The Role of Biotin-dependent Carboxylations in Biosynthetic Reactions," *Biochem. J.*, **102**, 381 (1967). This is the text of the Third Jubilee Lecture of the Biochemical Society. The enzymatic role of biotin and its mode of action are discussed and many references are given.

INDEX